INFRARED AND MILLIMETER WAVES

VOLUME 12 ELECTROMAGNETIC WAVES IN MATTER, PART II

CONTRIBUTORS

MOHAMMED NURUL AFSAR

KENNETH J. BUTTON

P. K. CHEO

B. G. DANLY

S. G. EVANGELIDES

GERT FINGER

FRITZ K. KNEUBÜHL

B. LAX

NOBORU MIURA

JOHN F. RABOLT

R. J. TEMKIN

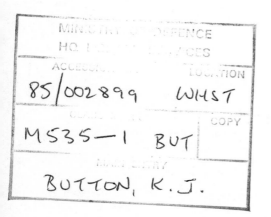

INFRARED AND MILLIMETER WAVES

VOLUME 12   ELECTROMAGNETIC WAVES IN MATTER, PART II

Edited by   **KENNETH J. BUTTON**

NATIONAL MAGNET LABORATORY
MASSACHUSETTS INSTITUTE OF TECHNOLOGY
CAMBRIDGE, MASSACHUSETTS

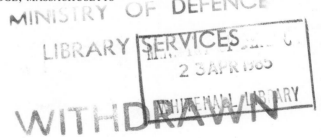
1984

ACADEMIC PRESS, INC.
(Harcourt Brace Jovanovich, Publishers)

Orlando   San Diego   New York   London
Toronto   Montreal   Sydney   Tokyo

ACADEMIC PRESS, INC.
Orlando, Florida 32887

United Kingdom Edition published by
ACADEMIC PRESS, INC. (LONDON) LTD.
24/28 Oval Road, London NW1 7DX

Library of Congress Cataloging in Publication Data

Main entry under title:

Infrared and millimeter waves.

    Includes bibliographies and indexes.
    Contents: v. 1. Sources of radiation. -- v. 2. Instrumen-
tation. -- [etc.] -- v. 12. Electromagnetic waves in matter,
part II.
    1.   Infra-red apparatus and appliances. 2. Millimeter
wave devices.   I. Button, Kenneth J.
TA1570.152   621.36'2   79-6949
ISBN 0-12-147712-6 (V.12)

CONTENTS

Chapter 4 **Spectral Thermal Infrared Emission
of the Terrestrial Atmosphere**
Gert Finger and Fritz K. Kneubühl

Chapter 5 **Frequency Tuning and Efficiency Enhancement of
High-Power Far-Infrared Lasers**
B. G. Danly, S. G. Evangelides, R. J. Temkin, and B. Lax

Chapter 6 **Far-Infrared Laser Scanner for
High-Voltage Cable Inspection**
P. K. Cheo

LIST OF CONTRIBUTORS

Numbers in parentheses indicate the pages on which the authors' contributions begin.

MOHAMMED NURUL AFSAR (1), *Massachusetts Institute of Technology, Francis Bitter National Magnet Laboratory, Cambridge, Massachusetts 02139*

KENNETH J. BUTTON (1), *Massachusetts Institute of Technology, Francis Bitter National Magnet Laboratory, Cambridge, Massachusetts 02139*

P. K. CHEO (279), *United Technologies Research Center, East Hartford, Connecticut 06108*

B. G. DANLY (195), *Plasma Fusion Center and National Magnet Laboratory, Massachusetts Institute of Technology, Cambridge, Massachusetts 02139*

S. G. EVANGELIDES (195), *Plasma Fusion Center and National Magnet Laboratory, Massachusetts Institute of Technology, Cambridge, Massachusetts 02139*

GERT FINGER (145), *Physics Department, ETH, CH-8093 Zürich, Switzerland*

FRITZ K. KNEUBÜHL (145), *Physics Department, ETH, CH-8093 Zürich, Switzerland*

B. LAX (195), *Plasma Fusion Center and National Magnet Laboratory, Massachusetts Institute of Technology, Cambridge, Massachusetts 02139*

NOBORU MIURA (73), *Institute for Solid State Physics, University of Tokyo, Roppongi, Minato-ku, Tokyo, Japan*

JOHN F. RABOLT (43), *IBM Research Laboratory, San Jose, California 95193*

R. J. TEMKIN (195), *Plasma Fusion Center and National Magnet Laboratory, Massachusetts Institute of Technology, Cambridge, Massachusetts 02139*

PREFACE

This is the second volume in this treatise to deal exclusively with the millimeter and submillimeter properties of materials and the methods of measuring and interpreting these properties. A third volume on this topic is now in preparation, and plans for a fourth are in progress.

The contents of Volumes 8 and 12 speak eloquently for the theme of this subseries on electromagnetic waves in matter. Each book opens with a chapter on dielectric materials studies. G. W. Chantry introduced the topic in Volume 8, and Mohammed Nurul Afsar has now provided us with his extensive dielectric measurements at millimeter wavelengths in Volume 12. In both volumes the second chapter deals with polymers. W. F. X. Frank and U. Leute described their far-infrared spectroscopy of high polymers in Volume 8. John F. Rabolt deals with low-frequency (far-infrared) vibrations of long-chain molecules and polymers in Volume 12. These chapters are followed by semiconductor topics. In Volume 8 S. Perkowitz gave a general treatment of the spectroscopy of semiconductors and B. Jensen gave a thorough treatment of free-carrier behavior in semiconductors (which is most important in the far infrared). In Volume 12 we have a review of the celebrated work of the Miura group, which uses extraordinarily high-intensity pulsed magnetic fields for magnetooptical spectroscopy of semiconductor phenomena.

Following the pathfinding work by Gert Finger and Fritz K. Kneubühl on spectral thermal infrared emission of the terrestrial atmosphere, we have two chapters on far-infrared lasers because lasers are of such great importance to the far-infrared spectroscopy of materials.

A third volume devoted to electromagnetic wave interactions in matter is being readied for press. Volumes 9, 10, 11, and 13 begin a subseries on millimeter-wave components and techniques; only the last of these four volumes is yet to be published.

CONTENTS OF OTHER VOLUMES

Volume 13: Millimeter Components and Techniques, Part IV
(In Press)

Volume 14: Millimeter Components and Techniques, Part V
(In Press)

CHAPTER 1

Millimeter-Wave Dielectric Properties of Materials*

Mohammed Nurul Afsar and Kenneth J. Button

Massachusetts Institute of Technology
Francis Bitter National Magnet Laboratory†
Cambridge, Massachusetts

I. Introduction

Almost no reliable data have been available in the millimeter and near-millimeter wavelength (60–600 GHz) range because measurements of the dielectric properties of materials at these wavelengths are extremely difficult

---

* Partially supported by the U. S. Army Research Office under Contract Number DAAG-29-81-K-0009 and the Office of Fusion Energy, U. S. Department of Energy, under Contract W-7405-eng-26 with Union Carbide Corporation and the Boeing Aerospace Corporation.
† Supported by the National Science Foundation.

1

to carry out accurately. The millimeter-wave region lies beyond conventional microwave techniques and forms a "bridge" to the optical techniques. In the past, one could rarely trust the millimeter-wave dielectric data for use in precision engineering design, because any extrapolated microwave method or extrapolated optical method that was used to make the measurements had many serious limitations and uncertainties. Until recently, engineers have been satisfied to know whether a material was "opaque" or "transparent" at millimeter wavelengths. More recently, a measurement good to 10% accuracy was considered to be better than nothing; after all, it is inconvenient and expensive to acquire and use precision-measurement facilities and sophisticated instrumentation. The real danger lies in the literature that is actually misleading. Most frequently the misleading data get into the literature when someone uses a familiar *microwave* instrument, such as waveguide interferometer or a cavity resonator or a Fabry–Perot open resonator, beyond the limit of its classical capabilities. For example, the millimeter wavelengths are too short for the practical use of a microwave single-mode resonant cavity. The millimeter wavelengths are too long at this extreme end of the *optical* spectrum for a familiar blackbody source such as a mercury-vapor lamp to be used; normally, it provides too little energy for millimeter-wave measurements with a Fourier spectrometer. Indeed, the use of a *conventional* plane-wave interference technique employing a mercury lamp to obtain millimeter-wave dielectric data is almost impossible. Nevertheless, the Fourier method has now been improved by Afsar to provide data from 5 mm (60 GHz) into the submillimeter range (Afsar and Button, 1983). New theories were also developed by Afsar giving a full treatment of all beams and interface effects (Afsar, 1977, 1984a; Afsar and Chantry, 1977; Afsar *et al.*, 1976b,c), and great care was taken to increase the efficiency of energy throughput and detection (Afsar and Button, 1981, 1983; Afsar, 1982). In such a special spectrometer, the phase determination, in particular, can be made very accurately when used in the asymmetric mode (dispersive Fourier transform spectroscopy), leading to the determination of the real part of the dielectric constant to five or six significant figures (Afsar and Button, 1983; Afsar, 1982). Because we employ a quasi-optical technique, we measure directly the optical parameters, namely, the absorption coefficient α and the refractive index n, simultaneously. Dielectric parameters ϵ', ϵ'' and loss tangent (tan δ) are easily calculated, as will be demonstrated. The present-day dispersive Fourier transform spectroscopic (DFTS) technique developed by Afsar, measures the refractive index spectrum and, simultaneously, the absorption coefficient spectrum from the analysis of the amplitude and phase information that the specimen has contributed to the output signal (Afsar, 1977, 1982, 1984a; Afsar and Button, 1981, 1983; Afsar and Chantry, 1977; Afsar and Hasted, 1977; Afsar *et al.*, 1976b,c). Although the phase information can be carried through to a determination of the refractive index (and the real part of the dielectric permittivity) to an accuracy of five or six significant figures for a low-loss material, the absorption coefficient (and loss tangent) can be determined only to about 1%, because the commercially available electronic amplifying equipment cannot ordinarily carry through amplitude information with higher precision and reproducibility (Afsar and Button, 1983; Afsar, 1982).

Several other classical methods are being improved in efforts to provide some kind of data, if not the best, to this barren region of the spectrum. The Fabry–Perot open-resonator method provides about an order-of-magnitude less accuracy in the measurement of loss tangent and only three significant figures in the dielectric constant, but in some ways it is more convenient to use (Cullen and Yu, 1971; 1979; Cook and Jones, 1976; Jones, 1976). Today, the most significant improvement in the Fabry–Perot system would be the use of a superheterodyne receiver with highly stabilized, phase-locked Gunn oscillators (Matsui et al., 1984). The Mach–Zehnder type of spectrometer used with Gunn or IMPATT sources also produces dielectric data at the typical IMPATT frequencies (Birch, 1980, 1981). Precision data in this case again is obtainable only by the use of a specially constructed, highly stable spectrometer system with a high degree of statistical fitting (Afsar, 1984b). Various other techniques such as rotation of a parallel-slab specimen with input and output devices (Shimabukuro et al., 1984), waveguide reflectometer (Vanloon and Finsy, 1973, 1974; Finsy and Vanloon, 1972), oversize cavity resonator (Stumper, 1972, 1973; Stumper and Frentrup, 1976), and oversize waveguide interferometer (Goulon et al., 1968, 1973; Butterweck, 1968) also produce dielectric data in the range 10–2 mm, but the accuracy is again limited to about 10% in most of these techniques. Among all of these methods, the DFTS is the best for the millimeter and submillimeter range. Other methods, such as the six mentioned previously, have their particular applications such as other wavelength ranges, odd specimen sizes, and different physical properties such as liquids and gases (Kolbe and Leskovar, 1982).

Why should we go to all of this trouble and expense just to get another order-of-magnitude, or even a factor of three, higher accuracy, reproducibility, and reliability? Why would a quick measurement providing "engineering values" be unsuitable for the purpose of exercising the trade-off process for the selection of materials for particular applications? The simple answer is that there are wide variations in the parameters of nominally identical specimens at millimeter wavelengths that microwave engineers rarely see at lower frequencies. When we are trying to determine the reasons for these variations so as to choose a "standard material" for our application, $\pm 10\%$ in reproducibility of measurement is just not good enough.

It is very important to have highly reproducible data, so that one would be able to distinguish the different dielectric properties among nominally identical specimens — dielectric properties that vary among specimens from different suppliers, among specimens prepared by somewhat different methods, or among specimens having physical properties that are not precisely controlled during preparation. In our recent dispersive Fourier transform spectroscopic dielectric measurement work, we have found significant variations in the dielectric properties of such common materials as SiO_2, fused silica glass (Afsar and Button, 1983; Afsar, 1982). There are notable differences in absorption coefficients in Al_2O_3, ceramic alumina, depending on the source of the alumina specimens. For example, hot-pressed ceramic beryllia, BeO, has much lower losses than cold-pressed beryllia.

We would expect to find differences in absorption among high-resistivity semiconductors such as semi-insulating GaAs, and large differences were found (Afsar and Button, 1983). Polymers (plastic) are well known to be very much in need of characterization before they can be used in engineering applications. Their degree of crystallinity must be controlled (Chantry *et al.,* 1971; Davies and Haigh, 1974; Konwerska-Hrabowska *et al.,* 1981). Therefore, it is now essential that a full description of a material be available along with accurate, reproducible measurements of its dielectric properties. Thus, it has been shown that not only is a microwave measurement of loss tangent untrustworthy at millimeter wavelengths but also traditional microwave methods used at millimeter wavelengths can be inaccurate and irreproducible.

The important differences in nominally identical specimens can be detected, verified, and understood only by using the most sophisticated, highly sensitive, and highly stable equipment backed by a most detailed evaluation of the theory of the technique. Therefore, it is important to rely on a "center of excellence" as a source of practical data.

This chapter will include a treatment of the relationship of measured and derived (calculated) quantities as well as brief descriptions and comparisons of some modern dielectric measurement techniques. Appendix A is a compendium of data, where illustrations will show a comparison of some common materials in terms of their spectra of absorption coefficient, refractive index, real and imaginary parts of the dielectric permittivity, and loss tangent. A discussion of differences among nominally identical specimens will also be given. These discussions will provide examples of the importance of "characterization of materials." Sometime in the near future, the community of millimeter-wave engineers should be provided with a "digest of millimeter-wave materials information and measurement." As more data can be collected and added to the illustrations in the appendix, the nucleus of the digest will be created. Up-to-date copies of this fledgling digest will be available from the Millimeter and Submillimeter-Wave Materials Information and Measurement Center at the MIT National Magnet Laboratory.

II. Electromagnetic Quantities

The complex refractive index $\hat{n}$ is derived from the complex dielectric permittivity $\hat{\epsilon}$ of Maxwell's equations so that

$$\hat{\epsilon} = (\hat{n})^2.$$

The real and imaginary parts of n are, by definition,

$$\hat{n} = n - ik = n - i(\alpha/4\pi\tilde{v}) = n - i(c\alpha/4\pi v),$$

where k is the absorption index $\alpha/4\pi\tilde{v} = c\alpha/4\pi v$, α the absorption coefficient (cm^{-1}), v the frequency in hertz, $\tilde{v}$ the wave number in cm^{-1}, c the velocity of light in vacuum; and it is convenient to note that 1 wave number (cm^{-1}) = 30 GHz.

The complex dielectric permittivity $\hat{\epsilon}$ has real (ϵ') and imaginary (ϵ'') parts defined as

$$\hat{\epsilon} = \epsilon' - j\epsilon'', \qquad i = j = \sqrt{-1}.$$

Then, our definitions provide us with the simple relationships between the fundamental optical quantities, α and n, and the dielectric quantities ϵ' and ϵ'', as follows:

$$\epsilon' = n^2 - k^2 = n^2 - (\alpha/4\pi\tilde{v})^2 = n^2 - (c\alpha/4\pi v)^2;$$

$$\epsilon'' = 2nk = (\alpha n)/2\pi\tilde{v} = (\alpha cn)/2\pi v.$$

The term loss tangent, or the popularly known "tan δ," is the ratio of the imaginary part (ϵ'') to the real part (ϵ') of the dielectric permittivity,

$$\tan \delta = \epsilon''/\epsilon'.$$

In the millimeter-wave region of the spectrum, we need all of these relationships, because we cannot simply extrapolate all of our microwave techniques into this nether region from the long-wavelength side, nor can we extrapolate all of our optical techniques from the high-frequency side. Both the microwave engineers and the optical engineers consider this nether world of millimeter waves by its strict Webster's definition as "world of the dead or of future punishment." The microwave engineer finds his millimeter wavelengths to be too small for fundamental-mode techniques for measurement of complex dielectric permittivity and loss tangent. The optical engineer, who typically uses free-space, plane-wave Michelson interference, finds his source of blackbody radiation (mercury-vapor lamp) to be too feeble.

This anticipation of future punishment can be ameliorated somewhat by the development of some "figures of merit" or means for evaluating trade-off selections of materials. One of these is the definition of a "millimeter-wave low-loss material" to eliminate the use of the loss tangent, tan δ, which is sometimes unreliable at millimeter wavelengths.

III. Definition of a Low-Loss Material

Note that the expression for the dielectric constant,

$$\epsilon' = n^2 - (\alpha/4\pi\tilde{v})^2,$$

is not a *constant* at all; it contains a term inversely dependent on frequency. It is not constant, that is, unless the second term that contains the frequency can be neglected. This is usually the case at optical frequencies where the denominator of the second term is very large. At millimeter-wave frequencies, however, we cannot drop the second term unless the absorption coefficient α is sufficiently small compared with the refractive index n to render the real part of the dielectric permittivity ϵ' independent of frequency. Specifically, if the absorption coefficient is less than unity at 300 GHz and less than 0.1 Np/cm at 30 GHz, then ϵ' is truly a dielectric *constant* to at least four significant figures, because the material is "low-loss" by definition. We shall show that SiO_2, fused-silica glass, is a low-loss millimeter-wave material. A number of other materials satisfy this criterion but most materials do not.

IV. Measurement Methods

Attempts have been made during the past decade to extend various classical cavity techniques and quasi-optical techniques toward the millimeter-wave region. As microwave methods are extended toward the millimeter wavelengths, Q values become very low, particularly for closed cavities. Therefore, it is practical to discuss only quasi-optical techniques here, namely, dispersive Fourier transform spectroscopy (DFTS), Mach–Zehnder–IMPATT spectrometer (MZI), and the open-resonator method. Current MIT National Magnet Laboratory facilities include all these techniques.

The highest-accuracy absorption coefficient and refractive index data can be obtained from dispersive Fourier transform spectroscopy when it employs a polarizing two-beam interferometer and a special detector consisting of a helium-cooled InSb hot-electron bolometer (Afsar, 1977, 1982, 1984a; Afsar and Button, 1981, 1983; Afsar and Chantry, 1977; Afsar *et al.,* 1976b,c). This system generates high-accuracy data in the range 3–0.25 mm and is limited only on the microwave end of its range (around 5 mm) by the weak radiative power from its mercury-vapor lamp. Therefore, at the microwave end of the range other methods can be used, such as Mach–Zehnder interferometer together with a 20-mW IMPATT (Birch, 1980, 1981) or Gunn oscillator or an open-resonator technique (Cook and Jones, 1976; Cullen and Yu, 1971, 1979; Jones, 1976a,b; Kolbe and Leskovar, 1982; Matsui *et al.,* 1984). In experienced hands, these methods can provide excellent, reliable data on both dielectric and magnetic parameters, ϵ', ϵ'', μ', and μ'' through the millimeter–submillimeter gap in the spectrum.

A. DISPERSIVE FOURIER TRANSFORM SPECTROSCOPY

The difference between the dispersive Fourier transform (DFTS) technique and the conventional Fourier transform technique lies in the simple fact that the specimen is placed in front of the fixed mirror *in an active arm* of the Michelson interferometer instead of in the conventional location in front of the detector, as shown in Fig. 1. When we use the DFTS configuration, we get both phase and amplitude information from the specimen into the interferogram. The Fourier transform of this interferogram contains a *complex* frequency spectrum, the real and imaginary parts of which provide the phase and amplitude spectra and, hence, the refractive index and the absorption coefficient spectra. This is the only *high precision* technique known that gives both the refractive index and the absorption coefficient simultaneously: the refractive index to six significant figures and the absorption to about one percent reproducibility. This method is not yet widely used because the complicated theory has only recently been developed and perfected by Afsar (Afsar, 1977, 1982, 1984a; Afsar and Button, 1981, 1983; Afsar and Chantry, 1977; Afsar and Hasted, 1977; Afsar *et al.,* 1976b,c) and adapted to the millimeter-wave range.

An interferogram is recorded by moving mirror M_2 in Fig. 1 so that the

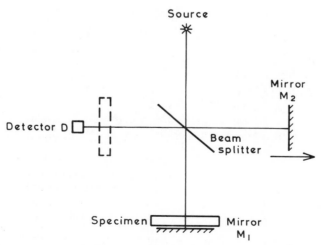

FIG. 1   Ray diagram of a two-beam Michelson interferometer.

path difference x creates constructive and destructive inferences at the beam splitter.

If the interferogram recorded with the specimen present is $F_s(x)$ and if the reference interferogram without the specimen is $F_0(x)$, where x is the path difference, the ratio of the full Fourier transform $\mathcal{F}$ gives the positional insertion loss $\hat{L}(\tilde{\nu})$:

$$\mathcal{F}\{F_s(x)\}/\mathcal{F}\{F_0(x)\} = \hat{L}(\tilde{\nu}) = \text{function } [\hat{n}(\tilde{\nu}); \text{ position}].$$

The relationship between the complex insertion loss $\hat{L}(\tilde{\nu})$ and the complex refractive index $\hat{n}(\tilde{\nu})$ depends on both the nature and the location of the specimen. The interferogram $F_s(x)$ may contain a number of interference signatures arising from reflections at various interfaces before and after transmitting through the specimen, and from multiple reflections between parallel surfaces within the specimen, as shown in Fig. 2. In general, $F_s(x)$ is the sum of $F_R(x)$, $F_T(x)$, and $F_M(x)$ so that $F_s(x) = F_R(x) + F_T(x) + F_M(x)$, where $F_R(x)$ represents signatures after reflections from front and rear surfaces of the specimen. The term $F_T(x)$ represents the interference signature arising after transmitting through the specimen, reflecting from the mirror M_1, and transmitting through the specimen again. The term $F_M(x)$ represents multiply reflected signatures. For a relatively low absorbing material, a thick specimen can be used that leads to a good separation of $F_R(x)$, $F_T(x)$, and $F_M(x)$ signatures, and any of these edited signatures could be used together with $F_0(x)$ to obtain the complex refractive index.† It is obviously preferable to use the $F_T(x)$ signature, because it contains the main transmission component. The insertion loss then becomes

$$\hat{L}_T(\tilde{\nu}) = \mathcal{F}\{F_T(x)\}/\mathcal{F}\{F_0(x)\}$$
$$\simeq \hat{S}(\tilde{\nu}) \exp\{-4\pi i \tilde{\nu}[\hat{n}(\tilde{\nu}) - 1]d\}\hat{M}(\tilde{\nu}),$$

where $\hat{S}(\tilde{\nu})$ and $\hat{M}(\tilde{\nu})$ represent respectively surface reflection loss and multiple reflection loss contributions of the incident beam. The real and the imaginary parts of the complex refractive index are then calculated via an iterative procedure. The full iterative procedure of Afsar (Afsar, 1977, 1984a; Afsar and Chantry, 1977) takes care of all interface effects and, theoretically, eliminates the systematic errors. It is therefore no longer

† Honijk et al. (1977) and Passchier et al. (1977) developed the theory for the full interferogram $F_s(x)$ method, which may be used only when interference signatures cannot be separated; otherwise, a reflection dispersive Fourier transform spectroscopic method that employs phase and amplitude change at reflection at the flat surface of a specimen, may be used (Afsar, 1979a; Afsar and Hasted, 1977, 1978; Birch et al., 1976; Mead and Genzel, 1978; Parker et al., 1976; Staal and Eldridge, 1977). A transmission method (transmission DFTS) is always preferred, and for a lossy liquid such as ethanol and methanol a subtraction procedure (Afsar et al., 1976a,b) can be used for isolation of the $F_T(x)$ signature from $F_R(x)$ and $F_M(x)$ signatures.

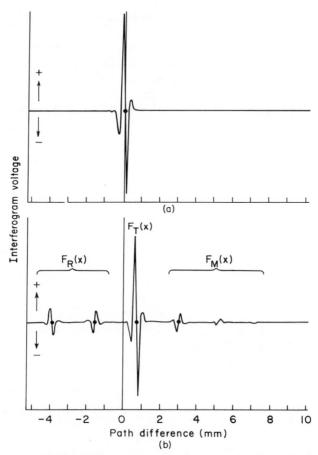

FIG. 2   Phase-modulated antisymmetric interferograms recorded using a polarizing interferometer for dispersive Fourier transform spectroscopy. (a) $F_0(x)$ is the reference intergerogram recorded without the specimen and (b) $F_s(x)$ is the interferogram recorded with a TPX specimen.

necessary to adjust the absolute level of absorption coefficient and refractive index with other methods, and these DFTS measurements can be considered the absolute standard.

The favorable kind of interferometric configuration in the millimeter-wave region is the polarization configuration. A "polarization" interferometer must be used in the millimeter-wave region. It uses two or three free-standing wire-grid polarizers (instead of plastic mylar beam splitters) to polarize the incident unpolarized beam, split, recombine, and analyze before reaching the detector.   In a Michelson-type configuration, one uses mylar as a beam-splitting material.   The effects of multiple reflections within the beam divider result in a strong frequency dependence of the transmissivity, thereby giving successively smaller lobes.   A polarizing interferometer is free of the multiple internal reflections within a mylar film and does not suffer the periodic zeros of response caused by multiple-beam interference in the beam splitter.   The transmissivity is flat up to a high cutoff frequency that is inversely proportional to the grid spacing.

Figure 3 shows the polarizing interferometric configuration used at the

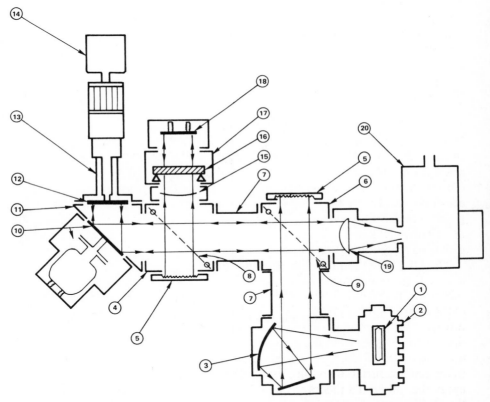

FIG. 3   Modular polarization interferometer for dispersive Fourier transform spectroscopy for millimeter-wave measurements of complex refractive index and complex dielectric permittivity of low-loss solids: (1) mercury lamp, (2) water-cooled lamp housing, (3) two-mirror collimator, (4) central cube, (5) radiation dump, (6) cube 2, (7) connectors (spacers), (8) wire grid beam splitter, (9) wire grid polarizer/analyzer, (10) 45° mirror and phase-modulation assembly, (11) half cube, (12) scanning mirror, (13) micrometer, (14) stepping motor, (15) mylar window, (16) solid specimen, (17) specimen holder, (18) adjustable fixed mirror, (19) focusing lens, (20) InSb Rollin detector.

MIT National Magnet Laboratory for millimeter-wave dielectric measurement of solids. Michelson interferometers have been used extensively in the submillimeter range, but this instrument had to be modified extensively to be useful in the millimeter range. The modification increased the energy throughput from the feeble mercury-vapor lamp and increased the signal-to-noise ratio (Afsar, 1982; Afsar and Button, 1983). For example, the phase modulation was provided by vibrating one of the mirrors in an active arm of the interferometer. The same mirror also bends the beam 90° toward the scanning mirror, mounted on the micrometer, that is coupled to a stepping motor. A pair of free-standing wire-grid polarizers made from 10-μm-diameter tungsten wire with wire spacing of 25 μm (center to center) was used. One of the grids acts as a polarizer and analyzer and the other as a beam splitter. The high wave-number cut off with 25-μm spacing is 200 per cm (50-μm wavelength). The orientation of wires of the beam-splitter grid is at 45° compared with the polarizer–analyzer grid, so that the perpendicular component of the polarized beam transmits to the scanning mirror arm and the parallel component reflects to the specimen arm. Beams recombine at the beam-splitter grid and get analyzed by the polarizer–analyzer grid before reaching the detector. A liquid-helium-cooled hot-electron-effect InSb Rollin-type (Kinch and Rollin, 1963) bolometer is found to be most suitable in the millimeter-wave region. It operates on the principle of free-carrier absorption. The working spectral range in wave number is from 1 to 40 per cm (10 to 0.25 mm). Liquid-helium-cooled germanium bolometers are also well known for their use in the infrared (10 μm) and far-infrared region. The recently developed composite type of Ge bolometer (Johnson *et al.*, 1979) works in the 2-mm region, but its response time is about 2 to 3 orders in magnitude less than a Rollin-type detector.

The DFTS method requires a highly stable reproducible interferometric system to generate precise phase information. The interferometer temperature is controlled by a fluid from a temperature-control bath. The main part of the interferometer is evacuated; the specimen chamber is flushed with dry nitrogen gas during measurement. The specimen rests on three balls. Specimen surfaces were flattened to one-quarter wavelength in the visible and made parallel to about one arc second of a degree. It is necessary to have the specimen surfaces and the adjustable fixed mirror optically parallel to apply the full DFTS theory for surface and interface reflection-loss corrections. The measurement of a low-loss material is equally difficult as the very absorbing–near opaque material (Afsar, 1979a; Afsar and Hasted, 1977, 1978; Birch *et al.*, 1976; Mead and Genzel, 1978; Parker *et al.*, 1976; Staal and Eldridge, 1977). The large specimen thickness has to be chosen so as to make the transmissive loss greater than the reflective loss (Afsar, 1982; Afsar and Button, 1983; Afsar and Chantry, 1977). It is also necessary to employ stable electronic components. A step change in the attenuator, or the gain knob in a lock-in amplifier, and the inclusion of the marked gain factor in the calculation can lead to a systematic change in absorption data. It is therefore recommended that the gain be kept unchanged during sets of measurements and the interference signal be resolved by higher digital discrimination.

For the measurement of a liquid, a slightly different configuration is employed, as if the interferometer in Fig. 3 were turned 180° vertically upward or downward. The source and collimation assembly are above the cube, and the specimen chamber is beneath the spectrometer. In this case the liquid rests directly on the fixed mirror (made from stainless steel) to form a plane-parallel, gravity-held layer.

B.   MACH – ZEHNDER – IMPATT SPECTROMETER

Because the polarizing dispersive Fourier transform spectrometer (DFTS) uses a blackbody continuum source of radiation, there are cases near microwave frequencies where the throughput of energy is too weak for any meaningful use. This occurs when one tries to measure highly absorbing materials at low frequencies. In that case one can expect equally high accuracy by using the Mach – Zehnder interferometer (MZI) with a high-power monochromatic source of radiation (Afsar, 1979b; Afsar *et al.*, 1976a, 1977). This method provides data only at a single frequency and at harmonic frequencies whenever these can be generated.

Until recently, MZI was not used in the *millimeter-wave* region until IMPATT and Gunn sources became available (Birch, 1980, 1981), although any stable source could have been used. Considerable past successes have been achieved with the MZI in the *submillimeter* range (Afsar, 1979b; Afsar *et al.*, 1976a, 1977), where molecular gas lasers provide an excellent stable, high-power source tunable to hundreds of different frequencies.

The MZI is compatible with the use of a monochromatic source because no beam returns to the source as happens, for example, in a Michelson interferometer. At low frequencies a ferrite isolator could protect the IMPATT, but it is well known that isolators do not perform well beyond 100 GHz.

Measurements can be made up to 400 GHz by using IMPATT harmonics. For example, Ino *et al.* (1976) and Ishibashi and Ohmori (1976) demonstrated high, in-waveguide, cw power levels at 200 and 300 GHz in the harmonics of IMPATT diode oscillators. An in-waveguide power level of about 2 mW was also achieved at 400 GHz (Ishibashi *et al.*, 1977). Birch (1980, 1981) measured harmonic contents at 200 and 300 GHz in the output of cw, 20-mW, and 100-GHz IMPATT oscillators by free-space methods. The oscillator employed by Birch was a silicon, double-drift,

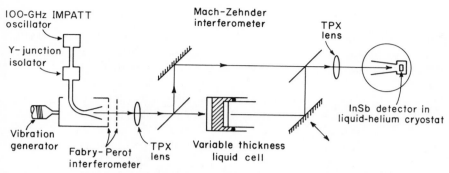

FIG. 4 Mach–Zehnder–IMPATT spectrometer for complex refractive index and complex dielectric measurements of liquids. The wire-mesh Fabry–Perot interferometer selects particular harmonics or the fundamental to be used for the measurement. The thickness of the liquid can be varied continuously or in steps by moving one of the windows of the liquid cell (Birch, 1981).

IMPATT diode in a high-Q waveguide cavity. Two interferometer systems were used to study harmonic content of the oscillator. Figure 4 illustrates the complete experimental arrangement for the liquid-dielectric measurement.† The radiation from the oscillator was fed to a wire-mesh Fabry–Perot interferometer. One mesh was moved (scanned) to change the spacing of the Fabry–Perot reflectors. The other was vibrated for phase modulation. Together, this selects or rejects different harmonics and the fundamental. Maximum power levels of 0.5 mW at 200 GHz and 4.5 μW at 300 GHz were measured, representing about 20-dB reduction per harmonic.

Figure 4 also shows the inclusion of a variable-thickness liquid cell in one arm of the Mach–Zehnder interferometer. The distance between windows in the cell could be varied from 0 to 20 mm by a stepping-motor-controlled micrometer drive to vary the liquid thickness.

The measurements of the absorption coefficient and the refractive index of liquid specimens were made by setting the Fabry–Perot interferometer to transmit the particular harmonic of interest and recording the detected interference pattern as the liquid thickness was decreased toward zero. The interferogram has the form of a damped cosine wave, and the optical parameters, the refractive index, and the absorption coefficient were determined from the period of the wave and its attenuation.

---

† It is essential when making dielectric measurements in this system to take data by varying the thicknesses of specimens. The simplest case is the liquid, because the thickness can easily be varied in steps or continuously.

C. Open-Resonator Method

Resonator methods make use of the change in Q and change in resonant frequency to provide ϵ'' and ϵ'. As we approach millimeter wavelengths, the *open resonator* method (Cook and Jones, 1976; Cullen and Yu, 1971, 1979; Culshaw and Anderson, 1962; Jones, 1976a,b) competes with the classical microwave fundamental-mode, resonant closed cavity. Obviously, at some short-millimeter wavelength, the microwave resonant closed cavity becomes too small. As we approach shorter-millimeter wavelengths, it first becomes inconvenient and eventually impractical to prepare specimens for the resonant closed cavity.

We shall now employ plausible arguments to show that the open resonator, particularly the *hemispherical* type, solves our problem at a few millimeter wavelengths. An open-resonator system uses two metal reflectors to form a resonant structure similar to that of a Fabry–Perot "etalon." One type of open resonator that is easy to understand is the *confocal* type. It provides high Q (small diffraction loss), because it has two large concave mirrors facing each other as shown in Fig. 5. The specimen must be placed in the center. At wavelengths longer than 1 mm (<300 GHz) the electromagnetic energy can be coupled in and out of the resonator by waveguides. A small coupling hole in each mirror is used to transmit the energy to and from the input and output waveguides. It is important to have the coupling hole at the center of each of the mirrors because an off-center hole is one of several ways whereby higher modes of oscillation of the cavity can be introduced (Dees and Sheppard, 1965). Theoretical analysis of the data is dependent on the fundamental mode; higher-mode degeneracy must be avoided (Cullen and Yu, 1971; Jones, 1976).

A convenient form of resonant structure is the *hemispherical* type (Jones, 1976) that employs one concave and one plane mirror, as shown in Fig. 6. The specimen rests on the plane mirror. Our problem is solved. The specimen can be a liquid or a flat solid of any size. The input and output

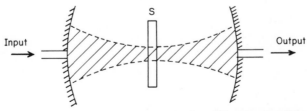

FIG. 5   The confocal Fabry–Perot open resonator for millimeter-wave dielectric measurement: S represents the specimen position. The shaded areas show the extent of the millimeter-wave beam in the resonator.

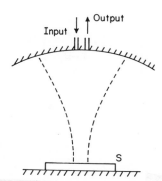

FIG. 6 The hemispherical type of open resonator for millimeter-wave dielectric measurement. The specimen in this case rests on the flat mirror of the resonator.

waveguides are attached at the back of the concave mirror. At frequencies greater than 100 GHz, the waveguides become more difficult to use. Certainly, beyond 300 GHz the energy must be focused by a lens through a small hole in a mirror (Batt and Luk, 1980), although sometimes a beam-splitter coupling (French and Arnold, 1967) can be used.

The theory of an open-resonator system for two different configurations has been developed by Cullen and Yu (1971, 1979), Jones (1976b), and Yu and Cullen (1982). The modern theory uses a Gaussian beam formulation rather than a plane-wave formulation (Cullen and Yu, 1971, 1979). Recently, Yu and Cullen (1982) used a vector field theory of the open resonator based on an exact solution of Maxwell's equations. They derived a simplified approximate formula for the electric and magnetic field vectors. A variational formula of the resonant frequency of an open resonator was then developed with the use of these simplified formulas. Lynch (1982) used the variational formula and measured the refractive index to within 0.5% (permittivity to 1%).

Jones (1976a,b) adapted the Gaussian beam formula for use with a *hemispherical* resonator where the specimen is placed on the plane mirror. The specimen can easily be positioned on the plane mirror perpendicular to the axis of the beam, and the specimen becomes automatically aligned in a hemispherical configuration. Moreover, if one wishes to introduce a liquid specimen, only one window is required above the plane mirror to form a plane-parallel layer. In a confocal configuration, the solid specimen must be located at the center and carefully aligned to be perpendicular to the axis (Cullen *et al.,* 1972).

Open resonators are very convenient for the measurement of low-loss materials because Q values can be made very much higher than a resonant closed cavity at these frequencies. The typical Q of the 13-cm-long, 15-cm-diameter, hemispherical resonator of Jones (1976a,b) and Clarke and Rosenberg (1982) at 144 GHz is 260,000. Cullen and Yu (1971, 1979), Jones

TABLE I

MILLIMETER-WAVE DATA OBTAINED WITH OPEN-RESONATOR SYSTEMS

Material	λ (mm)	Frequency (GHz)	n	ε'	tan δ (μrad)	References
Alumina	8.834	33.906	3.119 ± 0.015			Lynch (1982)
Polyethylene (high density)	0.87	344.83		2.31 ± 0.05		Degenford and Coleman (1966)
	2.1	142.86		2.31 ± 0.05		Degenford and Coleman (1966)
Polyethylebe (Rigidex 2000)	8.57	35	1.5218 ± 0.0015	2.316 ± 0.004	134 ± 7	Afsar and Jones (1978)
	8.6	34.88	1.5360 ± 0.0015	2.3593 ± 0.042	173 ± 9.9	Cook et al. (1974)
Polypropylene	8.718	34.38	1.5858 ± 0.0004			Lynch (1982)
	8.862	34.80	1.5012 ± 0.0004			Lynch (1982)
	8.57	35	1.5014 ± 0.002	2.254 ± 0.004	154 ± 8	Afsar and Jones (1978)
	0.87	344.83		2.07 ± 0.04		Lynch (1982)
	2.1	142.86		2.07 ± 0.04		Lynch (1982)
Teflon (sintered)	6	50	1.433 ± 0.007	2.052 ± 0.020	315 ± 26	Culshaw and Anderson (1962)
	8.6	34.88			250 ± 30	Degenford (1968)
	8.6	34.88		1.952 ± 0.007	48 ± 3.9	Cook et al. (1974)
Teflon (unsintered)	8.677	34.57	1.4589 ± 0.0013		427 ± 21	Lynch (1982)
TPX	8.508	35.26	1.458 ± 0.002	2.126 ± 0.004		Afsar and Jones (1978)
Polystyrene	0.87	344.83		2.57 ± 0.05		Degenford and Coleman (1976)
	2.1	142.86		2.56 ± 0.05		Degenford and Coleman (1976)
	6	50	1.590 ± 0.008	2.528 ± 0.025	721 ± 58	Culshaw and Anderson (1962)
	8.6	34.88			520 ± 60	Degenford (1968)
Plexiglas	0.87	344.83		2.61 ± 0.05		Degenford and Coleman (1976)
	2.1	142.86		2.60 ± 0.05		Degenford and Coleman (1976)
	6	50	1.599 ± 0.008	2.557 ± 0.026	3270 ± 262	Culshaw and Anderson (1962)
Crystal Quartz (ordinary ray)	8.57	35		4.430 ± 0.004		Jones (1976a)
(extraordinary ray)	8.57	35		4.633 ± 0.004		Jones (1976a)
Slip-cast fused silica	3.18	94.34	1.812 ± 0.003	3.285 ± 0.007	2640 ± 142	Breeden and Langley (1969)
Liquid cyclohexane	8.436	35.56		2.014 ± 0.003	79 ± 4	Cook and Jones (1978)
	4.167	72		2.018 ± 0.002	154 ± 8	Cook and Jones (1978)
	2.083	144		2.016 ± 0.004	294 ± 8	Cook and Jones (1978)

(1976a,b), and Clarke and Rosenberg (1982) have made measurements on solids and liquids at various frequencies in the range 10–150 GHz with an accuracy ranging from ± 0.2 to 1% for ϵ' and 2 to 10% for tan δ. These workers have employed phase-locked klystrons and frequency-multiplier combinations as frequency sources at 35, 72, and 144 GHz. Jones (1976a) had also measured anisotropic materials by exploiting the fact that the wave-guide feed provides a linearly polarized wave. Gases were earlier measured by French and Arnold (1967) (nitric oxide at 150 GHz), by Valkenberg and Derr (1966) (water vapor at several frequencies between 100 and 300 GHz), by Batt and Harries (1976), and by Batt and Luk (1980) (at 890 GHz). Batt and Harries (1976) and Batt and Luk (1980) have employed an HCN laser (890 GHz) together with a folded version of the confocal resonator by intro-ducing a third plane mirror in an intermediate position between two concave mirrors. Recently, Kolbe and Leskovar (1982, 1983) used a Fabry–Perot resonator with a Gunn oscillator and superheterodyne detection for the measurement of SiO_2 gas diluted in air at 70 and 140 GHz. Although the cavity Q was somewhat low (106,000 at 140 GHz), the use of the newer detection device permitted faster response time and greater sensitivity to be achieved (Kolbe and Leskovar, 1983).

At the MIT National Magnet Laboratory, a 60-GHz, hemispherical resonator–spectrometer system has been constructed (Matsui et al., 1984). This includes a high-Q resonator and two highly stable, phase-locked Gunn oscillators (a tunable-signal-frequency oscillator of 60 GHz, and a 56-GHz fixed-frequency local oscillator). Both oscillators together with a ring filter and an IF amplifier make the extremely low-noise-sensitive superhetero-dyne receiver. This could increase the accuracy of tan δ measurements by as much as two orders of magnitude. Table I compares some results on low-loss solids and liquids.

V. Discussion

When reliable and reproducible millimeter-wave data become available for a sufficiently large variety of materials, as in the case of submillimeter-wave data (Simonis, 1982), the origin of the losses can be discussed. In the meantime, we have been able to develop a few clues that, nevertheless, should be helpful in engineering design, because all of these materials (GaAs, Si, Al_2O_3, SiO_2, BeO, ZnSe, ZnS, spinel) are widely used in modern elec-tronic systems. For example, in some materials, high purity is not as impor-tant as the manufacturer's method of preparation. We also know that, for the semiconductors, the highest resistivity is needed to prevent free-carrier absorption and electron-plasma reflection.

A. Absorption Effects

The good news is that *all* of these are low-loss materials. The bad news is that the absorption coefficient varies among nominally identical specimens, depending on different manufacturer's methods of preparation. The fundamental parameter α, the absorption coefficient, is the reliable figure of merit for comparison of materials. Trying to use ϵ'' or tan δ as a comparative measure of dielectric losses may be misleading, because ϵ'' and tan δ are frequency dependent (see Section II).

It is surprising that such a common material as SiO_2 (fused silica glass) should be an excellent example of this problem. Figure 7 (Appendix) shows that the absorption coefficient profile of Corning U.V.-grade glass is nearly the same as that of SiO_2 deliberately contaminated with a heavy ion, 7% TiO_2. On the other hand, the lowest-loss material on the *entire list* is water-free SiO_2 from Thermal American Fused-Quartz Company. Before we all begin to think that some reliable rules have been developed, such as the insensitivity of the absorption to heavy ion contamination, we should note that MACOR, the Corning machinable ceramic, has ten times higher loss than any material in our list.

Hot-pressed ceramics have much lower losses than cold-pressed ceramics of the same chemical composition. For example, hot-pressed BeO containing $\frac{1}{4}$% lithia flux, made by Union Carbide, exhibits 42% less absorption loss at 300 GHz than the cold-pressed Ceradyne Cerraloy 418 S 99.5 beryllia.

Somewhat the same effect must occur in the case of ceramic Al_2O_3, because we were surprised again when AL 995 had *lower* losses than AL 999 despite the presumption that the latter has higher purity. Crystal sapphire (Al_2O_3) exhibited only half the absorption loss of any ceramic alumina, as expected, but was not better than Thermal American water-free fused silica.

The extrapolated microwave data that we have been depending on in the past for guidance to low-loss millimeter-wave materials was the most misleading in the case of magnesium–aluminum spinel. Having been led to expect that hot-pressed $MgAl_2O_4$ spinel would be an order of magnitude better than alumina, we found only a 17% advantage at 150 GHz.

The semiconductors such as GaAs and silicon are very special cases because of millimeter-wave free-carrier absorption and electron-plasma reflection. The highest possible room-temperature resistivity must be specified to assure the smallest number of electrons in the plasma. High resistivity, quoted in ohm-centimeters, is different for semiconductors having different electron effective masses and different dielectric constant. For example, a single crystal of GaAs having a low resistivity of the order of magnitude of $10^4 \ \Omega \cdot cm$ was opaque, but excellent low-loss measurements are shown in

Appendix G for $10^8 \ \Omega \cdot$ cm, high-purity single crystals. In the case of silicon, a high resistivity of $10^4 \ \Omega \cdot$ cm was satisfactory whereas $10^2 \ \Omega \cdot$ cm was a high-loss material.

In some cases the parameter tan $\delta = \epsilon''/\epsilon'$ can be misleading in milli-meter-wave technology. The measurements on low-loss liquids provide us with a convenient illustration of this, although the same phenomenon is present in solids to a lesser extent. Whereas the absorption coefficient increases with frequency, tan δ decreases for low-loss liquids such as fluoro-carbons and cyclohexane. Increasing absorption is fundamental to all liq-uids, polar and nonpolar, because we are on the tail of a broad submillimeter absorption band. Why, then, should tan δ *decrease* with frequency, giving us exactly the opposite impression of the losses? The trouble arises in $\epsilon'' = \alpha n/2\pi\tilde{\nu}$ when the product αn fails to increase with frequency as rapidly as $\tilde{\nu}$. Then the slope of ϵ'' is negative where the slope of α is positive.

As standard dielectric reference materials (Afsar *et al.*, 1980), the liquids surpass all of the solids because of the control over manufacturing processes and microqualitative chemical analysis. We have selected two groups of electronic coolant fluids to be included in the appendix. The Dow Corning dimethyl siloxanes have high cooling capacities but also higher dielectric loss than the 3M fluorocarbons. The fluorocarbons are available in several chemical compositions and generally exhibit dielectric loss as low as we have seen in the best solids.

B.   DISPERSION EFFECTS

Normally, the magnitude of the refractive index $n \simeq (\epsilon')^{1/2}$ is limited in accuracy and reproducibility to three, sometimes four, significant figures, as shown in Table I. This dielectric constant ϵ' is indeed nearly constant as a function of frequency and is adequate for all engineering applications. Nevertheless, serious problems arise when one must assess the engineering consequences of differences among siblings in a batch of material: different *sources* of the same material, different *methods* of preparation, different *aging* processes such as neutron irradiation or high-power electromagnetic radiation, and environmental changes in properties caused by assimilation of water vapor or chemical pollutants. These problems can be solved only by using the dispersive Fourier transform spectrometric method (Section IV), which uniquely provides the refractive index to six significant figures at millimeter wavelengths in low-loss materials. The refraction spectra (see Appendix A) show massive features as a function of frequency in the fifth significant figure and fine structure in the sixth figure. The differences can be determined simply by inspection.

Appendix: Compendium of Data

A. FUSED-SILICA GLASS (SiO_2)

Corning U.V.-grade fused silica #7940 and Corning titanium silicate #7971 were obtained from Dr. N. Borelli and Dr. J. W. Malmendier of Corning Glass Works, Corning, New York. Data are shown in Figs. 7–13, pp. 24–26.

The Corning 7940 U.V.-grade silica specimen is 50 mm in diameter and 17.068 mm thick. It is pure fused silica, SiO_2. The Corning 7971 titanium silicate specimen is 75 mm in diameter and 14.871 mm thick and is SiO_2 93%, TiO_2 7% wt.

The water-free fused silica "spectrasil WF" specimen was obtained from M. H. Robinson of Thermal American Fused Quartz Company, Montville, New Jersey. Its diameter and thickness are 75 and 21.6476 mm, respectively. The water content in this material is only five parts to a million compared with 1200 parts per million (0.18% wt) for non-water-free "spectrasil" fused silica.

B. ALUMINA (Al_2O_3)

This section compares two ceramic-alumina specimens obtained from different manufacturers with sapphire (single-crystal Al_2O_3) and spinel $MgAl_2O_4$) specimens. Data are shown in Figs. 14–20, pp. 26–28.

The alumina 995 specimen was obtained from W. S. Smith, WESGO Division of GTE Products Corp., Belmont, California. It was cold-pressed and sintered using a flux containing less than 0.5% $CaOMgSiO_2$. The specimen diameter and thickness are 65 and 49.942 mm, respectively. The alumina 999 was obtained from Jack Sibold of COORS Porcelain Co., Colorado. It was cold-pressed and sintered using a flux containing less than 0.1% MgO. The specimen diameter and thickness are 75 and 27.1248 mm, respectively.

The single-crystal Z-cut sapphire was obtained from Chandra Kattak and Fred Schmid of Crystal Systems Inc., Massachusetts. Its diameter and thickness are 65 and 31.7666 mm, respectively.

The hot-pressed $MgAl_2O_4$ spinel specimen of "COORS" Porcelain Co., Colorado, was obtained from Naresh Deo of Epsilon Lambda Electronics Corporation, Illinois. The specimen thickness and diameter are 10.1044 82 mm, respectively.

C. BERYLLIA (BeO)

The 99.5 beryllia specimen is a Ceralloy 418 S of Ceradyne, Inc., Santa Ana, California. Its diameter and thickness are 69.35 and 35.7788 mm, respectively. It was cold pressed and sintered using about 0.5% magnesium trisilicate flux. The density was about 2.88 g/cm^3.

The second BeO specimen was cut from a large disk that was hot-pressed at the Nuclear Division of the Union Carbide Corporation, Oak Ridge National Laboratory, using a flux containing 0.25% lithia. The thickness and diameter of the specimen are 10.509 and 75 mm, respectively. Both specimens were obtained through C. Marshall Loring at the Oak Ridge National Laboratory, Tennessee.

Data are shown in Figs. 21–25, pp. 28–30.

D. CORNING MACOR

Corning Macor™ Machinable Glass Ceramic is a product of Corning Glass Works, Corning, New York. The specimen was obtained from Dr. J. W. Malmendier and Dr. N. Borelli. Technical bulletins and information on the distributor–fabricator network can also be obtained through Corning France BP61, 77 Avon, France. Macor requires only ordinary metal-working tools for fabrication to any size and shape.

Data are shown in Figs. 26–28, pp. 30–31. Although the absorption losses are one order higher than low-loss fused-silica glass, Macor may now be used at millimeter wavelengths where precision machined elements are needed. The thickness and the diameter of the specimen are 18.6866 and 75 mm, respectively.

E. CORNING 9616 GREEN GLASS

A Corning 9616 green-glass specimen was obtained from Dr. N. Borelli and Dr. J. W. Malmendier of Corning Glass Works, Corning, New York. It is a lithium alumino–silicate green glass suitable for glass ceramic. Its absorption loss is one order higher than fused-silica glass. The specimen thickness and diameter are 13.6522 and 75 mm, respectively.

Data are shown in Figs. 29–31, pp. 31–32.

F. SILICON

Single-crystal, high-purity silicon specimens are now easily available from various manufacturers. For millimeter-wave application, it is important that the resistivity is higher than 1000 $\Omega \cdot$cm. Our specimen is an undoped monocrystal and was purchased from General Diode Corporation, Massachusetts. Its resistivity, thickness, and diameter are about 8000 $\Omega \cdot$cm, 11.8125 mm, and 75 mm, respectively.

Data are shown in Figs. 32–34, pp. 32–33.

G. GALLIUM ARSENIDE (GaAs)

Until recently, it was nearly impossible to obtain pure single-crystal high-resistivity gallium arsenide. We are reporting results on two specimens, one from each batch obtained from Marvin Klein of Hughes Research Lab., Malibu, California, and Robert Linares of MA/Com, Burlington, Massachusetts.

The Hughes specimen D3 is a Cr-doped single-crystal GaAs with Cr-concentration of $2 \times 10^{16}/cm^3$. Its resistivity is of the order of $7.8 \times 10^7 \Omega \cdot cm$. The specimen thickness and diameter are 14.5456 and 75 mm, respectively.

The MA/Com specimen 1089 is also a Cr-doped single-crystal GaAs with Cr-concentration of $5 \times 10^{15}/cm^3$. Its resistivity is greater than $5 \times 10^7 \Omega \cdot cm$, and the room-temperature mobility is 2500 cm^2/V·sec. The specimen thickness and diameter are 54.5154 and 75 mm, respectively.

Data are shown in Figs. 35–39, pp. 33–34.

H. ZINC SULPHIDE (ZnS)

Zinc sulphide is widely used in the infrared region as low-loss window material. Our zinc sulphide specimen was obtained from Dr. Theodore Kohane of Raytheon Research Division, 131 Spring Street, Lexington, Massachusetts. It was made by the chemical-vapor deposition process. The specimen thickness and diameter are 25.9943 and 75 mm, respectively.

Data are shown in Figs. 40–42, p. 35.

I. ZINC SELENIDE (ZnSe)

Like ZnS, zinc selenide is also used extensively in the infrared as low-loss-window material for applications such as the CO_2 laser. Our specimen was obtained from Dr. Theodore Kohane of Raytheon Research Division, 131 Spring Street, Lexington, Massachusetts. It was made by the chemical-vapor deposition process. The specimen thickness and diameter are 20.0628 and 75 mm, respectively.

Data are shown in Figs. 43–45, p. 36.

J. 3M FLUOROCARBON ELECTRONIC COOLING FLUIDS

Fluorinert™ electronic liquids are available from the Commercial Chemicals Division of the 3M Corporation, 3M Center, St. Paul, Minnesota, in viscosities from 0.4 to 73 cS (at 25°C) and has a dielectric strength on the order of 40 kV (0.1-inch gap). We obtained three types of Fluorinert fluids, the FC-75, FC-104, and FC-43, through C. Marshall Loring of Union Carbide Corporation, Oak Ridge National Laboratory. These coolant fluids are members of a family of completely fluorinated organic compounds that have a unique combination of properties. They are derived from common organic compounds by replacement of all carbon-bound hydrogen atoms with fluorine atoms.

These liquids are nonpolar and have low solvent action. They are colorless, odorless, low in toxicity, and nonflammable. They also have high thermal stability, low chemical reactivity, and leave essentially no residue. All Fluorinert-brand electronic liquids are a series of very efficient dielectric coolants. The cooling is effective in free or forced convection. Cooling by boiling is even more effective. FC-43 is a tertiary amine with a boiling point of 174°C; it exhibits very low absorption loss. FC-75 is a C_8 cyclic ether mixed with C_8 alkayne with a boiling point of 102°C. FC-104 is a C_8 alkane with a boiling point of 101°C.

Technical bulletins and information on these liquids can be obtained from the Commercial Chemical Division of the 3M Corporation. In the measurement, a liquid-thickness-difference combination of about 40 mm was chosen in each case. Data are shown in Figs. 46–52, pp. 37–39.

K. DOW CORNING DIMETHYL POLYSILOXANE

Dow Corning Corporation, Midland, Michigan, produces a wide variety of dimethyl, phenylmethyl, and trifluoropropyl silicone fluids for dielectric cooling, damping, impregnating, lubricating, coating, heat transfer, releasing, and defoaming. The Dow Corning 200 (DC 200) dimethyl polysiloxane fluid is available in viscosities from 0.65 to 60,000 cS (at 25°C) and has a dielectric strength greater than 30 V/mil. Three viscosities were chosen, 1, 5, and 50 cS. The 5-cS fluid exhibits twice the absorption loss of the 1-cS fluid, but 50-cS fluid shows only a 15% increase in absorption coefficient over the 5-cS fluid.

These fluids cannot be considered low-loss material. At 1 mm, the absorption coefficient of lowest-loss dimethyl-siloxane 1-cS fluid is about 30 times higher than 3M FC-43 Fluorinert fluid. These fluids were also obtained through C. Marshall Loring of Oak Ridge National Laboratory, Tennessee.

Data are shown in Figs. 53–57, pp. 39–40.

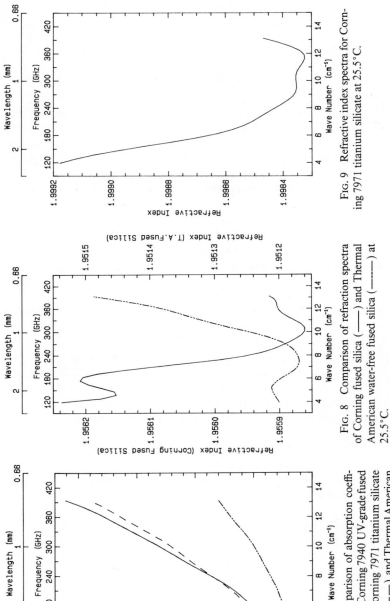

FIG. 9 Refractive index spectra for Corning 7971 titanium silicate at 25.5°C.

FIG. 8 Comparison of refraction spectra of Corning fused silica (———) and Thermal American water-free fused silica (———) at 25.5°C.

FIG. 7 Comparison of absorption coefficient spectra of Corning 7940 UV-grade fused silica (———), Corning 7971 titanium silicate (7% wt TiO$_2$)(— —), and Thermal American water-free fused silica Spectrasil WF (———) at 25.5°C.

24

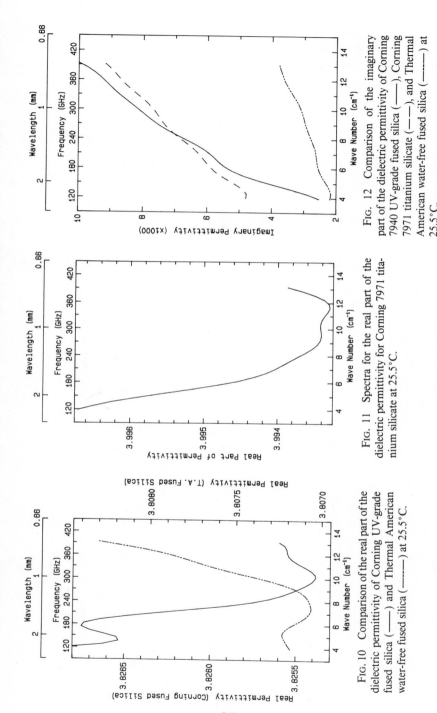

FIG. 10 Comparison of the real part of the dielectric permittivity of Corning UV-grade fused silica (———) and Thermal American water-free fused silica (———) at 25.5°C.

FIG. 11 Spectra for the real part of the dielectric permittivity for Corning 7971 titanium silicate at 25.5°C.

FIG. 12 Comparison of the imaginary part of the dielectric permittivity of Corning 7940 UV-grade fused silica (———), Corning 7971 titanium silicate (———), and Thermal American water-free fused silica (———) at 25.5°C.

25

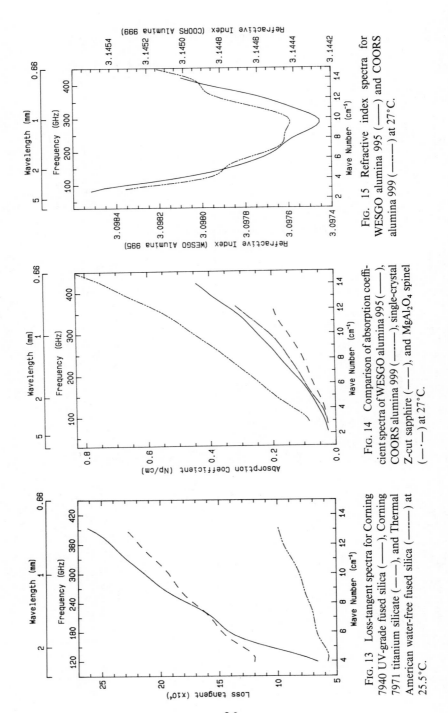

Fig. 15 Refractive index spectra for WESGO alumina 995 (——) and COORS alumina 999 (——·—) at 27°C.

Fig. 14 Comparison of absorption coefficient spectra of WESGO alumina 995 (——), COORS alumina 999 (——·—), single-crystal Z-cut sapphire (———), and MgAl₂O₄ spinel (——·—) at 27°C.

Fig. 13 Loss-tangent spectra for Corning 7940 UV-grade fused silica (——), Corning 7971 titanium silicate (———), and Thermal American water-free fused silica (——·—) at 25.5°C.

26

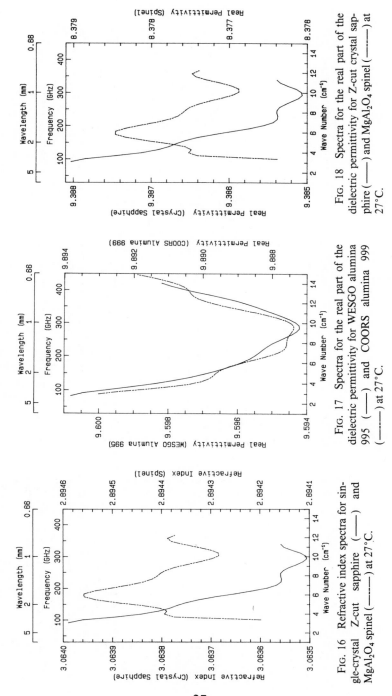

Fig. 16 Refractive index spectra for single-crystal Z-cut sapphire (———) and MgAl₂O₄ spinel (———) at 27°C.

Fig. 17 Spectra for the real part of the dielectric permittivity for WESGO alumina 995 (———) and COORS alumina 999 (———) at 27°C.

Fig. 18 Spectra for the real part of the dielectric permittivity for Z-cut crystal sapphire (———) and MgAl₂O₄ spinel (———) at 27°C.

27

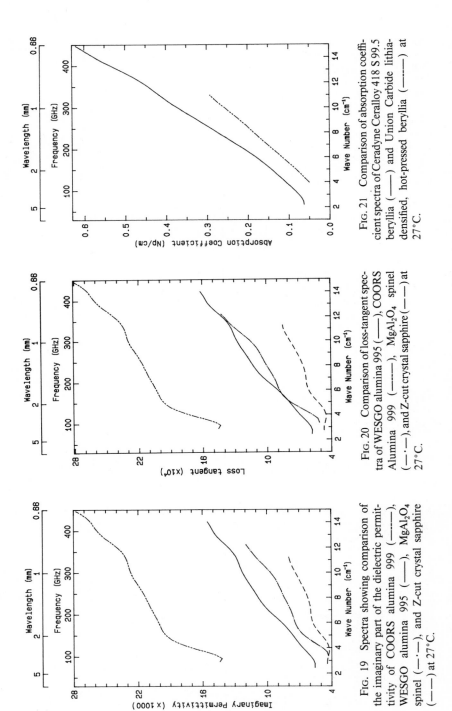

FIG. 19 Spectra showing comparison of the imaginary part of the dielectric permittivity of COORS alumina 999 (———), WESGO alumina 995 (———), MgAl₂O₄ spinel (—·—), and Z-cut crystal sapphire (———) at 27°C.

FIG. 20 Comparison of loss-tangent spectra of WESGO alumina 995 (———), COORS Alumina 999 (———), MgAl₂O₄ spinel (—·—), and Z-cut crystal sapphire (———) at 27°C.

FIG. 21 Comparison of absorption coefficient spectra of Ceradyne Ceralloy 418 S 99.5 beryllia (———) and Union Carbide lithia-densified, hot-pressed beryllia (———) at 27°C.

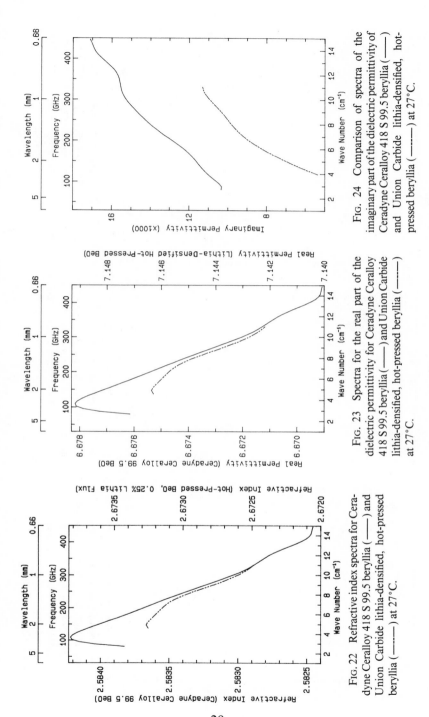

FIG. 24 Comparison of spectra of the imaginary part of the dielectric permittivity of Ceradyne Ceralloy 418 S 99.5 beryllia (——) and Union Carbide lithia-densified, hot-pressed beryllia (——) at 27°C.

FIG. 23 Spectra for the real part of the dielectric permittivity for Ceradyne Ceralloy 418 S 99.5 beryllia (——) and Union Carbide lithia-densified, hot-pressed beryllia (——) at 27°C.

FIG. 22 Refractive index spectra for Ceradyne Ceralloy 418 S 99.5 beryllia (——) and Union Carbide lithia-densified, hot-pressed beryllia (——) at 27°C.

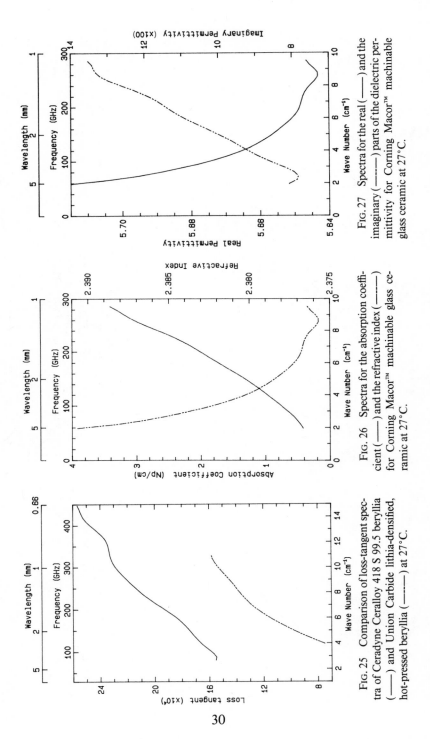

FIG. 25 Comparison of loss-tangent spectra of Ceradyne Ceralloy 418 S 99.5 beryllia (———) and Union Carbide lithia-densified, hot-pressed beryllia (———) at 27°C.

FIG. 26 Spectra for the absorption coefficient (———) and the refractive index (———) for Corning Macor™ machinable glass ceramic at 27°C.

FIG. 27 Spectra for the real (———) and the imaginary (———) parts of the dielectric permittivity for Corning Macor™ machinable glass ceramic at 27°C.

30

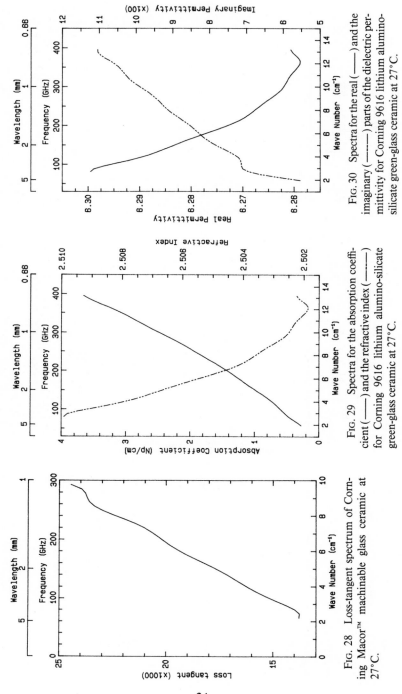

FIG. 30 Spectra for the real (——) and the imaginary (——) parts of the dielectric permittivity for Corning 9616 lithium alumino-silicate green-glass ceramic at 27°C.

FIG. 29 Spectra for the absorption coefficient (——) and the refractive index (——) for Corning 9616 lithium alumino-silicate green-glass ceramic at 27°C.

FIG. 28 Loss-tangent spectrum of Corning Macor™ machinable glass ceramic at 27°C.

31

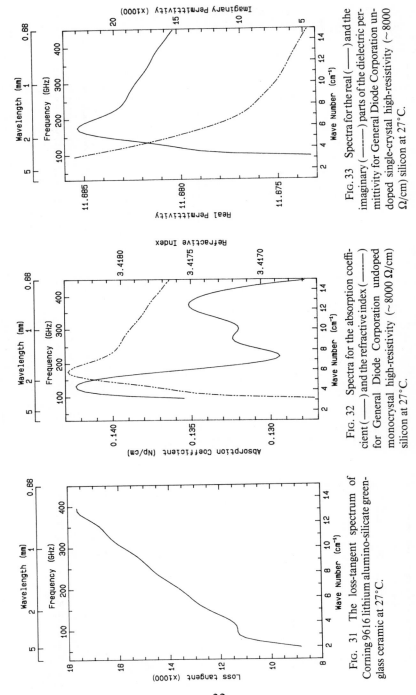

FIG. 33 Spectra for the real (——) and the imaginary (—·—·) parts of the dielectric permittivity for General Diode Corporation undoped single-crystal high-resistivity (~8000 Ω/cm) silicon at 27°C.

FIG. 32 Spectra for the absorption coefficient (——) and the refractive index (——) for General Diode Corporation undoped monocrystal high-resistivity (~8000 Ω/cm) silicon at 27°C.

FIG. 31 The loss-tangent spectrum of Corning 9616 lithium alumino-silicate green-glass ceramic at 27°C.

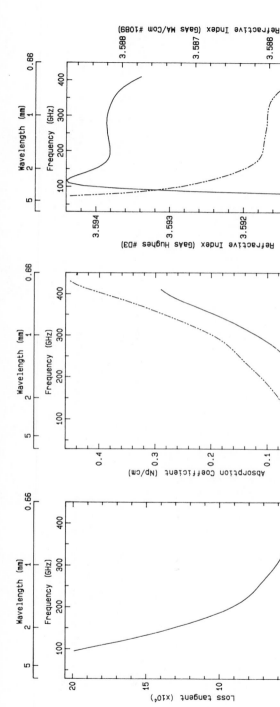

Fig. 36 Spectra for the refractive index for Hughes D3 (——) and MA/Com 1089 (——·——) Cr-doped single-crystal high-resistivity ($\rho > 5 \times 10^7$ Ω/cm) gallium arsenide specimens at 27°C.

Fig. 35 Absorption coefficient spectra for Hughes D3 (——) and MA/Com 1089 (——·——) Cr-doped single-crystal high-resistivity ($\rho > 5 \times 10^7$ Ω/cm) gallium arsenide specimens at 27°C. The increase in absorption with frequency is caused by a multiphonon absorption peak centering around 0.5 mm (600 GHz) (After Stolen, 1969).

Fig. 34 The loss-tangent spectrum for General Diode Corporation undoped monocrystal high-resistivity (~8000 Ω/cm) silicon at 27°C.

33

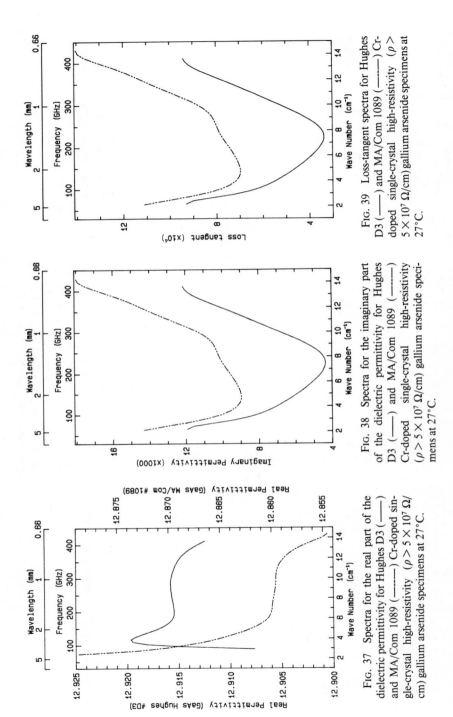

FIG. 37 Spectra for the real part of the dielectric permittivity for Hughes D3 (——) and MA/Com 1089 (———) Cr-doped single-crystal high-resistivity ($\rho > 5 \times 10^7$ Ω/cm) gallium arsenide specimens at 27°C.

FIG. 38 Spectra for the imaginary part of the dielectric permittivity for Hughes D3 (——) and MA/Com 1089 (———) Cr-doped single-crystal high-resistivity ($\rho > 5 \times 10^7$ Ω/cm) gallium arsenide specimens at 27°C.

FIG. 39 Loss-tangent spectra for Hughes D3 (——) and MA/Com 1089 (———) Cr-doped single-crystal high-resistivity ($\rho > 5 \times 10^7$ Ω/cm) gallium arsenide specimens at 27°C.

34

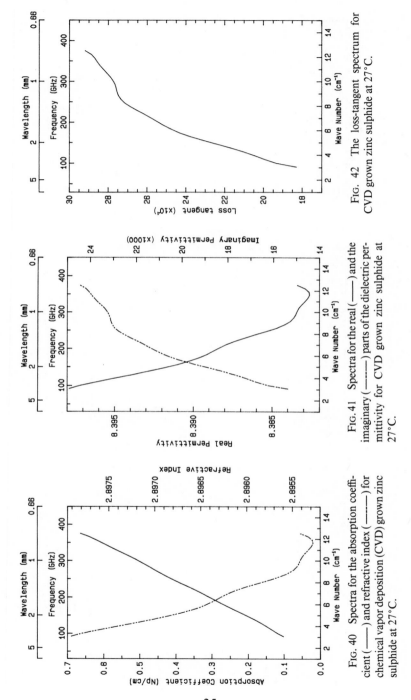

FIG. 40 Spectra for the absorption coefficient (——) and refractive index (——) for chemical vapor deposition (CVD) grown zinc sulphide at 27°C.

FIG. 41 Spectra for the real (——) and the imaginary (——) parts of the dielectric permittivity for CVD grown zinc sulphide at 27°C.

FIG. 42 The loss-tangent spectrum for CVD grown zinc sulphide at 27°C.

35

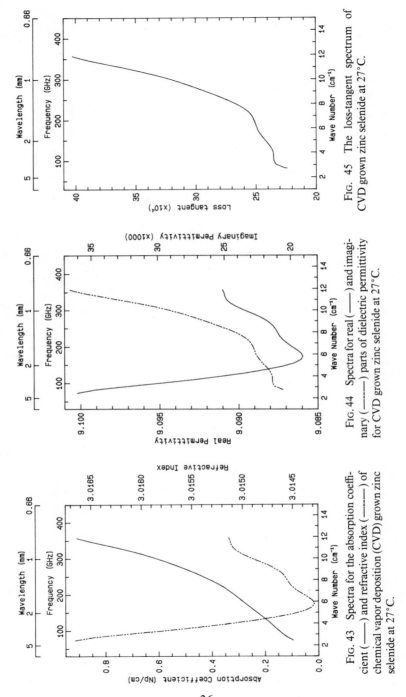

FIG. 45 The loss-tangent spectrum of CVD grown zinc selenide at 27°C.

FIG. 44 Spectra for real (——) and imaginary (——) parts of dielectric permittivity for CVD grown zinc selenide at 27°C.

FIG. 43 Spectra for the absorption coefficient (——) and refractive index (——) of chemical vapor deposition (CVD) grown zinc selenide at 27°C.

36

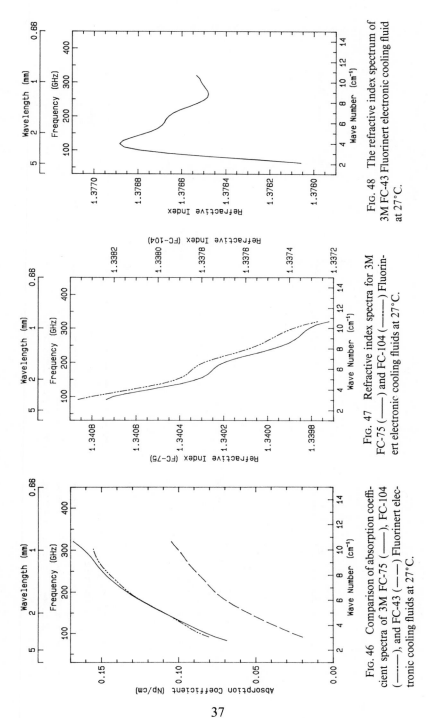

FIG. 46 Comparison of absorption coefficient spectra of 3M FC-75 (———), FC-104 (———·—·—), and FC-43 (— — —) Fluorinert electronic cooling fluids at 27°C.

FIG. 47 Refractive index spectra for 3M FC-75 (———) and FC-104 (———) Fluorinert electronic cooling fluids at 27°C.

FIG. 48 The refractive index spectrum of 3M FC-43 Fluorinert electronic cooling fluid at 27°C.

37

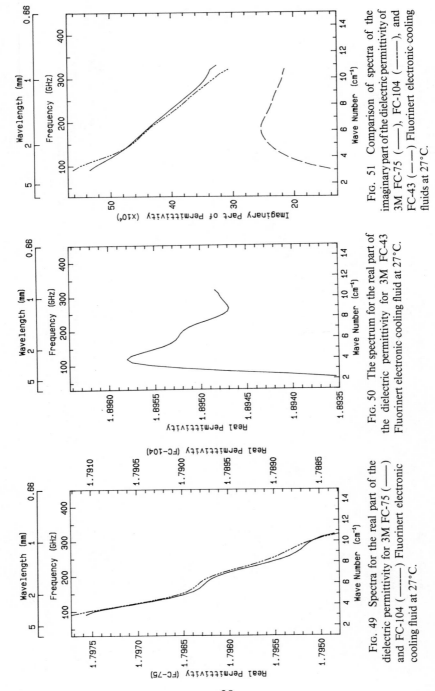

FIG. 51 Comparison of spectra of the imaginary part of the dielectric permittivity of 3M FC-75 (——), FC-104 (·······), and FC-43 (– – –) Fluorinert electronic cooling fluids at 27°C.

FIG. 50 The spectrum for the real part of the dielectric permittivity for 3M FC-43 Fluorinert electronic cooling fluid at 27°C.

FIG. 49 Spectra for the real part of the dielectric permittivity for 3M FC-75 (——) and FC-104 (·······) Fluorinert electronic cooling fluid at 27°C.

38

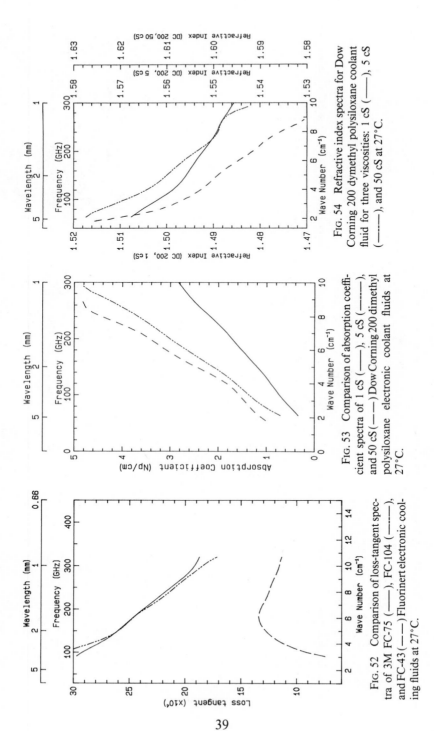

FIG. 54 Refractive index spectra for Dow Corning 200 dymethyl polysiloxane coolant fluid for three viscosities: 1 cS (——), 5 cS (——), and 50 cS at 27°C.

FIG. 53 Comparison of absorption coefficient spectra of 1 cS (——), 5 cS (——), and 50 cS (——) Dow Corning 200 dimethyl polysiloxane electronic coolant fluids at 27°C.

FIG. 52 Comparison of loss-tangent spectra of 3M FC-75 (——), FC-104 (——), and FC-43 (——) Fluorinert electronic cooling fluids at 27°C.

39

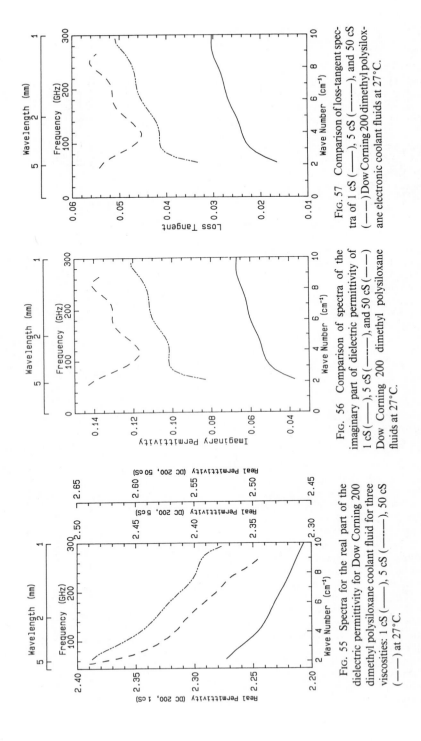

FIG. 55 Spectra for the real part of the dielectric permittivity for Dow Corning 200 dimethyl polysiloxane coolant fluid for three viscosities: 1 cS (———), 5 cS (–––––), 50 cS (— — —) at 27°C.

FIG. 56 Comparison of spectra of the imaginary part of dielectric permittivity of 1 cS (———), 5 cS (–––––), and 50 cS (— — —) Dow Corning 200 dimethyl polysiloxane fluids at 27°C.

FIG. 57 Comparison of loss-tangent spectra of 1 cS (———), 5 cS (–––––), and 50 cS (— — —) Dow Corning 200 dimethyl polysiloxane electronic coolant fluids at 27°C.

REFERENCES

Afsar, M. N. (1977). "The Measurement of Power Absorption Coefficients of Liquids and Solids by Transmission Dispersive Fourier Transform Spectrometry," Report DES 42. National Physical Laboratory, Teddington, Middlesex, United Kingdom.

Afsar, M. N. (1979a). *Conf. Digest. 4th. Int. Conf. IR and MM Waves*, IEEE Cat. No. 79 CH 1384-7 MTT, pp. 191–192.

Afsar, M. N. (1979b). *Proc. IEEE* **67**, 1460–1463.

Afsar, M. N. (1982). *NBS Conf. Precision Electromagnetic Measurements Dig.*, IEEE Cat. No. 82 CH1737-6, Boulder, Colorado, pp. c-12 to c-14.

Afsar, M. N. (1984a). *In* "Infrared and Millimeter Waves," vol. 14 (K. J. Button, ed.). Academic Press, New York, forthcoming.

Afsar, M. N. (1984b). *Int. J. Infrared and Millimeter Waves*, forthcoming.

Afsar, M. N., and Button, K. J. (1981). *Int. J. Infrared and Millimeter Waves* **2**, 1029–1044.

Afsar, M. N., and Button, K. J. (1983). *IEEE Trans. Microwave Theory Tech.* **MTT-31**, 217–223.

Afsar, M. N., and Chantry, G. W. (1977). *IEEE Trans. Microwave Theory Tech.* **MTT-25**, 509–511.

Afsar, M. N., and Hasted, J. B. (1977). *J. Opt. Soc. Am.* **67**, 902–904.

Afsar, M. N., and Hasted, J. B. (1978). *Infrared Phys.* **18**, 835–841.

Afsar, M. N., and Jones, R. G. (1978). Unpublished work and the Ph.D. Thesis of M. N. Afsar, University of London, London, United Kingdom.

Afsar, M. N., Chamberlain, J., and Chantry, G. W. (1976a). *IEEE Trans. Instrum. Meas.* **IM-25**, 290–294.

Afsar, M. N., Chamberlain, J., and Hasted, J. B. (1976b). *Infrared Phys.* **16**, 587–599.

Afsar, M. N., Hasted, J. B., and Chamberlain, J. (1976c). *Infrared Phys.* **16**, 301–310.

Afsar, M. N., Chamberlain, J., Chantry, G. W., Finsy, R., and Van Loon, R. (1977). *Proc. IEEE* **124**, 575–577.

Afsar, M. N., *et al.* (1980). *IEEE Trans. Instrum. Meas.* **IM-29**, 283–288.

Batt, R. J., and Harries, D. J. (1976). *Infrared Phys.* **16**, 325–327.

Batt, R. J., and Luk, S. C. (1980). *Int. J. Infrared Millimeter Waves* **1**, 333–338.

Birch, J. R. (1980). "*Electron Lett.* **16**, 799–800.

Birch, J. R. (1981). "Dielectric Reference Materials for the High Frequency Region: Measurement at 100, 200 and 300 GHz." Final report of National Physical Laboratory, United Kingdom to Bureau Communautaire de Reference (BCR), Brussels, BCR Contract No. 382/1/4/148/78/2. BCR Brussels.

Birch, J. R., Price, G. D., and Chamberlain, J. (1976). *Infrared Phys.* **16**, 311–315.

Breeden, K. H., and Langley, J. B. (1969). *Rev. Sci. Instrum.* **40**, 1162–1163.

Butterweck, H. J. (1968). *IEEE Trans. Microwave Theory Tech.* **MTT-16**, 274–281.

Chantry, G. W., Fleming, J. W., Smith, P. M., Cudby, M., and Willis, H. A. (1971). *Chem. Phys. Lett.* **10**, 473–477.

Clarke, R. N., and Rosenberg, C. B. (1982). *J. Phys. E.* **15**, 9–24.

Cook, R. J., and Jones, R. G. (1976). *Electron. Lett.* **12**, 1–2.

Cook, R. J., and Jones, R. G. (1978). *Proc. 8th. European Microwave Conf.*, Paris, France, pp. 528–532.

Cook, R. J., Jones, R. G., and Rosenberg, C. B. (1974). *IEEE Trans. Instrum. Meas.* **IM-23**, 438–442.

Cullen, A. L., and Yu, P. K. (1971). *Proc. Roy. Soc. London, Ser. A* **325**, 493–509.

Cullen, A. L., and Yu, P. K. (1979). *Proc. Roy. Soc. London, Ser. A* **366**, 165–171.

Cullen, A. L., Nagenthiram, P., and Williams, A. D. (1972). *Electron Lett.* **8**, 577–579.

Culshaw, W., and Anderson, M. V. (1962). *Proc. IEEE* **109**, part B, suppl. 23, 820–826.

Davies, J. G., and Haigh, J. (1974).   *Infrared Phys.* **4**, 183–188.

Dees, J. W., and Sheppard, A. P. (1965).   *IEEE Trans. Instrum. Meas.* **IM-11**, 52–58.

Degenford, J. E. (1968).   *IEEE Trans. Instrum. Meas.* **IM-17**, 413–417.

Degenford, J. E., and Coleman, P. D. (1966).   *Proc. IEEE* **54**, 520–522.

Finsy, R., and VanLoon, R. (1972).   *In* "High Frequency Dielectric Measurement" (J. Chamberlain and G. W. Chantry, eds.) pp. 42–46.   IPC Science and Technology Press, Guildford, Surry, United Kingdom.

French, I. P., and Arnold, T. E. (1967).   *Rev. Sci. Instrum.* **38**, 1604–1607.

Goulon, J., Rossy, G., and Rivail, J. L. (1968).   *Rev. Phys. Appl.* **3**, 231–236.

Goulon, J., Sarteaux, J. P., Brondeau, J., and Roussy, G. (1973).   *Rev. Phys. Appl.* **8**, 165–174.

Honijk, D. D., Passchier, W. F., Mandel, M., and Afsar, M. N., (1977).   *Infrared Phys.* **17**, 9–24.

Ino, M., Ishibashi, T., and Ohmori, M. (1976).   *Electron. Lett.* **12**, 148–149.

Ishibashi, T., and Ohmori, M. (1976).   *IEEE Trans. Microwave Theory Tech.* **MTT-24**, 858–859.

Ishibashi, T., Ino, M., Makimura, T., and Ohmori, M. (1977).   *Electron. Lett.* **13**, 299–300.

Johnson, C., Low, F. J., and Davidson, A. W. (1979).   *SPIE Instrum. in Astronomy III* **172**, 178–183.

Jones, R. G. (1976a).   *J. Phys. D.* **9**, 819–827.

Jones, R. G. (1976b).   *Proc. IEEE* **123**, 285–290.

Kinch, M. A., and Rollin, B. V. (1963).   *Br. J. Appl. Phys.* **14**, 672–676.

Kolbe, W. F., and Leskovar, B. (1982).   *Rev. Sci. Instrum.* **53**, 769–775.

Kolbe, W. F., and Leskovar, B. (1983).   *Int. J. Infrared Millimeter Waves* **4**, 733–749.

Konwerska-Hrabowska, J., Chantry, G. W., and Nicol, E. A. (1981).   *Int. J. Infrared Millimeter Waves* **2**, 1135–1150.

Lynch, A. C. (1982).   *Proc. R. Soc. London, Ser. A* **380**, 73–76.

Matsui, T., Afsar, M. N., and Button, K. J. (1984).   *Int. J. Infrared Millimeter Waves,* forthcoming.

Mead, D. G., and Genzel, L. (1978).   *Infrared Phys.* **18**, 555–564.

Parker, T. J., Ledsham, D. A., and Chambers W. G. (1976).   *Infrared Phys.* **16**, 293–297.

Passchier, W. F., Honijk, D. D., Mandel, M., and Afsar, M. N. (1977).   *Infrared Phys.* **17**, 381–391.

Shimabukuro, F. I., Lazar, S., Chernick, M. R., and Dyson, H. B. (1984).   *IEEE Trans. Microwave Theory Tech.* **MTT-32**, 659–665.

Simonis, G. J. (1982).   *Int. J. Infrared Millimeter Waves* **3**, 439–469.

Staal, P. R., and Eldridge, J. E. (1977).   *Infrared Phys.* **17**, 299–303.

Stolen, R. H. (1969).   *Appl. Phys. Lett.* **15**, 74–75.

Stumper, U. (1972).   *In* "High Frequency Dielectric Measurement" (J. Chamberlain and G. W. Chantry, eds), pp. 51–55.   IPC Press, Guildford, United Kingdom.

Stumper, U., (1973).   *Rev. Sci. Instrum.* **44**, 165–169.

Stumper, U., and Frentrup, K. P. (1976).   *Rev. Sci. Instrum.* **47**, 1196–1200.

Valkenberg, E. P., and Derr, V. E. (1966).   *Proc. IEEE* **54**, 493–498.

Vanloon, R., and Finsy, R. (1973).   *Rev. Sci. Instrum.* **44**, 1204–1208.

Vanloon, R., and Finsy, R. (1974).   *Rev. Sci. Instrum.* **45**, 523–525.

Yu, P. K., and Cullen, A. L. (1982).   *Proc. Roy. Soc. London, Ser. A* **380**, 49–71.

CHAPTER 2

Low-Frequency Vibrations in Long-Chain Molecules and Polymers by Far-Infrared Spectroscopy

John F. Rabolt

IBM Research Laboratory
San Jose, California

I. Introduction

Initial interest in polymeric materials in the far infrared was generated not by a curiosity about their molecular structure but by their attractive optical properties that made them ideal for spectral-filter applications (Bell and Goldman, 1967; Fateley, 1966; McCarthy, 1967; McKnight and Moller, 1964; Mon and Sievers, 1975; Ulrich, 1969). These optical properties were further enhanced by unique structural integrity, low vapor pressure, and ease of processing, which made these materials an obvious choice for vacuum-cell components (Johnson and Rabolt, 1973), infrared windows (Willis *et al.*,

43

1963), and sample-support matrices (Brasch and Jacobsen, 1964; Decamps *et al.,* 1967; May and Schwing, 1963).

At the same time, there was a growing cognizance that polymers crystallized in three-dimensional structures under suitable processing conditions. As the number of x-ray diffraction studies on polymers increased, it became apparent (Krimm, 1960) that the unit cell of such crystalline structures often accommodated more than one polymer chain and that the resulting intermolecular forces could affect the observed vibrational spectrum. If these forces were strong, each band of the isolated chain could be observed to split into a number of closely spaced components equal to the number of polymer chains in the unit cell. This splitting was caused by an interaction with the crystal field, because similar vibrating atoms on each polymer chain in the unit cell were in a different surrounding environment, and hence the vibrational frequencies of each chain would be slightly perturbed. Experimentally, this splitting has been observed in the infrared and Raman spectra of only a few polymers, either because the splitting is too small, the intermolecular forces are too weak, or the symmetry-determined optical-selection rules prohibit the observation of all components spectroscopically.

However, when crystal field splitting is observed, it is generally indicative of strong intermolecular interactions that can also give rise to low-frequency lattice modes involving the relative motion of polymer chains within the unit cell. The number and spectral activity of these modes can be predicted from a knowledge of the space-group symmetry of the unit cell and the number of polymer chains it contains.

In addition to low-frequency intermolecular modes, nonlocalized intramolecular modes characteristic of a polymer's skeletal structure can also be found in the far-infrared region of the spectrum. In contrast to localized modes such as C – H stretch, in which the displacements of most atoms from their equilibrium positions are negligible except for those atoms that contribute significantly to the potential-energy distribution, nonlocalized modes involve significant displacements of all atoms from equilibrium during the normal mode. Thus, these low-frequency modes in long-chain molecules and polymers are generally observed below 400 cm^{-1}; their frequencies are very sensitive to the conformational structure of the backbone.

The far-infrared region of the spectrum can therefore provide a wealth of structural information about both the spatial arrangement of atoms in a polymer chain and the intermolecular environment and packing symmetry of the chain in the crystal lattice.

In the following discussion, an attempt will be made to assess the impact of far-infrared spectroscopy on our understanding of the conformational and crystal structure of a number of well-characterized semicrystalline polymers and copolymers. A discussion will follow on the use of far-infrared spectroscopy to explore secondary and tertiary structure in model polypeptides, proteins, and polynucleotides. Finally, several new research areas will be highlighted in an attempt to bring into focus the future role of far-infrared spectroscopy in the structural investigation of polymeric materials.

II. Homopolymers

A. POLYETHYLENE

Crystalline polyethylene, $(-CH_2-)_n$, has been the most thoroughly investigated polymer and is one of the few to exhibit crystal field splitting of the $-CH_2-$ rocking (720–730 cm^{-2}) and bending (1460–1470 cm^{-1}) vibrations in the mid-infrared (Krimm, 1954). An examination of the region below 100 cm^{-2} (Bertie and Whalley, 1964; Frenzel and Butler, 1964; Krimm and Bank, 1965) revealed the presence of a strong band at 71 cm^{-1} whose intensity was correlated with the crystalline content (Bertie and Whalley, 1964) and whose low-temperature shift to higher frequency was attributed to a contraction of the crystalline lattice with a subsequent increase or "stiffening" of the intermolecular force constants (Krimm and Bank, 1965). Recently, the peak position of this 70-cm^{-1} band has been shown to vary linearly with crystallinity (Zirke and Meissner, 1978), after having been initially observed to shift to 66.5 cm^{-1} in perdeutero-polyethylene (Krimm and Bank, 1965) in accordance with the mass effect expected for a translatory lattice mode. Final assignment of this band to a B_{1u} lattice mode of the orthorhombic unit cell was assured when infrared measurements of orthorhombic (2 molecules per unit cell) $C_{36}H_{74}$ indicated the presence of a band at 70 cm^{-1} whereas identical measurements on triclinic (1 molecule per unit cell) $C_{20}H_{42}$ showed the absence of any such feature (Krimm and Bank, 1965). Polarized far-infrared measurements on machine-oriented polyethylene films (Bank and Krimm, 1968) revealed that the 70-cm^{-1} band exhibited an a-axis dichroism, thus confirming its assignment to the B_{1u} symmetry species.

The effects of chain branching, crystallization habit (end-group packing), and morphology on the far-infrared spectrum of polyethylene (PE) were studied by Bank and Krimm (1968), and the results helped elucidate the nature of the packing of chains into the crystallographic unit cell. Their observations are shown in Table I. It is interesting to note that, as the a-axis lattice dimension increases, the band position of this lattice vibration is found at lower frequencies, reflecting the decrease in intermolecular forces. Identical behavior has been observed in linear PE by following the peak position of this band as a function of increasing temperature from 14 to 413 K (Frank, 1977; Frank et al., 1977), because the lattice expands with increasing temperature. In the extreme case of highly branched PE (37.9

TABLE I

B_{1u}-LATTICE FREQUENCY OF POLYETHYLENE

	Lattice dimensions (Å)		Observed frequency $(cm^{-1})^a$	
	a	b	$T = 30°C$	$T \approx -170°C$
Branched (37.9 CH_3/1000C)	7.529	4.993	70.4	76.9
Cast film	7.445	4.975	71.7	76.4
Single crystal	7.435	4.970	71.8	76.6
Melt crystallized	7.414	4.947	72.3	78.1
High-pressure crystallized	—	—	72.7	78.4

[a] The setting angle θ of the zigzag plane relative to the crystallographic a axis can vary slightly and also affect the lattice frequency by $1-2$ cm^{-1}.

CH_3/1000C), the observed frequency is consideraly lowered and is convincing evidence for incorporation of branches into the lattice.

Observation of a second peak at 109 cm^{-1} in the far-infrared spectrum of PE at 2 K was first reported by Dean and Martin (1967); this band was subsequently assigned to the B_{2u} translatory lattice mode predicted to occur at 109 cm^{-1} by the normal coordinate calculations of Tasumi and Krimm (1967). Because the band did not appear in the room-temperature spectrum (Amrhein and Heil, 1971), it was not possible to determine its frequency position as a function of temperature. However, development of high radiometric precision in the far infrared as reported by Fleming et al. (1972), allowed the observation of this weak band at 94 cm^{-1} in the room-temperature spectrum of a highly crystalline sample of polyethylene. Its shift to 107 cm^{-1} at 125 K was consistent with the earlier assignment to a B_{2u} lattice mode. Recent far-infrared studies (Zirke and Meissner, 1978) at room temperature on a series of commercially available PE samples of medium crystallinity revealed the presence of the B_{2u} lattice mode at 96 cm^{-1}. When these materials were annealed close to the melting point, an increase in the background absorption continuum was observed that obscured this band. Thus it appears that early attempts to observe weak lattice bands in the far infrared of PE may have been hampered not by instrumental sensitivity but by the presence in PE of a broad absorption continuum (Davies and Haigh, 1974) whose origin has been attributed (Fleming et al., 1972) to "liquid lattice" (poley) bands, i.e., broad bands reflecting a disordering of the lattice spectrum due to a breakdown of the optical selection rules.

When PE is heated to ~ 500 K under moderate pressure (4–5 kbar), a new hexagonal phase results that contains a considerable amount of disorder (Bassett et al., 1974). If the PE is annealed (5–40 h) then slow-cooled while the pressure is maintained, a very highly crystalline (>95%) material results. Its far-infrared spectrum (Schlotter and Rabolt, 1984) at room temperature and at 10 K is shown in Fig. 1. In the room-temperature spectrum, only the B_{1u} lattice mode is present at 72.5 cm^{-1}. On lowering the temperature to 10 K, the B_{2u} mode at 110 cm^{-1} is observed, but, in addition, a new band appears at 39 cm^{-1}. The appearance of this band was initially reported by Frank *et al.* (1981a), who also showed that it first appeared below 170 K. An explanation for the existence of the new band was given, based on the possible presence of a monoclinic modification in low concentration. However, recent measurements (Schlotter and Rabolt, 1984) of monoclinic and orthorhombic PE (>10% monoclinic) at 10 K did not indicate the presence of any bands below 50 cm^{-1}, and an assignment to the optically inactive A_u out-of-phase *c*-axis translatory mode [calculated at 51 cm^{-1} (Tasumi and Shimanouchi, 1965)] was proposed instead. In fact, incoherent, inelastic neutron scattering studies (Berghmans *et al.,* 1971; Stafford *et al.,* 1966), in which peaks representing maxima in the density of vibrational states are observed independent of optical-selection rules, indicated the presence of a medium-intensity band at 39 cm^{-1} that was assigned to the A_u lattice mode. Further studies in the low-frequency infrared and Raman spectra of high-pressure crystallized PE are still needed before definite assignment of the 39-cm^{-1} band can be made.

More recent studies on polyethylene in the far infrared have focused on

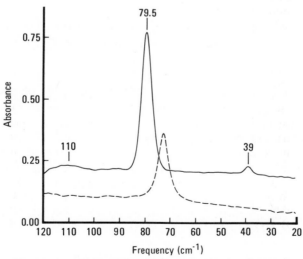

FIG. 1   Far-infrared spectra of high-pressure crystallized polyethylene at room (– – –) and liquid helium (——) temperatures: 298 K and 10 K, respectively.

the effects of irradiation (Frank, 1977), high temperature (Prettl and Frank, 1979), and high pressure (Leute, 1981).

B. POLYTETRAFLUOROETHYLENE

Polytetrafluoroethylene (PTFE) initially received only minor attention as a room-temperature infrared optical material (Chamberlain and Gebbie, 1966) prior to 1970. However, considerable interest was generated within the polymer spectroscopy community by the first reports of crystal field splitting (Boerio and Koenig, 1971) in the Raman spectrum of the low-temperature ($<19°C$) crystal phase. This 19°C phase transition, involving primarily a change in conformation, was not believed to involve substantial changes in the crystal packing of the unit cell (Clark and Muus, 1962). The spectroscopic results, however, suggested that the unit cell of the low-temperature phase contained more than one molecule, contrary to the results of preliminary x-ray measurements (Clark and Muus, 1962; Kilian, 1962). Additional support for the existence of more than one molecule per unit cell came from far-infrared measurements reported by Chantry et al. (1972b) and Johnson and Rabolt (1973), although the first published low-temperature far-infrared spectrum had appeared, without analysis, in earlier work on the surface resistance of lead supported on PTFE substrates (Brandli and Sievers, 1972). The appearance of four or five new infrared bands, whose increase in intensity and frequency position as a function of decreasing temperature was analogous to that expected for intermolecular or lattice vibrations, could not be explained with a crystallographic unit cell containing only one chain stem.

Additional evidence for two molecules per unit cell was presented by Jones et al. (1976) and Piseri et al. (1975) using oriented PTFE specimens for far-infrared dichroic measurements. At low temperatures, three of the strong bands in the 50-cm^{-1} region (see Fig. 2) showed strong polarization properties, with the transition dipole moments lying perpendicular to the chain axis. Jones et al. (1976) concluded that the two components of the $54-57$-cm^{-1} doublet (at $-150°C$) were the in-phase and out-of-phase rotatory (hindered rotation about the chain axis) lattice modes of a crystallographic unit cell containing two chain stems, whereas that at 45 cm^{-1} was one of the three infrared-active translatory lattice modes expected for such a structure. Furthermore, they also found that assignment of the two additional lattice modes, one polarized perpendicular and one polarized parallel, was not unambiguous, because four weak bands at 34, 70, 76, and 84 cm^{-1} still remained unassigned.

Support for more than one molecule per unit cell continued to accumulate from studies of highly crystalline (Chantry et al., 1974) and highly irradiated (Rabolt, 1983) PTFE. In the latter case, low-temperature spectroscopic

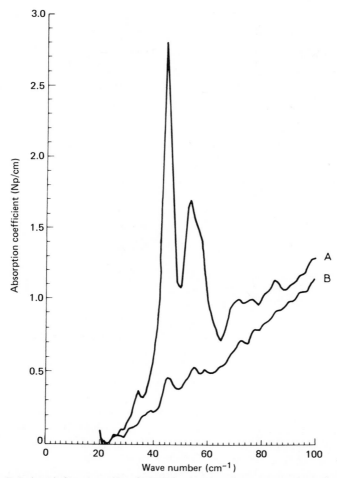

FIG. 2 Polarized infrared spectra of PTFE in the low-temperature crystal phase: curve A, radiation polarized perpendicular to the orientation direction; curve B, radiation polarized parallel to the orientation direction. (Reproduced from Jones *et al.*, 1976, *Polymer* **17**, 153, by permission of the publishers, Butterworth and Co., Ltd.)

studies of progressively irradiated PTFE showed a two-fold effect in the far-infrared region of the spectrum. As seen in Fig. 3, with increasing radiation dosage, there is an apparent increase in the background absorbance level toward the high-frequency region of the spectrum that is accompanied by an obvious disappearance of the intense features ascribed to lattice modes.

The change in background absorbance, similar to that observed in *n*-al-

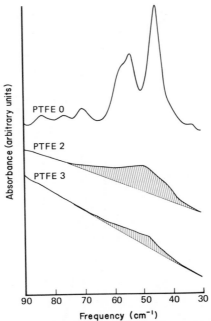

FIG. 3   Low-temperature far-infrared spectra of progressively irradiated PTFE.   Irradiation dose increases from PTFE 0 to PTFE 3.   (From Rabolt, 1983.   © 1983 John Wiley and Sons, Inc.)

kanes, has been the subject of much discussion (Chantry, 1971).   Generally, the broad continuum is believed to originate from a disorder-induced smearing out of the crystal lattice spectrum.   However, another contributing factor in these irradiation experiments may come from the creation of voids containing small molecular segments, caused by chain scission, that can act as scattering centers for the incoming infrared beam.

The disappearance of the strong features with increasing irradiation and decreasing crystallinity supported earlier results indicating that these vibrations were intermolecular in origin.   The optical activity of more than one lattice mode in the far infrared was taken as an indication that the crystallographic unit cell contained more than one chain stem.   Recent x-ray studies (Weeks et al., 1981) bore out this conjecture.

Further studies (Willis et al., 1975) of PTFE in the submillimeter-wave region of the spectrum revealed the presence of a broad weak feature at 15 cm^{-1} that was subsequently assigned to an intermolecular mode belonging to the ν_9 dispersion curve.   Its counterpart in the Raman was found at 13 cm^{-1}, but no explanation for the apparent discrepancy in frequency was given.

Interesting spectroscopic results were recently reported by Chantry et al.

(1981) when attempts were made to degrade the molecular weight of PTFE by high-temperature oxidation with potassium nitrate. Finding that the low-temperature far-infrared spectrum contained many sharp bands led to the conclusion that potassium fluoride dihydrate, a crystalline substance, had formed in the polymer. Furthermore, evidence was obtained to suggest that the interaction between the inorganic dihydrate and PTFE was sufficiently strong to cause a modification of its low-frequency infrared spectrum.

C. POLYPROPYLENE

Polypropylene, $(-CH_2-C(CH_3)H-)_n$, exists in three configurations: two stereoregular (isotactic and syndiotactic) and one random or irregular (atactic), depending on the polymerization scheme and the substance used to catalyze the reaction. The two stereoregular forms crystallize in highly ordered structures, whereas the latter is amorphous. Isotactic polypropylene (IPP) has been the subject of considerable study since its commercial introduction in the mid-1950s. Syndiotactic polypropylene, on the other hand, has been available only in small quantities, via custom synthesis, and thus has evaded the close scrutiny of the vibrational spectroscopist.

Structural studies of IPP indicated that a 3/1 (three chemical repeat units in one turn) helical conformation existed in the crystalline state. The factor group of the line group is isomorphic to the $C(2\pi/3)$ point group. Infrared-active modes belong to either the A (parallel polarized) or E (perpendicularly polarized) symmetry species. Early investigations on uniaxially oriented specimens in the far infrared indicated the presence of many bands in the region below 400 cm^{-1} (Miyazawa et al., 1963). Polarization measurements were obtained, and both parallel (398, 251, 200, and 155 cm^{-1}) and perpendicular (321, 210, and 169 cm^{-1}) bands were identified. A medium feature at 106 cm^{-1} was also observed, and although its dichroic behavior was not determined, it was assigned to a torsional vibration about the $C-C$ bond through normal coordinate calculations using a modified Urey–Bradley force field.

Additional investigations of IPP (Chantry et al., 1971) in the submilli-meter-wave region revealed the splitting of the 106-cm^{-1} band at liquid-nitrogen temperatures into at least two components at 98 and 110 cm^{-1}, as shown in Fig. 4. The mean frequency of the two components did not shift in position with temperature, thus supporting an assignment of the doublet to intramolecular vibrations. Below 100 cm^{-1} a very broad absorption band was observed whose peak was located in the vicinity of 55 cm^{-1}. The origin of this broad feature was attributed to the superposition of a sharp torsional vibration (predicted to lie at 63 cm^{-1}) and a broad underlying continuum arising from "liquid-lattice" disordered contributions (Chantry et al., 1971).

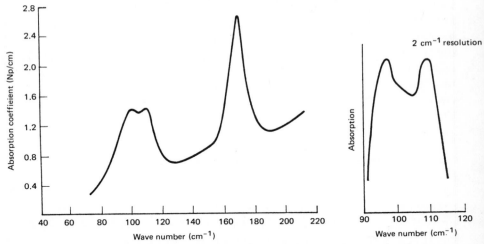

FIG. 4   The low-frequency infrared spectrum of isotactic polypropylene at reduced temperatures. (From Chantry *et al.*, 1971. © 1971 Pergamon Press, Ltd.)

An alternative explanation of this 55-cm⁻¹ band was later provided by the work of Goldstein *et al.* (1973). By suitably controlling the thermal processing history, a series of samples containing varying degrees of crystalline content were prepared and investigated in the far infrared. The intensity and sharpness of the 55-cm⁻¹ absorption was observed to increase with increasing crystallinity, whereas its frequency position was observed to shift from 48 to 55 cm⁻¹ on lowering the temperature to 100 K. Based on these observations, this low-frequency mode was reassigned to an intermolecular vibration of the crystal lattice, with this conclusion being reconfirmed by Haigh *et al.* (1975).

In addition, dichroic studies of oriented, highly crystalline IPP specimens revealed that both components of the low-temperature doublet at 98 and 110 cm⁻¹ exhibited perpendicular dichroism, invalidating a previous assignment (Chantry *et al.*, 1971) of the band at 98 cm⁻¹ to the A symmetry species (which would exhibit parallel polarization). Instead, these authors favored an assignment of this doublet and the one at 135–145 cm⁻¹ to correlation field splitting of E (former) and A (latter) line-group modes. With these assignments, all internal modes of the isolated chain had been accounted for with the three additional bands (135, 98, and 55 cm⁻¹) being attributed to crystal lattice effects.

D.   POLYVINYLIDENE FLUORIDE

Polyvinylidene fluoride (PVF₂), (—CH₂CF₂—)ₙ, has long been recognized as being polymorphic, existing in at least three crystalline modifications (Hasegawa *et al.*, 1972) depending on thermal and mechanical processing history. Kondrashov (1960) observed that x-ray diffraction patterns of drawn PVF_2 exhibited two identity periods, indicating the presence of two crystalline forms. Lando *et al.* (1966) used both NMR and x-ray diffraction studies to elucidate the crystal structures of the two forms, after first postulating that the unit cell of Form I (also referred to as the β phase) contained two planar zigzag chains. Galperin *et al.* (1965) had suggested a TGTG' (T, trans; G, gauche) conformation for the chains in Form II (also referred to as α phase) that was later confirmed by the results of Doll and Lando (1970). Evidence for yet a third form (Form III) was presented by Cortili and Zerbi (1967) and Natta *et al.* (1965), who suggested a conformational structure similar to Form I. Elucidation of all three crystal modifications was later provided by the x-ray studies of Hasegawa *et al.* (1972).

Polarized far-infrared measurements of Forms I and II were first reported by Luongo (1972). Bands unique to Form II were found at 360, 220, 110, 55, and 48 cm^{-1}, whereas a broad feature at 85 cm^{-1} was found to be characteristic of Form I. Although no specific band assignments were made, it was noted that on poling a uniaxially oriented film a conversion from Form II to Form I was observed spectroscopically.

A more detailed study of Form II in the low-frequency region was reported by Rabolt and Johnson (1973) with the resulting spectrum shown in Fig. 5. In addition to the strong bands noted by Luongo (1972), a medium band at 53 cm^{-1} was observed at room temperature and found to shift to 60 cm^{-1} at liquid nitrogen temperatures. Its intensity was correlated with the degree of crystallinity, thus prompting its assignment to a vibration of the crystal

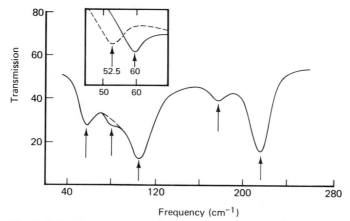

FIG. 5 The far-infrared spectrum of PVF_2 (Form II). Inset indicates the frequency shift of the 52.5-cm^{-1} lattice mode upon lowering the sample temperature to 77 K. (From Rabolt and Johnson, 1973.)

TABLE II

FAR-INFRARED (400–10 CM⁻¹) BANDS OF THE
THREE FORMS OF POLYVINYLIDENE FLUORIDE[a]

Form I	Form II	Form III
		380 m
	372 vw	
	355 m	
350 m		
		334 m
	325 vw	
		300 m
	285 w	
	215 m	
		204 w
200 w		
	175 w	175 w
	102 s	
		90 s
	85 w	
70 m		
	53 m	

[a] vw, very weak; w, weak; m, medium; s, strong.

lattice. Factor-group analysis of the monoclinic unit cell first proposed by Hasegawa *et al.* (1972) predicted the presence of a single infrared-active rotatory lattice mode, to which the 53 cm⁻¹ was ultimately attributed.

Spectroscopic studies of all three forms in the far infrared were reported by Kobayashi *et al.* (1975) and Latour *et al.* (1981). Although there is some difficulty in obaining samples that contain only a single-crystal modifica-

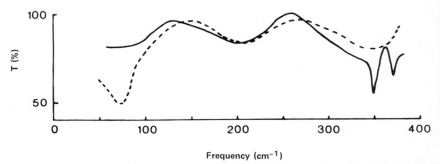

FIG. 6   Spectra of β PVF$_2$ prepared by anisotropic stretching of film: ——, light polarized parallel to the orientation direction; (– –), light polarized perpendicular to the orientation direction. (From Latour *et al.*, 1981. © 1981 John Wiley and Sons, Inc.)

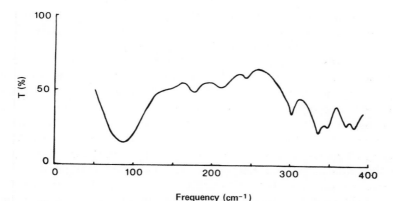

FIG. 7 The far-infrared spectrum of γ PVF$_2$ obtained from powder solution in dimethyl-sulfoxide at 85° C. (From Latour *et al.*, 1981. © 1981 John Wiley and Sons, Inc.)

tion, there are a number of strong – medium bands that can be identified with each crystal structure. These are listed in Table II. The reader is referred to Fig. 6 (β phase) and Fig. 7 (γ phase), to the aforementioned references, and to Tashiro *et al.* (1981) for a more detailed comparison of the weaker bands of Forms I, II, and III.

E. POLYETHYLENE OXIDE

Of the various models proposed for polyethylene oxide (PEO), $(-CH_2CH_2-O-)_n$, the 7/2 helix has been the one most compatible with x-ray measurements and spectroscopic data (Tadokoro, 1979). Four helical molecules have been found to pack into the unit cell that belongs to the monoclinic crystal system with space group P2$_1$/a (Takahashi and Tadokoro, 1973). The results of factor-group analysis shown in Table III indicate that, in addition to low-frequency intramolecular modes, five lattice modes should appear in the far infrared.

TABLE III

FACTOR GROUP ANALYSIS OF MONOCLINIC PEO (P2$_1$/a)[a]

Symmetry species	E	C$_2$	i	σ	T	T′	R′	Activity
A$_g$	1	1	1	1	0	3	1	R
B$_g$	1	−1	1	−1	0	3	1	R
A$_u$	1	1	−1	−1	1	2	1	IR
B$_u$	1	−1	−1	1	2	1	1	IR

[a] T, translations; T′, translatory lattice modes; R′, rotatory lattice modes; R, Raman active; IR, infrared active.

Low-frequency (400–100 cm⁻¹) infrared studies by Yoshihara *et al.*
(1964) and Miyazawa *et al.* (1962) revealed the presence of polarized bands
at 216 ($\perp$), 165 ($\perp$), and 107 (∥) cm⁻¹. All three were assigned (Matsuura and
Miyazawa, 1968) to vibrational modes involving torsional motion of the
CCO backbone. The 107 cm⁻¹ is particularly interesting because it was
subsequently shown (Rabolt *et al.*, 1974) to shift 7 cm⁻¹ to higher frequency
(see Fig. 8) on lowering the sample temperature to 110 K. This behavior is
usually associated with lattice vibrational modes indicating that, although
there is considerable backbone motion involved in the normal mode, there is
a significant intermolecular component that contributes to the potential-energy distribution. In fact, because no normal mode analysis of the PEO
crystal has appeared in the literature, the reassignment of the 107 cm⁻¹ band
to one of the predicted infrared-active lattice modes cannot be ruled out.

In addition to the one at 107 cm⁻¹, three additional bands were observed at
81, 52, and 37 cm⁻¹ by Rabolt *et al.* (1974). No significant frequency shift or
intensity change was found on lowering the sample temperature or its crystalline content, respectively. The 81- and 52-cm⁻¹ infrared modes shown in
Fig. 8 were thought to correspond to those reported at 80 and 51 cm⁻¹ by
Trevino and Boutin (1967) from neutron-scattering measurements. The
lowest calculated normal mode for an isolated PEO chain was found to be at
92 cm⁻¹ by Matsuura and Miyazawa (1968) and at 94 cm⁻¹ by Yokoyama *et
al.* (1972), and thus the band observed at 80 cm⁻¹ was tentatively assigned to
this vibration. No specific assignment of the 52- and 37-cm⁻¹ absorptions

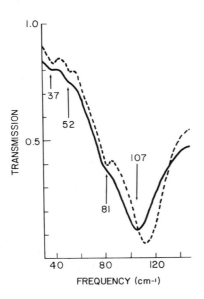

FIG. 8 Low-frequency infrared spectra of
polyethylene oxide 20,000 at 110 (- - -) and
300 K (——). (From Rabolt *et al.*, 1974.)

was made because of a lack of available normal-mode calculations incorporating the crystal field.

F. POLYOXYMETHYLENE

Polyoxymethylene, $(-CH_2-O-)_n$, the simplest of the polyester structures, has been shown to exhibit both an orthorhombic (Carazollo and Mammi, 1963) and a hexagonal (Huggins, 1945) crystal modification. POM assumes a 2/1 helix in the orthorhombic crystal structure, whereas it adopts a 9/5 helical conformation in the hexagonal form. Because the orthorhombic form contains more than one molecule in the unit cell, it was not surprising that crystal field splitting (434–428 cm⁻¹) was observed and reported by Zerbi and Massetti (1967). In addition to the strong mode observed at 304 cm⁻¹, assigned to a torsional vibration of the backbone, several medium–strong bands were also observed at 130, 89, and 83 cm⁻¹. As seen in Fig. 9, when the temperature is lowered to 120 K, these absorptions shift to 143, 97, and 92 cm⁻¹, respectively, characteristic of intermolecular vibrations. Based on this behavior, all three modes were assigned to either translatory (130 and 89 cm⁻¹) or rotatory (83 cm⁻¹) lattice modes. Similar observations with decreasing temperature were made for the 71 cm⁻¹

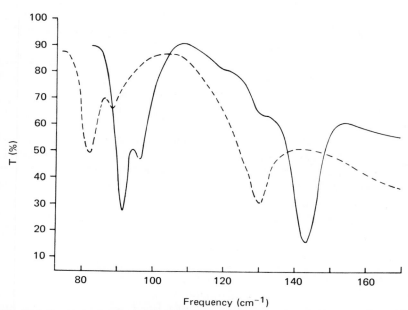

FIG. 9 Infrared spectra of orthorhombic polyoxymethylene at room (– – –) and liquid nitrogen (——) temperature. (From Zerbi and Massetti, 1967.)

band of PE by Bertie and Whalley (1964), whereupon an assignment of this feature to a lattice mode was also made.

The hexagonal form of POM contains only one molecule per unit cell, and thus the potential for only one spectroscopically active (infrared or Raman) lattice vibration exists. The lowest calculated torsional mode of the isolated chain (Matsui *et al.*, 1965) was assigned to the band observed at 230 cm^{-1} with no further reports of any infrared band observed in the region below 200 cm^{-1}.

III. Polypeptides

Polypeptides, $(-CRH-CONH-)_n$, generally assume either α helical (3.6 residues per helical turn), β pleated-sheet (Pauling and Corey, 1951), or β rippled-sheet (Pauling and Corey, 1953) structures. Although much of the earlier work concentrated on the localized vibrations of the amide group, a considerable amount of information concerning both secondary (conformation) and tertiary structure eventually became accessible in the far-infrared region where intramolecular skeletal-angle-bending modes and interchain vibrations involving motion of the hydrogen bond were expected to appear.

A. POLYGLYCINE

Polyglycine, $(-CH_2-CONH-)_n$, is the simplest polypeptide that has been studied extensively in the far infrared. The structure of polyglycine I (PGI) was initially thought to be a β pleated sheet, but a slightly better agreement between the vibrational modes observed and those calculated was obtained by Moore and Krimm (1976b) when a β rippled sheet was considered. Low-frequency infrared bands have been observed at 320, 285, 215, and 133 cm^{-1} by Fanconi (1973), and all but the one at lowest frequency was attributed to skeletal bending and torsional vibrations of the polymer backbone. The feature at 215 cm^{-1} was previously reported by Miyazawa (1961) and was also assigned to a torsional mode about the peptide group corresponding to the observed vibration at 206 cm^{-1} in N-methylacetamide. The lowest observed infrared band at 133 cm^{-1} (Fanconi, 1973) was initially assigned to a translational lattice mode involving a significant contribution (92%) to the potential-energy distribution caused by stretching of the NH $\cdots$ O intermolecular hydrogen bond. More recent calculations (Moore and Krimm, 1976b) indicate that the mode is a combination of CN torsion and NH out-of-plane bending influenced to a large degree by NH $\cdots$ O stretching (49%). Perhaps low-temperature far-infrared data could further elucidate the precise origin of the 133-cm^{-1} band, because the subsequent lattice contraction should result in significant frequency shifts for bands with intermolecular character.

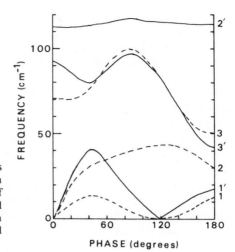

FIG. 10 Low-frequency dispersion curves of PGII. Phase angle is the difference in phase between the atomic displacements of adjacent chemical repeat units: – – –, isolated chain; ——— parallel chain lattice. (From Fanconi, 1973. © 1973 John Wiley and Sons, Inc.)

PGII is a threefold helical chain (Crick and Rich, 1955) that is hydrogen-bonded via the peptide group to adjacent parallel chains in the crystalline lattice. The vibrational spectrum of PGII down to a low-frequency limit of 300 cm⁻¹ was first reported by Suzuki *et al.* (1966) with only one vibration at 363 cm⁻¹ observed in the far infrared. Two additional bands at 267 and 115 cm⁻¹ were reported by Fanconi (1973), who also determined their shift on N-deuteration. Although assignment of both the 363- and 267-cm⁻¹ bands to skeletal-bending vibrations was straightforward using the isolated chain calculation of Abe and Krimm (1972), the assignment of the 115-cm⁻¹ mode required a normal-coordinate calculation of the parallel-chain lattice incorporating interchain interactions. The major effect of the inclusion of such interactions is to change the shape and intersection of the PGII dispersion curves, as shown in Fig. 10. Curves 1, 2, and 3 result from a calculation of the isolated chain, whereas these become 1′, 2′, and 3′ when lattice interactions are introduced. Fanconi (1973) assigned the 115 cm⁻¹ far-infrared band to the *E*-mode intersection (120°) of the 2′ dispersion curve at 116 cm⁻¹, corresponding to the CN + CC torsional mode at 42 cm⁻¹ calculated by Abe and Krimm (1972) in the isolated chain approximation. Although the parallel-lattice calculation also predicts an additional low-frequency vibration below 90 cm⁻¹, no spectroscopic evidence for such a mode has appeared in the literature.

B. POLY-L-ALANINE

Poly-L-alanine, $(-CH(CH_3)-CONH-)_n$, in both the α helical (α PLA) and extended (β PLA) form have been investigated in the far infrared by a number of research groups. The α helix is stabilized by strong intrachain hydrogen bonds between third neighboring peptide groups, thus enhancing their interaction. Miyazawa *et al.* (1967) initially reported the far-infrared spectrum of α PLA, finding strong features at 375, 329, and 295 cm^{-1} characteristic of the helical structure. Dichroic measurements on oriented α PLA were subsequently published (Fig. 11) by Itoh and Shimanouchi (1970) after they initially reported the low-frequency spectrum of unoriented α PLA below 250 cm^{-1} (Itoh *et al.*, 1968, 1969). In the oriented sample, a series of bands were observed at 371, 324, 284, 185, 163, 120, and 113 cm^{-1}, and all were attributed to hydrogen-bond stretching and bending vibrations coupled with backbone torsional motions, as shown in Table IV. Recent studies of α PLA at ambient and low temperatures (Shotts and Sievers, 1973, 1974; Rabolt and Rein, 1983; Rabolt *et al.*, 1977) tend to confirm these assignments. It is interesting to note that the infrared band at 115 cm^{-1}, assigned (Rabolt *et al.*, 1977) to a skeletal torsional mode with a contribution from NH out-of-plane bending, shifts to 122 cm^{-1} at 77 K, perhaps reflecting the contraction of the α helix (Rabolt *et al.*, 1977) at low temperatures with a subsequent increase in the strength of intramolecular hydrogen bonds. A similar shortening of the intramolecular hydrogen bond in poly γ-benzyl-L-glutamate on N-deuteration was noted by Tomita *et al.* (1962).

Two additional vibrations in α PLA were observed by Rabolt *et al.* (1977) to lie below 100 cm^{-1}. A medium band at 84 cm^{-1}, thought to correspond to the Raman band at 87 cm^{-1}, was assigned to the calculated A mode at 89 cm^{-1} involving torsional motion about the backbone CC nd CN bonds. The remaining weak band at 46 cm^{-1} was tentatively assigned to backbone deformation coupled with hydrogen-bond stretching.

In contrast to α PLA, the far-infrared spectrum of β PLA is considerably simplified. Strong features at 442 and 247 cm^{-1}, characteristic of the β

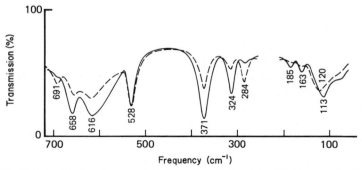

FIG. 11   Polarized infrared transmission spectra of oriented α PLA: $---$, electric vector parallel to the direction of orientation; ——, electric vector perpendicular to the direction of orientation. (From Itoh and Shimanouchi, 1970. © 1970 John Wiley and Sons, Inc.)

TABLE IV

ASSIGNMENTS OF α PLA WITH THE RIGHT-HANDED
HELICAL CONFORMATION[a]

Calculated		Observed		
A	E	A	E	Assignment
	368		371	Helix deformation
353				
	333		324	$C_\alpha C_\beta$ bending[b]
295		284		
259				Helix deformation
	203		185	Helix deformation[b]
	162		163	$C_\alpha C_\beta$ bending
141		120		
	93		113	Helix deformation
80		89[c]		
	44		46[c]	

[a] Itoh and Shimanouchi (1970).
[b] C_α: carbon adjacent to peptide group; C_β: methyl carbon.
[c] Rabolt *et al.* (1977).

pleated-sheet structure, were observed by Itoh and Katabuchi (1972) and Itoh *et al.* (1968). In addition, a rather broad band was found at 122 cm^{-1} and subsequently assigned to a skeletal-bending mode of the peptide group and the adjacent α carbon atom by Moore and Krimm (1976a). Rabolt *et al.* (1977) reported two additional bonds at 97 and 46 cm^{-1}, but no normal mode assignment was given.

An interesting result from a series of oligopeptides of L-alanine was reported by Shotts and Sievers (1974), who investigated the low-frequency infrared spectra of oligomers containing 2–6 alanine residues. They concluded that in the solid state these compounds do not assume the α helical form. Extension of this work to longer chain lengths (10 and 40 residues) by Rabolt and Rein (1983) yielded the spectra of PLA (10) and PLA (40), shown in Fig. 12, recorded at low temperature. As seen in this figure and in Table V, PLA (10) exhibits the strong β peaks at 250 and 440 cm^{-1}; in comparison, in the PLA (40) spectrum there are additional bands observed at 370, 330, and 117 cm^{-1}, indicative of α helical conformation. This then suggests that at least ten L-alanine residues are required before an α helical conformation in PLA is adopted. The implication of these results on the stability of secondary structure in proteins is significant, because their func-

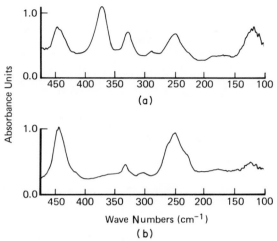

FIG. 12   Low-temperature far-infrared spectra of L-alanine oligomers: (a) 40, (b) 10 at 77 K. The numbers 40 and 10 refer to the number of L-alanine residues in the chain.

tion or dysfunction is highly correlated with the distribution of amino acid residues along the backbone, which, in turn, determines the conformational structure.

C.   POLY-L-VALINE

Itoh *et al.* (1969) were the first to report the far-infrared spectrum of poly-L-valine (PLV), ($-CH(CH_3CHCH_3) -CONH-)_n$, cast from trifluoroacetic acid. Infrared bands were observed at 360, 324, 261, and 120 cm⁻¹, with the latter band taken as indicative of a β pleated-sheet structure. On the other hand, Fanconi (1973) has reported the spectrum shown in Fig. 13 for a similarly prepared sample. In addition to the features at 305 and 240 cm⁻¹, there is a sharp band present at 150 cm⁻¹ with a weaker feature present at

TABLE V

OBSERVED FAR-INFRARED BANDS (500–50 CM⁻¹)
OF α POLY-L-ALANINE AND L-ALANINE OLIGOMERS

Poly-L-alanine (10)	Poly-L-alanine (40)	α Poly-L-alanine (∞)	Conformation
	117	117	α
250	250		β
	330	330	α
	370	370	α
440	440		β

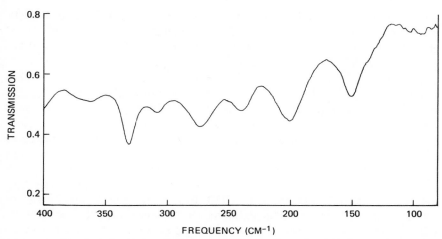

FIG. 13 The far-infrared spectrum of poly-L-valine. (From Fanconi, 1973. © 1973 John Wiley and Sons, Inc.)

100 cm⁻¹. It is this lowest frequency band that was assigned to the hydrogen-bond-stretching translatory lattice mode of the β pleated-sheet structure (Fanconi, 1973). No apparent evidence of the broad 120-cm⁻¹ band reported by Itoh *et al.* (1969) was observed (Fig. 13). To date, no infrared data on PLV below 100 cm⁻¹ has been reported in the literature.

D. POLYAMINO ACIDS CONTAINING LONG ALKYL SIDE CHAINS

Polypeptides containing long alkyl side chains have been the subject of a number of far-infrared studies. These include poly-L-α-amino-*n*-butyric acid (Itoh *et al.,* 1969; Itoh and Shimanouchi, 1970), poly-L-proline (Isemura *et al.,* 1968; Rabolt *et al.,* 1975; Shotts and Sievers, 1974), poly-L-leucine (Shotts and Sievers, 1974), poly-L-norvaline, and poly-L-norleucine (Itoh and Shimanouchi, 1970). Unfortunately, normal coordinate calculations for polypeptides containing long alkyl side chains have lagged behind those for synthetic macromolecules and, thus, information concerning the effect of lattice (intermolecular) interactions on the low-frequency vibrational (infrared and Raman) spectrum is unavailable.

IV. Copolypeptides

With a general understanding of the conformational and crystal structures of simple polypeptides, the natural extension toward understanding protein structure resulted in investigations of copolypeptides. Polyalanylglycine was the first to be studied both experimentally (Itoh *et al.,* 1968) and theoretically (Moore and Krimm, 1976a). Far-infrared studies by Itoh *et al.* (1968) showed, quite nicely, the presence of bands at 440 and 250 cm^{-1}, indicative of the β aniparallel pleated-sheet structure. An additional medium band was observed at 332 cm^{-1} and assigned to backbone deformation mixed with a bending motion of the C_β atom (in the L-alanine residue) relative to the backbone. Although Moore and Krimm (1976a) generally agreed with these assignments, they reassigned the band at 250 cm^{-1} to a methyl torsion, since a mode with such character was confidently predicted by normal-coordinate calculations. Subsequent support for this reassignment came from neutron-scattering experiments (Drexel and Peticolas, 1975).

In related copolypeptides, low-frequency infrared studies of copoly-D,L-alanines (Itoh and Shimanouchi, 1970; Itoh *et al.,* 1968) and a series of copoly-tripeptides (Itoh and Katabuchi, 1972) have served to elucidate the effect of side-chain interactions on backbone conformation and lattice interactions.

V. Proteins

As infrared and Raman data provided a general understanding of the effect of the nature and distribution of amino acid side chains on polymer conformation, studies of small and intermediate-size proteins began to appear in the literature. In an investigation of seven globular proteins (including lysozyme, myoglobin, hemoglobin), Buontempo *et al.* (1971) found the far-infrared spectra to be remarkably similar, containing only a strong, broad band in the $100-200$-cm^{-1} region, which was attributed to the short sequences of α and β structure determined by the distribution of amino acid side chains of different length. This conclusion was supported by the work of Shotts and Sievers (1974) on myoglobin, casein, and egg albumin. In their infrared spectra at 4 K, only a broad continuum absorption in the $100-200$-cm^{-1} region was observed.

In a far-infrared study of sperm whale myoglobin, on the other hand, Chirgadze and Ovsepyan (1973) observed a series of weak absorption bands superimposed on the broad continuum (see Fig. 14). These bands, attributed to vibrations of amino acid residues of various length were observed to shift or disappear upon denaturation, indicating a conformational change. Similar weak features were observed in the far-infrared spectrum (Ataka and Tanaka, 1979) of single-crystal lysozyme supported on a silicon substrate. Bands at 387, 90, and 56 cm^{-1} were attributed to skeletal deformation vibrations, whereas those at 484 and 322 cm^{-1} were assigned to side-chain vibrations.

In a related study, far-infrared magnetic resonance (FIRMR) was used by

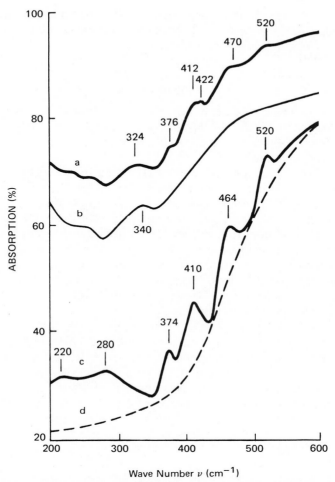

FIG. 14   Infrared spectra of sperm-whale myoglobin: (a) native protein, pH 6.6, T = 20°C;
(b) denatured protein, pH 9.5, T = 90°C; (c) difference spectrum, a − b; (d) background continuum component of spectrum. (From Chirgadze and Ovsepyan, 1973. © 1973 John Wiley
and Sons, Inc.)

Champion and Sievers (1980) to detect a magnetic-field-induced optical
absorption band at 3.5 cm⁻¹ in deoxymyoglobin and deoxyhemoglobin at
1.2 K. The very-low-frequency position of this band was used to determine
the energy level structure of the low-lying electronic states of high-spin
ferrous iron; its broad bandwidth (2–3 cm⁻¹) was thought to reflect the large
number of electronic-state configurations, corresponding to slightly different conformational structures of the protein.

Most recently, a preliminary investigation of collagen and egg-white albumin was reported by Evans *et al.* (1982), but no attempt at interpreting the low-frequency-infrared spectrum was made.

VI. Polynucleotides

The importance of low-frequency vibrations in determining the structure of such complex molecules as DNA has been acknowledged; however, studies of polynucleotides in the far infrared are rare, although Raman measurements have been reported.

Motivated by the theoretical prediction (Eyster and Prohofsky, 1974) that large-amplitude breathing and rocking modes of double-helical poly(I) · poly(C) RNA should appear in the far infrared, Beetz and Ascarelli (1976) were the first to measure its spectrum below 100 cm^{-1}. They observed a broad band at 40 cm^{-1} that was attributed to backbone motion of the A conformation (11-fold helices). By changing the NaCl concentration of the sample (at constant relative humidity), a structural phase transition to the A′ conformation (12-fold helices) was brought about, and the low-frequency band was observed to shift to 70 cm^{-1}, consistent with what would be expected for a change in conformation.

In a recent infrared study, Beetz and Ascarelli (1982) investigated the 500–25-cm^{-1} region of several nucleotides and poly(I)·poly(C) RNA at cryogenic temperatures. Changes in the infrared spectrum between 100–30 cm^{-1}, noted for the poly(I)·poly(C) RNA as a function of relative humidity, were correlated with changes in backbone conformation. Some sharpening of the spectral features of the polynucleotide were observed at 80 K and subsequently assigned to ribose-ring modes and out-of-plane base-ring vibrations.

VII. Conducting Polymers

A. POLYACETYLENE

Highly conducting polymers have been the subject of an increasing number of studies because of their attractive combination of electrical properties and potentially unique processing characteristics (Street and Clarke, 1980). Polyacetylene, $(-CH=CH-)_n$, has been by far the most widely studied conducting polymer, primarily because of its simple molecular structure and high conductivity (> 500 $\Omega \cdot$ cm^{-1}) after chemical doping.

B. THE DOPING MECHANISM

Numerous infrared and Raman studies of polyacetylene before, during, and after the doping process have been reported in an attempt to elucidate the nature and dynamics of the doping mechanism. Although there is general agreement that the doping procedure involves charge transfer between the dopant molecule and the polymer, there is some controversy (Clarke *et al.,* 1982) over the nature and extent of charge distribution remaining on the polymer chain.

The recent far-infrared measurements of Rabolt *et al.* (1982) and Hoffman *et al.* (1982) have provided useful information concerning the applicability of the "phase-kink" soliton model (Su *et al.,* 1979) to lightly doped polyacetylene. The soliton model predicted a strong far-infrared absorption by a pinned, charged soliton at 300 cm^{-1} in doped polyacetylene (Rice and Mele, 1980). Studies down to 20 cm^{-1} over a range of dopant concentrations using either AsF$_5$ (Rabolt *et al.,* 1982) or I$_2$ (Hoffman *et al.,* 1982) provided no evidence of any such band in the far infrared. Far-infrared spectroscopy should play an increasing role in understanding both the nature of doping and the extent of modification of the polymer conformation and crystal structure that results.

VIII.   Other Polymers of Interest

Although the polymers and copolymers treated thus far have attracted significant interest among far-infrared spectroscopists, there are a number of additional macromolecular systems that are equally as interesting but that have commanded less attention. Among those studied in detail are polybutene (Goldstein *et al.,* 1978), polyamides (Frank and Fiedler, 1979; Matsubara and Magill, 1973), polyethylene terephthalate (Manley and Williams, 1969; Frank *et al.,* 1978, 1981), polyphenylene terephthalamide (Kevlar) (Wong *et al.,* 1980), poly-*p*-xylylene (Mathur and Tabisz, 1974), polystyrene (Jasse and Monnerie, 1975; Spells *et al.,* 1977), polymethylmethacrylate (Chantry and Fleming, 1972), poly-bis (*p*-toluene sulfonate) diacetylene (Bloor and Kennedy, 1980), and ethylene-vinylacetate copolymers (Chantry *et al.,* 1972, 1973).

An area in which far-infrared spectroscopy may be particularly fruitful in elucidating molecular structure is that of ionomers. Rouse *et al.* (1979) studied ion clustering in polystyrene – methacrylic acid ionomers in the low-frequency infrared and located a vibration at 170 cm^{-1} that was subsequently assigned to the vibrations of aggregates involving many cations and anions. The potential for studying ionic interactions in polyelectrolytes is exciting and should be ideally suited for investigation in the far infrared.

IX.   Conclusion

The far-infrared region of the vibrational spectrum contains a wealth of information about the conformational and crystal structures of polymeric materials. In spite of the 15 or so years that Michelson interferometry has been used to study long-chain molecules, only the simplest structures have been fully understood. Perhaps with the resurgence of interest in the far infrared and the concomitant commercial availability of rapid-scan far-infrared interferometers, we are poised on the horizon of a commitment to renewed excitement and activity in both biological and synthetic macromolecules.

As evidenced by the discussion of the last several sections, there are a number of intriguing areas of study for the infrared and Raman spectroscopist that remain relatively untouched. Certainly the lack of availability of previously characterized materials has played a major role in limiting spectroscopic activities in many research areas. However, this has begun to change through collaborative efforts between researchers in spectroscopy and their materials science colleagues. The development of new product markets has brought about the commercial production and processing of new materials providing a ready supply of many novel compounds for characterization.

Normal coordinate analyses, on the other hand, have also reached a level of moderate sophistication whereby attempts to understand some of the new, more complex systems has become feasible. With all these tools in place it only remains for the spectroscopist to venture out onto these new frontiers of basic science.

ACKNOWLEDGMENTS

The assistance and expertise of Karen Bryan and Sylvia Fujii in the preparation of this chapter are very much appreciated and gratefully acknowledged.

REFERENCES

Abe, Y., and Krimm, S. (1972). *Biopolymers* **11**, 1841–1853.
Amrhein, E. M. (1972). *Ann. N.Y. Acad. Sci.* **196**, 179–194.
Amrhein, E. M., and Heil, H. (1971). *J. Phys. Chem. Solids* **32**, 1925–1933.
Ataka, M., and Tanaka, S. (1979). *Biopolymers* **18**, 507–516.
Bank, M., and Krimm, S. (1968). *J. Appl. Phys.* **39**, 4951–4958.
Bassett, D. C., Block, S., and Piermarini, G. J. (1974). *J. Appl. Phys.* **45**, 4146–4150.
Beetz, C., Jr., and Ascarelli, G. (1976). *Biopolymers* **15**, 2299–2301.
Beetz, C., Jr., and Ascarelli, G. (1982). *Biopolymers* **21**, 1569–1586.
Bell, R. J., and Goldman, G. M. (1967). *J. Opt. Soc. Am.* **57**, 1552–1553.
Berghmans, H., Stafford, G. J., and Leung, P. S. (1971). *J. Polym. Sci.-Polym. Phys. Ed.* **9**, 1219–1234.
Bertie, J. E., and Whalley, E. (1964). *J. Chem. Phys.* **41**, 575–576.
Bloor, D., and Kennedy, R. J. (1980). *Chem. Phys.* **47**, 1–7.
Boerio, F. J., and Koenig, J. L. (1971). *J. Chem. Phys.* **54**, 3667–3669.
Brandli, G., and Sievers, A. J. (1972). *Phys. Rev.* **B 5**, 3550–3557.
Brasch, J. W., and Jacobsen, R. J. (1964). *Spectrochim. Acta* **20**, 1644–1646.

Buontempo, U., Careri, G., Fasella, P., and Ferraro, A. (1971). *Biopolymers* **10**, 2377–2386.
Carazzolo, G., and Mammi, M. (1963). *J. Polym. Sci.* **1**, 965–983.
Chamberlain, J. E., and Gebbie, H. A. (1966). *Appl. Opt.* **5**, 393–396.
Champion, P. M., and Sievers, A. J. (1980). *J. Chem. Phys.* **72**, 1569–1582.
Chantry, G. W. (1971). "Submillimetre Spectroscopy." Academic Press, New York.
Chantry, G. W., and Fleming, J. W. (1972). *Brit. Polym. J.* **4**, 279–290.
Chantry, G. W., Fleming, J. W., Pardoe, G. W. F., Reddish, W., and Willis, H. A. (1971). *Infrared Phys.* **11**, 109–118.
Chantry, G. W., Fleming, J. W., and Nicol, E. A. (1972a). *Infrared Phys.* **12**, 101–107.
Chantry, G. W., Fleming, J. W., Nichol E. A., Willis, H. A., and Cudby, M. E. A. (1972b). *Chem. Phys. Lett.* **16**, 141–144.
Chantry, G. W., *et al.* (1973). *Infrared Phys.* **13**, 157–160.
Chantry, G. W., *et al.* (1974). *Polymer* **15**, 69–73.
Chantry, G. W., Nichol, E. A., Willis, H. A., and Cudby, M. E. A. (1981). *Int. J. Infrared Millimeter Waves* **2**, 97–105.
Chirgadze, Y. N., and Ovsepyan, A. M. (1973). *Biopolymers* **12**, 637–645.
Clark, E. S., and Muus, L. T. (1962). *Z. Kristallogr.* **117**, 119–126.
Clarke, T. C., McQuillan, B. W., Rabolt, J. F., Scott, J. C., and Street, G. B. (1982). *Mol. Cryst. Liq. Cryst.* **83**, 1033–1048.
Cortili, G., and Zerbi, G. (1967). *Spectrochim. Acta* **23A**, 285–299.
Crick, F. H. C., and Rich, A. (1955). *Nature* **176**, 780–781.
Davies, G. J., and Haigh, J. (1974). *Infrared Phys.* **14**, 183–188.
Dean, G. D., and Martin, D. H. (1967). *Chem. Phys. Lett.* **1**, 415–416.
Decamps E., Chanal, D., and Hadni, A. (1967). *J. Phys.* **28**, 120–122.
Doll, W. W., and Lando, J. B. (1970). *J. Macromol. Sci-Phys.* **4**, 309–329.
Drexel, W., and Peticolas, W. L. (1975). *Biopolymers* **14**, 715–721.
Evans, G. L., Evans, M. W., and Pethig, R. (1982). *Spectrochim. Acta* **38A**, 421–422.
Eyster, J. M., and Prohofsky, E. W. (1974). *Biopolymers* **13**, 2527–2543.
Fanconi, B. (1973). *Biopolymers* **12**, 2759–2776.
Fateley, W. (1966). *Appl. Spectrosc.* **20**, 190–196.
Fleming, J. W., *et al.* (1972). *Chem. Phys. Lett.* **17**, 84–85.
Frank, W. F. X. (1977). *J. Polym. Sci.-Polym. Lett. Ed.* **15**, 679–682.
Frank, W. F. X., and Fiedler, H. (1979). *Infrared Phys.* **19**, 481–489.
Frank, W. F. X., Schmidt, H., and Wulff, W. (1977). *J. Polym. Sci., Polym. Symp.* **61**, 317–326.
Frank, W. F. X., Fiedler, H., and Strohmeier, W. (1978). *J. Appl. Polym. Sci.: Appl. Polym. Symp.* **34**, 75–87.
Frank, W. F. X., *et al.* (1981a). *Polymer* **22**, 17–19.
Frank, W. F. X., Strohmeier, W., and Hallensleben, M. L. (1981b). *Polymer* **22**, 615–618.
Frenzel, A. O., and Butler, J. P. (1964). *J. Opt. Soc. Am.* **54**, 1059–1060.
Galperin, Y., Strogalin, L. V., and Mlenik, M. P. (1965). *Vysokomol. Soedin* **7**, 933–936.
Goldstein, M., Seeley, M. E., and Willis, H. (1978). *Polymer* **19**, 1118–1122.
Goldstein, M., Seeley, M. E., Willis, H. A., and Zichy, V. J. I. (1973). *Polymer* **14**, 530–534.
Haigh, J., Ali, A. S. M., and Davies, G. J. (1975). *Polymer* **16**, 714–716.
Hasegawa, R., Takahashi, Y., Chatani, Y., and Tadokoro, H. (1972). *Polym. J.* **3**, 600–610.
Hoffman, D. M., Tanner, D. B., Epstein, A. J., and Gibson, H. W. (1982). *Mol. Cryst. Liq. Cryst.* **83**, 143–150.
Huggins, M. L. (1945). *J. Chem. Phys.* **13**, 37–42.
Isemura, T., Okabayashi, H., and Sakakibara, S. (1968). *Biopolymers* **6**, 307–321.
Itoh, K., and Katabuchi, H. (1972). *Biopolymers* **11**, 1593–1605.

Itoh, K., and Katabuchi, H. (1973). *Biopolymers* **12**, 921–929.

Itoh, K., and Shimanouchi, T. (1970). *Biopolymers* **9**, 383–399.

Itoh, K., Oya, M., and Shimanouchi, T. (1972). *Biopolymers* **11**, 1137–1148.

Itoh, K., Nakahara, T., Shimanouchi, T., Oya, M., Uno, K., and Iwakura, Y. (1968). *Biopolymers* **6**, 1759–1766.

Itoh, K., Katabuchi, H., and Shimanouchi, T. (1972). *Nature (London), New Biol.* **89**, 42–43.

Itoh, K., Shimanouchi, T., and Oya, M. (1969). *Biopolymers* **7**, 649–658.

Jasse, B., and Monnerie, L. (1975). *J. Phys. D.* **8**, 863–871.

Johnson, K. W., and Rabolt, J. F. (1973). *J. Chem. Phys.* **58**, 4536–4538.

Jones, R. G., *et al.* (1976). *Polymer* **17**, 153–154.

Kilian, H. G. (1962). *Kolloid Z. Z. Polym.* **185**, 13–16.

Kobayashi, M., Tashiro, K., and Tadokoro, H. (1975). *Macromolecules* **8**, 158–171.

Kondrashov, Y. D. (1960). *Tr. Gipkh'a* **46**, 166–168.

Krimm, S. (1954). *J. Chem. Phys.* **22**, 567–568.

Krimm, S. (1960). *Fortschr. Hochpolym.-Forsch.* **2**, 51–172.

Krimm, S., and Bank, M. (1965). *J. Chem. Phys.* **42**, 4059–4060.

Lando, J. B., Olf, H. G., and Peterlin, A. (1966). *J. Polym. Sci.-Polym. Chem. Ed.* **4**, 941–951.

Latour, M., Montaner, A., Galtier, M., and Geneves, G. (1981). *J. Polym. Sci.-Polym. Phys. Ed.* **19**, 1121–1129.

Leute, U. (1981). *Polym. Bull.* **4**, 89–96.

Luongo, J. P. (1972). *J. Polym. Sci.-Polym. Phys. Ed.* **10**, 1119–1123.

Manley, T. R., and Williams, D. A. (1969). *J. Polym. Sci.-Polym. Symp.* 1009–1018.

Mather, M. S., and Tabisz, G. C. (1974). *J. Cryst. Mol. Struct.* **4**, 23–29.

Matsubara, I., and Magill, J. H. (1973). *J. Polym. Sci.-Polym. Phys. Ed.* **11**, 1173–1187.

Matsui, Y., Kubota, T., Tadokoro, H., and Yoshihara, T. (1965). *J. Poly. Sci. Part A* **3**, 2275–2288.

Matsuura, H., and Miyazawa, T. (1968). *Bull. Chem. Soc. Japan* **41**, 1798–1808.

May, L., and Schwing, K. J. (1963). *Appl. Spectrosc.* **17**, 166–168.

McCarthy, D. E. (1967). *J. Opt. Soc. Amer.* **57**, 699–700.

McKnight, R. V., and Möller, K. D. (1964). *J. Opt. Soc. Am.* **54**, 132–133.

Miyazawa, T. (1961). *Bull. Chem. Soc. Japan* **34**, 691–696.

Miyazawa, T., Fukushima, K., and Ideguchi, Y. (1962). *J. Chem. Phys.* **37**, 2764–2776.

Miyazawa, T., Fukushima, K., and Ideguchi, Y. (1963). *J. Polym. Sci. Part B* **1**, 385–387.

Miyazawa, T., Fukushima, K., Sugano, S., and Masuda, Y. (1967). *In* "Conformation of Biopolymers." (G. N. Ramachandran, ed.), pp. 557–568. Academic Press, New York.

Mon, K. K., and Sievers, A. J. (1975). *Appl. Opt.* **14**, 1054–1055.

Moore, W. H., and Krimm, S. (1976a). *Biopolymers* **15**, 2465–2483.

Moore, W. H., and Krimm, S. (1976b). *Biopolymers* **15**, 2439–2464.

Natta, G., *et al.* (1965). *J. Polym. Sci. Part A* **3**, 4263–4278.

Pauling, L., and Corey, R. B. (1951). *Proc. Natl. Acad. Sci. U.S.* **37**, 729.

Pauling, L., and Corey, R. B. (1953). *Proc. Natl. Acad. Sci. U.S.* **39**, 253–256.

Piseri, L., Cabassi, F., and Masetti, G. (1975). *Chem. Phys. Lett.* **33**, 378–380.

Prettl, W., and Frank, W. (1979). *Proc. 2nd Int. Conf. Infrared Phys.,* 409–411.

Rabolt, J. F. (1983). *J. Polym. Sci.-Polym. Phys. Ed.* **21**, 1797.

Rabolt, J. F., and Johnson, K. W. (1973). *J. Chem. Phys.* **59**, 3710–3712.

Rabolt, J. F., and Rein, A. (1983). IBM Instruments Inc., Application Note. Danbury, Connecticut.

Rabolt, J. F., Johnson, K. W., and Zitter, R. N. (1974). *J. Chem. Phys.* **61**, 504–506.

Rabolt, J. F., Wedding, W., and Johnson, K. W. (1975). *Biopolymers* **14**, 1615–1622.

Rabolt, J. F., Moore, W. H., and Krimm, S. (1977). *Macromolecules* **10**, 1065–1074.

Rabolt, J. F., Clarke, T. C., and Street, G. B. (1982). *J. Chem. Phys.* **76,** 5781.

Rice, M. J., and Mele, E. J. (1980). *Solid State Commun.* **35,** 487–492.

Rouse, G. B., Risen, W. M., Jr., Tsatsas, A. T., and Eisenberg, A. (1979). *J. Polym. Sci.-Polym. Phys. Ed.* **17,** 81–85.

Schlotter, N. E., and Rabolt, J. F. (1984). *Macromolecules* (in press).

Shotts, W. J., and Sievers, A. J. (1973). *Chem. Phys. Lett.* **21,** 586–588.

Shotts, W. J., and Sievers, A. J. (1974). *Biopolymers* **13,** 2593–2614.

Spells, S. J., Shepherd, I. W., and Wright, C. J. (1977). *Polymer* **18,** 905–912.

Stafford, G. J., Naumann, A. W., and Simon, F. T. (1966). *J. Chem. Phys.* **45,** 3787–3794.

Street, G. B., and Clarke, T. C. (1980). *ACS Adv. in Chem.* **186,** 177.

Su, W. P., Schrieffer, J. R., and Heeger, A. J. (1979). *Phys Rev. Lett.* **42,** 1698–1701.

Suzuki, S, Iwashita, Y., Shimanouchi, T., and Tsuboi, M. (1966). *Biopolymers* **4,** 337–350.

Tadokoro, H. (1979). "Structure of Crystalline Polymers." Wiley (Interscience), New York.

Takahashi, Y., and Tadokoro, H. (1973). *Macromolecules* **6,** 672–675.

Tashiro, K., Kobayashi, M., and Tadokoro, H. (1981). *Macromolecules* **14,** 1757–1764.

Tasumi, M., and Krimm, S. (1967). *J. Chem. Phys.* **46,** 755–766.

Tasumi, M., and Shimanouchi, T. (1965). *J. Chem. Phys.* **43,** 1245–1258.

Tomita, K., Rich, A., de Lozé, C., and Blout, E. R. (1962). *J. Mol. Biol.* **4,** 83.

Trevino, H., and Boutin, S. (1967). *J. Macromol. Sci.* **A1,** 723–746.

Ulrich, R. (1969). *Appl. Opt.* **8,** 319–322.

Weeks, J. J., Clark, E. S., and Eby, R. K. (1981). *Polymer* **22,** 1480–1486.

Willis, H. A., Miller, R. G. J., Adams, D. M., and Gebbie, H. A. (1963). *Spectrochim. Acta* **19,** 1457–1461.

Willis, H. A., *et al.* (1975). *Chem. Phys. Lett.* **33,** 381–383.

Wong, P. T. T., Garton, A., Carlsson, D. J., and Wiles, D. M. (1980). *J. Macromol. Sci., Phys.* **B 18,** 313–324.

Yokoyama, M., Ochi, H., Tadokoro, H., and Price C. (1972). *Macromolecules* **5,** 690–698.

Yoshihara, T., Tadokoro, H., and Murahashi, S. (1964). *J. Chem. Phys.* **40,** 2902–2911.

Zerbi, G., and Massetti, G. (1967). *J. Mol. Spectrosc.* **22,** 284–289.

Zirke, J., and Meissner, M. (1978). *Infrared Phys.* **18,** 871–876.

CHAPTER 3

Infrared Magnetooptical Spectroscopy in Semiconductors and Magnetic Materials in High Pulsed Magnetic Fields

Noboru Miura

Institute for Solid State Physics
University of Tokyo
Roppongi, Minato-ku, Tokyo, Japan

I. Introduction

Infrared and far-infrared magnetooptical spectroscopy, which employs a combination of molecular gas lasers with a high-field magnet, is a powerful tool for the investigation of various branches of solid-state physics. In semiconductors, semimetals, and magnetic substances the motion of con-

73

duction electrons and electron spins are greatly influenced by high magnetic fields. The infrared and far-infrared electromagnetic radiation that interacts with electrons can be used to investigate the electronic states that are perturbed by high magnetic fields.

In the last two decades much progress has been made in the development of infrared and far-infrared lasers and also in the generation of high magnetic fields. Button et al. (1966) first succeeded in observing the quantum cyclotron resonance in p-Ge in the far-infrared range using an HCN laser and water-cooled high-field magnets. Since then much work has been done on far-infrared magnetooptical spectroscopy in high magnetic fields. With the widespread use of superconducting magnets, this sort of measurement has become popular in many laboratories. The stationary magnetic fields produced by either water-cooled or superconducting magnets are presently limited to 25 T. Even using the most advanced hybrid magnets the highest available field is 30 T.

The recent development of techniques for generating high-pulsed magnetic fields has greatly extended the magnetic field range in which measurements can be done. Kapitza was the first to produce a high-pulsed magnetic field (Kapitza, 1924). Much progress has been achieved since then for pulsed-magnet technology. At the Institute for Solid State Physics (ISSP) of the University of Tokyo, high pulsed fields up to 280 T (2.8 megagauss) were generated and used conveniently for various solid-state experiments (Miura et al., 1979a). In addition, nondestructive pulsed fields up to 45 T were generated with a very long-duration time (half-period of the pulse is 23 msec) (Miura et al., 1981). Such long-pulsed fields can be used to obtain measurements with a higher accuracy.

A pulsed magnetic field can be generated basically by apparatus that is compact and inexpensive in comparison with that used for steady high fields. Although the duration of the field is short, we can now perform various solid-state measurements in pulsed fields using recently developed electronic instruments such as transient recorders, detectors of infrared and far-infrared radiation with a fast response, fast preamplifiers, etc. Laser spectroscopy is one of the most suitable experimental methods for use in pulsed fields, and high-precision experiments in laser spectroscopy have now become possible using high pulsed fields. The pulsed-field measurement has some advantages over the steady-field measurements. For example, because the measurement is accomplished instantaneously, we need not worry about the fluctuation of the output power of the lasers or the temperature of the samples. Sometimes the experimental accuracy of the pulsed-field measurement is comparable to that of the steady-field measurement. The duration of the pulsed field is generally much longer than the relaxation times of various electronic processes in solids, so that the measurements can be made when the electronic system is in an equilibrium state.

Very high magnetic fields produced by pulsed magnets open up a variety of new possibilities. Electronic states in solids are so much affected by such high fields that we can expect to observe new nonlinear or striking effects. Let us consider the motion of the conduction electrons, for example. In high magnetic fields, the energy of the cyclotron orbital motion $\hbar\omega_c = \hbar eH/m^*c$ becomes very high and may even exceed various excitation energies in solids such as bandgap, optical-phonon energy, exciton, or impurity-binding energy, etc. Figure 1 shows the magnitude of $\hbar\omega_c$ for the case of the effective mass $m^* = 0.1m_0$ (m_0 is the free-electron mass) and the radius of the cyclotron orbit $r_c = \sqrt{\hbar c/eH}$ in high magnetic fields. At 100 T, for instance, $\hbar\omega_c$ reaches 116 meV (1340 K). Thus the cyclotron-resonance experiments in such high fields enable us to investigate the energy-band structure far from the band extremes, the polaron states near the polaron pinning region, the relaxation processes of electrons in the extreme quantum limit, and so on. Another advantage of the high magnetic fields is that we can study the electronic states in low-mobility materials, because the condition $\omega_c\tau > 1$ can be readily fulfilled even with small τ. Figure 1 also shows the

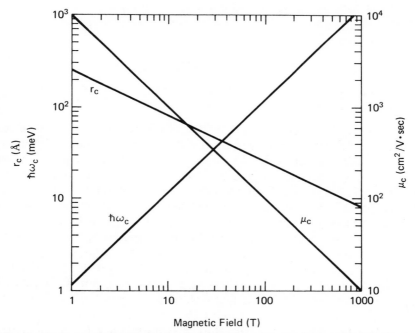

FIG. 1   The magnetic-field dependence of the radius of the cyclotron orbital motion $r_c = (\hbar c/eH)^{1/2}$, the energy of the cyclotron energy $\hbar\omega_c = \hbar eH/m^*c$, and the minimum mobility μ_c required for carriers to satisfy the condition $\omega_c\tau > 1$. The line for $\hbar\omega_c$ is shown for $m^* = 0.1m_0$.

value of the minimum mobility of samples μ_c (cm²/V·sec) for which the condition $\omega_c \tau > 1$ can be satisfied. It is seen that at 100 T, μ_c has only to be larger than 100 cm²/V·sec for the condition $\omega_c \tau > 1$. If we can employ fields up to 1000 T, even the cyclotron resonance in amorphous semiconductors may become observable. As is shown in Fig. 1, the radius of the cyclotron orbital motion r_c becomes 26 Å at 100 T. At 1000 T, it is reduced to 8 Å, which is almost comparable with the lattice spacing in most crystals. Consequently, the application of very high magnetic fields may bring about various electronic phase transitions.

On the other hand, the electron-spin states are also greatly altered by the high magnetic fields. The Zeeman splitting energy $g\mu_B H$ can become extremely large in high magnetic fields and may exceed the internal fields in solids. For example, fields of several megagauss are comparable with the exchange field of iron. Therefore, various phase transitions in spin states are expected to occur in very high magnetic fields.

In the following, we will describe recent achievements of infrared and far-infrared magnetooptical spectroscopy in high pulsed-magnetic fields at ISSP. Many new nonlinear phenomena characteristic of the high magnetic fields have been found in semiconductors, semimetals, and magnetic substances.

II. Experimental Techniques

A. GENERATION OF NONDESTRUCTIVE PULSED MAGNETIC FIELDS

A steady magnetic field higher than 30 T is difficult to produce. By using a pulsed magnet, however, we can easily obtain a field beyond this limit. In producing high magnetic fields we are faced with two major problems. One is the large power consumption in magnets. For steady-field magnets using resistive coils, a large power supply is required. In addition, the large Joule heat produced in the magnet should be taken off by a large cooling system (Landwehr, 1981). There are several high-magnetic-field laboratories where steady high fields are produced by employing water-cooled resistive magnets combined with a large power supply and a large cooling system (Kurti, 1975). To produce fields up to 20 to 25 T, about 10 MW of power and several hundred tons of cooling water per hour are required. Superconducting magnets can generate a field without producing Joule heat, but the generated fields are limited to below 17.5 T because of the presence of the critical field H_{c2} (Tachikawa, 1981). Unless entirely new superconducting materials with high H_{c2} are developed, it will be difficult to produce fields much above this value.

In the pulsed magnets, the Joule heat produced is absorbed by the heat capacity of the magnet, which is usually precooled by liquid nitrogen and allowed to heat up to about 100°C during the pulse. The magnet can absorb energy of about 100 kJ/kg copper. The difficulty arising from the Joule heat can be reduced by shortening the pulse duration at the expense of the convenience of the measurements. When the field becomes very high, however, we must consider the second problem: the large electromagnetic force exerted on the coil. This force, called Maxwell stress, is proportional to the square of the field, so that it becomes a serious problem at high fields. In fact, the magnets will be destroyed by the electromagnetic force if fields exceeding some limitation are generated.

At ISSP, pulsed magnetic fields up to 45 T are generated by copper-wire multiturned magnets (Miura *et al.*, 1981). Figure 2 shows a cross section of the pulsed magnets employed at ISSP. The coil is wound with rectangular copper wire 1.8×3.5 mm^2 in cross section. The wire has three-fold insulation layers of polyvinyl formyl, polyimid tape, and glass-fiber tape. Specially strengthened copper wire containing Zr and Cr is used for a stronger coil. This special wire has a much higher yield strength than ordinary copper and only 160% larger resistivity at 77 K. The wire is wound on a bobbin made of glass epoxy (FRP). After the winding of the wire, many

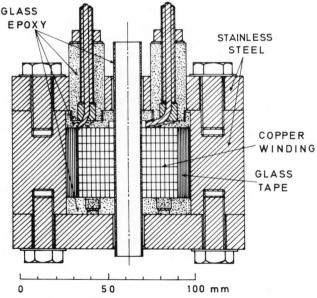

FIG. 2 Cross-sectional view of a nondestructive pulse magnet. The number of windings and the actual sizes differ from magnet to magnet. This figure shows a relatively small magnet.

layers of glass-fiber tape are tightly wound on it, and the coil is impregnated with epoxy resin. The expoxy resin is solidified under a high pressure of about 50 bar at 140°C in oil using a high-pressure apparatus. This method of solidification of the epoxy resin increases the strength of the magnet. After the solidification, the outer surface of the coil is machined with a lathe, and the coil is pushed tightly into a cylinder of stainless steel.

The magnet is immersed in a liquid-nitrogen bath for precooling to 77 K. As a current source, a condenser bank with a total storing energy of 200 kJ is used. The condenser bank consists of two blocks, and we can select a series or a parallel connection simply by a control switch (Miura et al., 1981). Thus the bank can be used either as 10-kV, 4-mF condenser or a 5-kV, 16-mF condenser. This changeable function of the bank is useful to change the duration of the pulsed field. Four ignitrons are used to switch the current.

The magnet shown in Fig. 2 is one of the smallest models we are using. This type of magnet can produce a field up to 45 T quite safely. The rise time of the field (a quarter period, $T/4$) for this magnet is about 2 msec. With a larger magnet having an outermost diameter of 240 mm, the rise time of 11 msec is obtained, but the maximum field in this case is 37 T.

The pulsed magnets and the condenser bank as mentioned previously are very convenient for solid-state experiments, because of a fairly long pulse duration and readiness for the repetition of experiments. Another type of magnet made of a beryllium–copper helix can be also used for solid-state experiments (Suzuki and Miura, 1975). The coil is machined by a lathe from a solid Be–Cu rod into the shape of a helix. This type of magnet was first developed by Foner and Kolm (1957). The helix-type coil is capable of generating higher fields than the multiturned coil, but has the disadvantage of a shorter duration of the field pulse. In addition, it needs elaborate machining.

B.   GENERATION OF MEGAGAUSS FIELDS

The generation of very high magnetic fields in the megagauss range (higher than 1 MG = 100 T) is a difficult task because of the large electromagnetic force acting on the coil. The Maxwell stress reaches 200 kg/mm^2 at 70 T and 400 kg/mm^2 at 100 T. Therefore it becomes larger than the yield strength of any materials from which we can construct the magnet. Accordingly, any magnets will likely be broken in fields higher than 70 T.

To produce magnetic fields higher than 70 T, various techniques have been developed. Fowler et al. (1960) generated fields up to 1400 T by an explosive-driven magnetic-flux-compression technique. Because chemical explosives are used as an energy source, this is a very destructive method. Fowler et al. (1976) also generated fields up to about 250 T by using a bellows-type explosive-driven flux compression. Although this technique utilizes chemical explosives, it is less destructive and the generated fields were actually applied in magnetooptical measurements.

Electromagnetic flux compression is the most suitable technique for application to solid-state experiments because of the less destructive characteristics and the easy controllability. Figure 3 shows the schematic diagram of the coil system of electromagnetic flux compression. The principle of the megagauss field generation is as follows.

Inside a primary coil made of an iron plate, a liner made of a copper ring is set coaxially. When a large primary current is supplied from a condenser bank to the primary coil, a secondary current is induced in the liner to shield the magnetic field. The magnitude of the secondary current is almost the same as that of the primary current, but its direction is opposite. A large electromagnetic repulsive force rapidly squeezes the liner inwards. During this pinch process, if we have injected a seed field in the inner area of the liner by a pair of injection coils in advance, the magnetic flux in the inner area of the liner is compressed by the motion of the liner. Thus the field at the center is increased almost in proportion to the inverse of the inner area, and very high fields can be obtained when the diameter of the liner is sufficiently decreased.

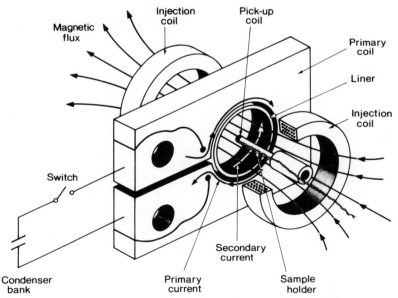

FIG. 3 A coil system for generating megagauss magnetic fields by electromagnetic flux compression.

By using this technique, megagauss fields up to 280 T have been generated (Miura *et al.,* 1979a). For supplying the primary current, a main condenser bank with a stored energy of 285 kJ (30 kV) is used. Twenty ignitron switches are employed to switch the primary current. For injecting the seed fields, a subcondenser bank of 72 kJ (3.3 kV) is used. Liners are machined by a lathe from copper pipes into 66 mm in diameter, 20 mm in length, and 1 – 1.5 mm in thickness. A phenolic pipe is inserted between the primary coil and the liner, for insulation and for vacuum seal. An example of the observed waveforms of the magnetic-field pulse is shown in Fig. 4. The field rises from a few tens of tesla to the megagauss range (> 100 T) in several microseconds. After the field reaches the maximum value the field probe is destroyed. At this instant the sample is destroyed together with the liner, the insulating phenolic pipe, and other surrounding materials. Therefore, care must be taken to obtain as much information as possible in every experiment.

The entire process of electromagnetic-flux compression can be well represented by a computer simulation because it is described by a combination of the kinetic equation and classical electromagnetic dynamics. Figure 5 shows an example of the results of the computer simulation (Miura and Chikazumi, 1979). Parameters necessary for the calculation are chosen in accordance with the actual experiment. The temperature rise of the liner, the nonuniform field distribution in the liner are taken into account. The primary current, the secondary current, the radius of the liner, the magnetic

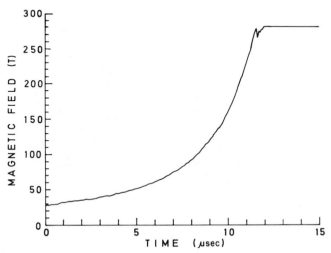

FIG. 4   An example of the experimental trace of a generated megagauss field pulse. (From Miura *et al.,* 1981.)

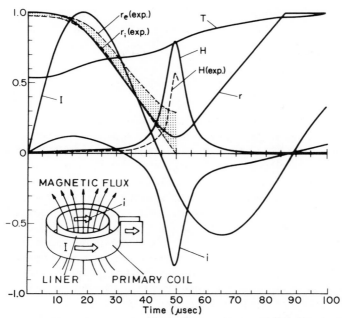

FIG. 5 The result of a computer simulation of the electromagnetic flux compression compared with the experimental curves: I is the primary current, i the secondary current in the liner, r the radius of the liner, T the temperature of the liner, and H the magnetic field; r_e(exp.) and r_i(exp.) are the experimental curves for the outer and the inner radius of the liner, and H(exp.) the experimental curve for the magnetic field. The scale of the vertical axis is normalized with respect to the maximum value of each quantity. (From Miura *et al.*, 1979a.)

field, the temperature of the liner are calculated as a function of time. Experimentally observed curves for the radius of the liner and the magnetic field are shown for comparison. Agreement between experiment and calculation is reasonably good.

At ISSP, a new project is being conducted to generate even higher fields by installing a main condenser bank of 5 MJ (40 kV) and a subcondenser bank of 1.5 MJ (10 kV). The 5-MJ bank uses 240 gap switches and can supply 6 MA of current. It is expected that very high fields up to 500 to 1000 T will be generated (Kido *et al.*, 1983; Miura *et al.*, 1981).

A direct discharge into a one-turn coil is an alternative method for generating megagauss fields suitable for solid-state experiments. A large-enough current pulse supplied to a small one-turn coil with a bore diameter of 5 to 10 mm can produce very high fields. Although the coil is broken by the large electromagnetic force, fields of the megagauss range are obtained before the destruction of the coil if the current pulse is sufficiently fast. Herlach *et al.* (1974) employed this technique for cyclotron resonance in the megagauss range. We have also installed at ISSP an ultrafast condenser bank for megagauss-field generation by this technique, and fields higher than 200 T have been obtained.

C. INFRARED AND FAR-INFRARED SPECTROSCOPY IN PULSED MAGNETIC FIELDS

As radiation sources for the infrared and far-infrared magnetooptical spectroscopy in pulsed magnetic fields, molecular-gas lasers are operated at various wavelengths: 337 and 311 μm with an HCN laser, 119, 28, and 16.9 μm with an H_2O laser, and 9.4 – 11.9 μm (selective) with a CO_2 laser. A molecular gas laser pumped optically by a CO_2 laser is also useful to obtain many lines with different wavelengths. For the nondestructive fields with rise time 1 – 10 msec, the lasers are used in continuous operation. From an H_2O laser and an HCN laser with resonator lengths of 4 and 3 m, respectively, about 1 mW of far-infrared radiation can be obtained. In the case of the H_2O laser, a circularly polarized radiation at 119 μm is obtained, whereas from the HCN laser, a random polarization is obtained. Figure 6 shows a block diagram of the system for measuring cyclotron resonance. Light pipes are employed to guide the laser radiation. For magnetoreflection measurements in pulsed fields where the available space is limited, a light pipe can also be used for the propagation of both the incident and the reflected radiation by a combination with a Ge half-mirror (Hiruma *et al.*, 1981; Suzuki *et al.*, 1976).

When it is necessary to keep the polarization of the laser radiation either circular or linear, a focusing system using mirrors is also employed. A block

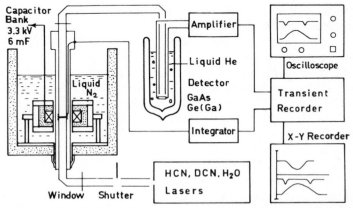

FIG. 6   Block diagram of a light-pipe system for measuring the infrared cyclotron resonance in nondestructive pulsed magnetic fields. (From Kido *et al.*, 1981.)

diagram of such a system for measuring magnetoreflection is shown in Fig. 7 (Kido *et al.,* 1983). In this case the pulsed magnet is set with a horizontal axis. In both systems, as shown in Fig. 6 and Fig. 7, the temperature of a sample is cooled by flowing liquid helium or liquid nitrogen through the narrow path surrounding the sample. The temperature is monitored by a thermocouple. The lowest temperature attained is about 6 K.

For measurements in megagauss fields, pulsed laser radiation is employed. An HCN laser, an H_2O laser, and a CO_2 laser with resonator lengths of about 2 m are energized by a small condenser bank. For operating an HCN laser, a bias current of about 1 A is supplied for stabilizing the laser oscillation. The laser lines of 119, 36.6, 28.0, and 16.9 μm from an H_2O laser are convenient for cyclotron resonance in the megagauss range. When we use an H_2O laser in a pulsed operation, however, the duration of the pulse is usually very short (of the order of 10^{-6} sec), irrespective of the long duration of the exciting discharge pulse. Because the rise time of the megagauss fields is several microseconds, laser pulses with a sufficiently long flat-top part (of the order of 10^{-5} sec) is necessary. The admixture of a large amount of helium gas to the discharge not only makes the pulse duration longer, but also enhances the peak power of the output (Kido and Miura, 1978). The effect on the 16.9-μm line is of particular importance because the oscillation of this line is usually weak and short. Waveforms of the prolonged laser pulses are shown in Fig. 8. For the 28.0-μm line, a maximum power of 380 mW is obtained with a duration of about 100 μsec by mixing He gas with H_2O gas.

Figure 9 shows a block diagram of the system for measuring cyclotron resonance in megagauss fields. The coil system is enclosed in a protecting box to save the measuring apparatus from the scattering fragments of the liner and to shield the electromagnetic noise. Because of this protecting

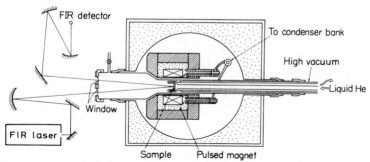

FIG. 7 Block diagram of a system for measuring the infrared magnetoreflection in nondestructive pulsed magnetic fields. The mirror system is used for preserving the polarization of the radiation. (From Kido *et al.,* 1983.)

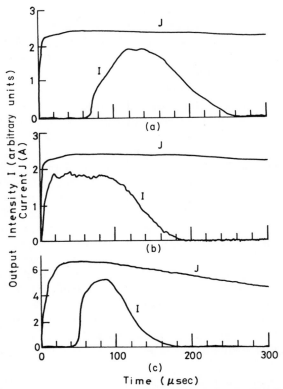

FIG. 8 Elongated pulses of the radiation from an H_2O laser at various wavelengths λ: (a) 16.93 μm, (b) 27.97 μm, and (c) 36.60 μm; I and J represent the intensity of the current and the radiation, respectively. (From Kido and Miura, 1978.)

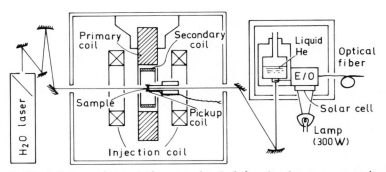

FIG. 9 Block diagram of a system for measuring the infrared cyclotron resonance in mega-gauss fields. The H_2O laser can be replaced by others such as a CO_2 or an HCN laser for different wavelengths of radiation. The detector is cooled to liquid-helium temperature in a cryostat set in a shielded box.

box, we can set the measuring apparatus close to the coil system. The radiation of the lasers is focused on the sample by a mirror system. The radiation transmitted by the sample is also focused by a concave mirror on a detector. To refrigerate the sample in megagauss fields, specially designed sample holders such as shown in Fig. 10 are employed (Kido *et al.*, 1981). The holder is made of double glass pipes sealed at the edge. By flowing a large amount of liquid helium, the temperature of the samples can be lowered to about 4 K. For the liquid-nitrogen temperature range, sample holders with a simpler structure are employed with a flow of liquid nitrogen.

As fast infrared detectors for the magnetooptical measurements in pulsed fields, extrinsic photoconductivity of Ga-doped Ge and GaAs is used. The detectors are cooled by liquid helium. These photoconductive detectors have sufficiently fast response time and high sensitivity. The signals are amplified by a high-speed preamplifier (Kido *et al.*, 1976).

The magnetic field is measured by integrating a voltage induced in a pickup coil wound around the sample. The sensitivity of the pickup coil is calibrated by various means: by a comparison with the Faraday rotation (Kido *et al.*, 1976), with the electron spin resonance (ESR) in ruby (Kido and Miura, 1982), and with peaks of the Shubnikov–de Haas or the magneto–phonon resonance (Hiruma *et al.*, 1977). The overall accuracy of the measurement of the field is estimated to be within ± 3% for megagauss fields and within ± 1% for nondestructive fields below 45 T.

The signals of the magnetic field and the radiation intensity are digitized and memorized by a transient recorder connected to a computer. As an example of the experimental recordings in nondestructive pulsed fields, the results of the electron-spin resonance in ruby are shown in Fig. 11. The identical resonances appear twice on the rising and falling slope of the magnetic field at the same positions. Three peaks are resolved showing good resolution of the measurement. Figure 12 shows an experimental recording of cyclotron resonance in *n*-type Ge as an example of the measurement in

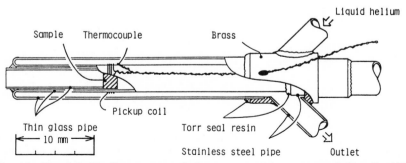

FIG. 10 Sample holder for magnetooptical measurements in megagauss fields at liquid-helium temperatures. (From Kido *et al.*, 1981.)

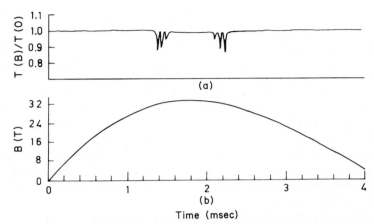

FIG. 11   Experimental recording of the electron-spin resonance in ruby (0.01% Cr^{3+}), where $v = 0.89096$ THz, B ∥ c, and temperature is 300 K: (a) transmission of far-infrared radiation at wavelength $\lambda = 337$ μm; (b) magnetic-field pulse.   (From Kido and Miura, 1982.)

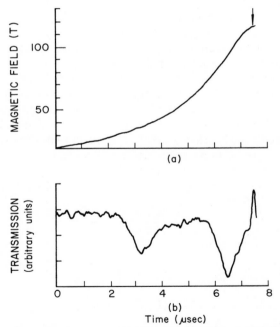

FIG. 12   Experimental recording of the cyclotron resonance in n-type Ge in megagauss fields at 300 K: (a) magnetic-field pulse; (b) transmission of infrared light at $\lambda = 28.0$ μm, **H** ∥ ⟨111⟩. At the time indicated by the arrow, the sample is destroyed.   Two resonance peaks are observed.   (From Miura et al., 1976a.)

megagauss fields. The two resonance peaks are well resolved. The figure shows a good signal-to-noise ratio and fast-enough response time of the detecting system. If we replot the data shown in Figs. 11 and 12 as a function magnetic field by using a computer treatment, we can obtain the usual curves of the magnetooptical spectra, as shown in the following sections.

III. Cyclotron Resonance and Electron Spin Resonance in Ultrahigh Magnetic Fields

A. GENERAL REMARKS

Since the first successful experiments for Ge and Si at microwave frequencies (Dresselhaus et al., 1955; Lax et al., 1954), cyclotron resonance has served as a powerful tool to investigate semiconductor physics. It is the most direct method for determining effective mass and band parameters. In addition, the mechanism of the carrier scattering can be studied from the linewidth.

In high magnetic fields such as we are discussing here, the resonant frequency is in the infrared and the far-infrared range. As the energy of the cyclotron orbital motion $\hbar\omega_c$ increases, various new effects are expected to occur under the high-magnetic fields. The nonparabolicity should become extremely large in narrow-gap materials such as InSb, in which $\hbar\omega_c$ becomes comparable with the bandgap. The high fields are also effective for a detailed study of rather unusual nonparabolicity that exists in materials such as GaP or Te. The quantum cyclotron resonance of degenerate valence bands can be observed in high fields where the condition $\hbar\omega_c \gtrsim kT$ is fulfilled. This enables us to determine a set of valence-band parameters.

The most striking effect of the high-magnetic fields appears only when $\hbar\omega_c$ exceeds some critical value. The polaron effect in polar materials is an example of such an effect. Usually at low-field regions the effective mass involved in the cyclotron resonance is the polaron mass. However, in the high-field region where $\hbar\omega_c$ is larger than the longitudinal optical-phonon energy $\hbar\omega_0$, the bare mass is observed in the cyclotron resonance. In the field region in the vicinity of $\hbar\omega_c \simeq \hbar\omega_0$, a large polaron pinning effect is observed. It is of interest to investigate the polaron effect in fairly polar semiconductors such as CdS and CdSe, for which very high fields are required.

As was described in Section I, measurement of the cyclotron resonance can be performed even for low-mobility carriers in high fields. In the same way, experiments can be carried out at relatively high temperatures if the fields are high. This is a great advantage because we need not be limited to low temperatures.

Studies of carrier-scattering mechanisms through linewidth measurement is another vast branch of high-field cyclotron resonance. In megagauss fields the condition $\hbar\omega_c \gtrsim kT$ may be satisfied even at room temperature. Therefore we can study the quantum transport phenomena at room temperature from the cyclotron resonance linewidth.

The electron-spin resonance in high fields is of interest in the sense that the Zeeman splitting of the spin states can exceed other energies in solids: the crystal field splitting or the spin–orbit interaction.

B. NONPARABOLIC ENERGY BANDS

In the two-band model, the energy of the nth Landau level is represented by

$$E_n = \tfrac{1}{2}E_g + \tfrac{1}{2}E_g[1 + (4/E_g)(n + \tfrac{1}{2})\hbar\omega_c(0)]^{1/2}, \tag{1}$$

where E_g is the energy gap between the conduction and the valence band, $\hbar\omega_c(0) = \hbar eH/m_c^*(0)c$, and $m_c^*(0)$ denotes the band-edge mass. When $\hbar\omega_c(0) \ll E_g$ at low magnetic fields, Eq. (1) tends to

$$E_n \simeq (n + \tfrac{1}{2})\hbar\omega_c(0), \tag{2}$$

as in the parabolic band. When $\hbar\omega_c(0)$ approaches E_g, however, the effect of the nonparabolicity becomes prominent. For the conduction band in crystals with Ge or zinc–blende-type structure having degenerate valence bands, Lax et al. (1961) gave an expression

$$E_n^{\pm} = (n + \tfrac{1}{2})2\mu_B H \frac{m_0}{m_c^*(0)} \frac{E_g(E_g + \Delta)}{3E_g + 2\Delta}$$

$$\times \left(\frac{2}{E_n^{\pm} + E_g} + \frac{1}{E_n^{\pm} + E_g + \Delta}\right) \pm \tfrac{1}{2}g^*\mu_B H \frac{E_g(E_g + \Delta)}{\Delta}$$

$$\times [(E_n^{\pm} + E_g)^{-1} - (E_n^{\pm} + E_g + \Delta)^{-1}], \tag{3}$$

where Δ is the spin–orbit splitting, μ_B the Bohr magneton, and g^* the g factor at the band edge. Figure 13 shows the Landau level energies of the conduction band in InSb. Reflecting the small effective mass and the small band gap, the nonparabolicity is prominent in high fields. For the CO_2 laser lines, the photon energy is more than half of E_g, so that the transition between different sets of levels should occur at different fields.

Figure 14 shows experimental recordings of cyclotron resonance in various semiconductors (Miura et al., 1979a). The data for InSb, GaAs, and Ge where $\lambda = 9.5–10.8$ μm were obtained at room temperature. Figure 15 shows the resonant photon energy as a function of the magnetic field for InSb and GaAs (Miura et al., 1976). As for InSb, two resonance peaks are ob-

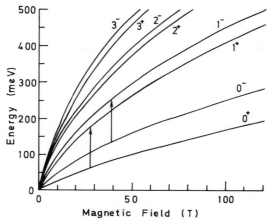

FIG. 13 Landau levels in the conduction band of InSb. The following parameters were assumed in Eq. (3): $E_g = 170$ meV, $\Delta = 810$ meV, $m_c^*(0) = 0.0127 \, m_0$, $g^* = -70$. The arrows indicate the cyclotron resonance transition for the CO_2 laser radiation with a wavelength of 10.6 μm.

served corresponding to $(0^+ \rightarrow 1^+)$ and $(0^- \rightarrow 1^-)$ transitions. This can be regarded as a kind of quantum cyclotron resonance that is observed at room temperature for the first time. Assuming $E_g = 170$ meV (Mooradian and Fan, 1966) and $\Delta = 810$ meV (Pidgeon et al., 1967), $m_c^*(0)$ and g^* are adjusted as fitting parameters. The best fit values are $m_c^*(0) = 0.0127 m_0$ and $g^* = -70$. Theoretically, Roth's relation should hold between $m_c^*(0)$ and g^* (Roth et al., 1959), as shown here:

$$g^* = 2\left[1 + \left(1 - \frac{m_0}{m_c^*(0)}\right)\Delta/(3E_g + 2\Delta)\right]. \qquad (4)$$

With $m_c^*(0) = 0.0127 m_0$, Eq. (4) leads to $g^* = -57$, in conflict with the experimental value of -70. The discrepancy indicates that a more rigorous treatment including the effect of the remote bands (Pidgeon and Brown, 1966) is necessary for better agreement, because the photon energy is comparable to E_g.

As for GaAs, the effect of spin splitting is negligible because $|g^*|$ is very small. For the conduction band in Ge, the nonparabolicity is represented by Eq. (1), because the band minimum is located at the L point where the valence band is not degenerate. Then the observed value of the effective mass $m_c^*(0) = 0.0960 m_0$ at $\hbar\omega = 115.3$ meV is well explained by $m_c^*(0) = 0.086 m_0$.

In contrast to the nonparabolicity of ordinary conduction bands as in InSb, GaAs, and Ge, the conduction band in GaP exhibits an unusual

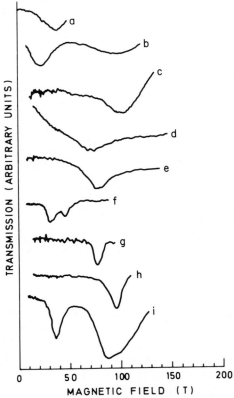

FIG. 14 Experimental recordings of the infrared cyclotron resonance spectra in various semiconductors. Curve a, GaP, H ∥ ⟨111⟩, 119 μm; curve b, GaP, H ∥ ⟨100⟩, 119 μm; curve c, CdS, H ∥ c, 16.9 μm; curve d, CdSe, H ∥ c, 16.9 μm; curve e, Te, H ∥ c, 16.9 μm; curve f, InSb, 9.5 μm; curve g, GaAs, 10.7 μm; curve h, Ge, H ∥ ⟨111⟩, 10.8 μm; curve i, Ge, H ∥ ⟨111⟩, 28.0 μm.

nonparabolicity. Although this substance is of importance in its application to optical devices, little has been known of the details of the band structure, because of the low mobility of carriers. Cyclotron resonance can be observed only in the pulsed-field range. Leotin *et al.* (1975) first observed cyclotron resonance of the conduction band in GaP at a wavelength of 337 μm in pulsed magnetic fields. They obtained the transverse mass $m_t^* = (0.25 \pm 0.01)m_0$ and the mass anisotropy factor $K^* = m_l^*/m_t^* = 20_{-6}^{+10}$ in the framework of the ellipsoidal surface energy model. Suzuki and Miura (1976) obtained $m_t^* = (0.254 \pm 0.004)m_0$ and $K^* = 7.9_{-2.0}^{+3.2}$ from the cyclotron resonance at 119-μm wavelength. There is a large discrepancy in the value of K^* between the two data obtained at different

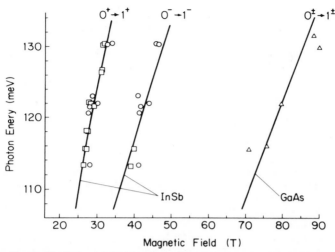

FIG. 15 Resonant photon energy as a function of the magnetic field for n-type InSb and GaAs. For InSb, the data were taken in magnetic fields generated by both the Cnare effect (electromagnetic flux compression) (○) and a one-turn coil (□); for GaAs, △. The lines were calculated from Eq. (3) using the following band parameters: for InSb, $E_g = 170$ meV, $\Delta = 810$ meV, $m_c^*(0) = 0.0127m_0$, $g^* = -70$; for GaAs, $E_g = 1430$ meV, $\Delta = 340$ meV, $m_c^*(0) = 0.065m_0$. (From Miura $et\ al.$, 1976b).

wavelengths. However, in both of these experiments the second resonance peak corresponding to $\sqrt{m_t^* m_l^*}$ that should appear for $\boldsymbol{H} \parallel \langle 100 \rangle$ was not observed within the available field range because of the large anisotropy. So the value of K^* was estimated from the dependence of the resonance field position of the first peak in the direction against the $\langle 100 \rangle$ axis. As a result, a large uncertainty was involved in the estimation of K^*. The cyclotron resonance spectra at $\lambda = 337\ \mu$m for $\boldsymbol{H} \parallel \langle 100 \rangle$ is shown in Fig. 16 (Kido $et\ al.$, 1981). But for the first peak at 8.0 T, the second peak is not discernible up to the maximum field of 37 T. This implies that K^* should be larger than 21.3 at least.

In a recent experiment in megagauss fields, both the first and the second resonance peaks were observed for $\boldsymbol{H} \parallel \langle 100 \rangle$ (Miura $et\ al.$, 1983a). Figure 17 shows the cyclotron resonance absorption spectra at $\lambda = 119\ \mu$m. The second peak, corresponding to $\sqrt{m_t^* m_l^*}$, was observed at 100 T in addition to the first peak at 23 T, corresponding to m_t^*. At $\lambda = 337\ \mu$m, the second peak was observed as a shoulder on the slope of the absorption curve centered at 42 T, as shown in Fig. 18. From these measurements, m_t^* was determined to be $0.25m_0$, but K^* was found to depend strongly on the photon energy. At 10.45 meV ($\lambda = 119\ \mu$m), $K^* = 19 \pm 2$, whereas at 3.68 meV ($\lambda = 337\ \mu$m), $K^* = 28 \pm 7$. That is to say, the longitudinal mass m_l^* de-

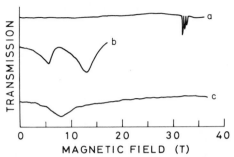

FIG. 16   Curve a, electron spin resonance in ruby for **H** ∥ *c*; cyclotron resonance in the conduction bands of Si (curve b) and GaP (curve c) for **H** ∥ ⟨100⟩. $\lambda = 337\ \mu$m, $T = 80$ K. While two resonance peaks were observed in Si, only a single peak was observed in GaP up to 37 T. (From Miura *et al.*, 1983a.)

creases as the energy is increased.   This dependence is contrary to the usual nonparabolicity.   The experimentally observed effective mass and the anisotropy parameter of the conduction band in GaP are listed in Table I.

Lawaetz (1975) proposed a camel's-back model for the conduction band in GaP based on a comparison between Si and GaP.   As is shown in Fig. 19, Si has a conduction band with a minimum close to the X point, where there is a double degeneracy.   In GaP, on the other hand, the double degeneracy at X is lifted because of the lack of inversion symmetry, and there exist two split bands X_1 and X_3.   The effective Hamiltonian for the conduction band in GaP is given in the following form:

$$
\mathcal{H} =
\begin{matrix}
& X_1 & X_3 \\
\end{matrix}
\left[
\begin{matrix}
Ak_\perp^2 + Bk_\parallel^2 - \Delta/2 & Pk_\parallel \\
Pk_\parallel & Ak_\perp^2 + Bk_\parallel^2 + \Delta/2
\end{matrix}
\right]
\tag{5}
$$

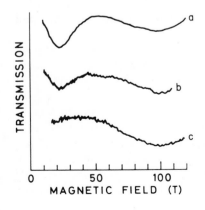

FIG. 17   Cyclotron resonance-absorption spectra in *n*-type GaP, **H** ∥ ⟨100⟩, at $\lambda = 119\ \mu$m.   The magnetic field is in the ⟨100⟩ direction.   Radiation is polarized left-handed circularly (curve a), linearly (curve b), and right-handed circularly (curve c); $T = 106$–152 K.   (From Miura *et al.*, 1983a.)

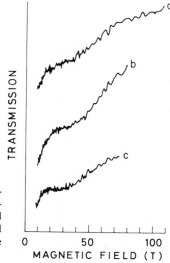

FIG. 18 Cyclotron resonance-absorption spectra in n-type GaP at $\lambda = 337\ \mu m$, $H \parallel \langle 100 \rangle$. Curve a is the unpolarized and curve b the left-handed circularly polarized radiation; curve c, with left-handed circularly polarized radiation, was obtained for a magnetic field $5°$ off the $\langle 100 \rangle$ axis; $T = 80-100$ K. (From Miura et al., 1983.)

where $k_{\parallel}$ and $k_{\perp}$ are the wave vector components along the $\langle 100 \rangle$ axis and the direction perpendicular to it, respectively; Δ the energy splitting at X; $A = \hbar^2/2m_t$ and $B = \hbar^2/2m_l$ the effective mass parameters; and P the band parameter representing the magnitude of the k-linear term. It should be noted that the parameter m_l is completely different from the apparent effective mass m_l^*. A straightforward calculation gives the energy dispersion of the X_1 band as follows:

$$E(k) = Ak_{\perp}^2 + Bk_{\parallel}^2 - [(\tfrac{1}{2}\Delta)^2 + P^2k_{\parallel}^2]^{1/2}. \tag{6}$$

TABLE I

EFFECTIVE MASS AND THE ANISOTROPY PARAMETER OF THE CONDUCTION BAND IN GaP

Experiment	Wavelength (μm)	m_t^*/m_0	$K^* = m_l^*/m_t^*$	m_l^*/m_0	Reference
Cyclotron resonance (angular dependence)	337	0.25 ± 0.01	20^{+10}_{-6}	—	Leotin et al. (1975)
	119	0.254 ± 0.04	$7.9^{+3.2}_{-2.0}$	—	Suzuki and Miura (1976)
Infrared absorption	—	—	29.0	—	Carter et al. (1977)
Luminescence	—	—	28.5	—	Bimberg, Skolnick, and Sanders (1979)
Cyclotron resonance (angular dependence)	337	0.252 ± 0.003	20 ± 10	—	Kido et al. (1981)
Cyclotron resonance	337	0.252 ± 0.003	28 ± 7	6.9 ± 1.7	
Cyclotron resonance	119	0.254 ± 0.004	19 ± 2	4.8 ± 0.5	Miura et al. (1983a)

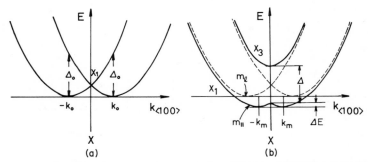

FIG. 19   The structure of the conduction-band edge of Si (a) and GaP (b) near the X point of the Brillouin zone.   The abscissa represents the $\mathbf{k}$ vector in the $\langle 100 \rangle$ direction.   (From Miura *et al.*, 1983b.)

Thus the bottom of the conduction band has the shape of a camel's back when $\Delta \leq P^2/B$.   This structure is similar to the valence band of tellurium (Nakao *et al.*, 1971).   Measurements of the indirect exciton absorption spectra verified the validity of the camel's-back model (Dean and Thomas, 1966; Humphreys *et al.*, 1978).

Landau levels in such a conduction band can be calculated using the effective Hamiltonian [Eq. (5)] (Miura *et al.*, 1983b).   The result of the calculation is shown in Fig. 20.   Landau levels are classified into two series

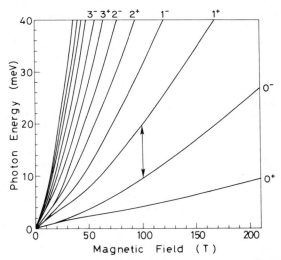

FIG. 20   Landau-level energies as a function of the magnetic field in the conduction band of GaP: $\mathbf{H} \parallel \langle 100 \rangle$.   The vertical bar with arrows indicates the transition for the radiation with a wavelength of 119 μm.   (From Miura *et al.*, 1983b.)

N^+ and N^- having different symmetry. The levels $N^\pm$ are doubly degenerate at low fields, but split in two at higher fields. This corresponds to the magnetic breakdown phenomenon. The selection rule for cyclotron resolance is such that the electronic transition takes place between the levels with the different symmetries. Moreover, the transition is not limited to $\Delta N = 1$ but is allowed between states with many different ΔN. The result shown in Fig. 20 was calculated to assign the second peak as arising mainly from the $0^- \rightarrow 1^+$ transition. Taking into account the intensity of each line, the experimental results were well explained by the calculated result shown in Fig. 20, and the parameters were determined. Another notable point in the cyclotron resonance in GaP is that the second resonance peak is observed not only for the electron-active polarization (left-handed circularly polarized radiation) but also for the electron-inactive polarization, as shown in Fig. 17. This is a common characteristic of an energy band with a large anisotropy (Narita *et al.*, 1972).

Similar to GaP, the valence band in tellurium also has a camel's-back structure. Many investigations have been made on the valence band structure of Te (Miura and Tanaka, 1970; Nakao *et al.*, 1971). However, there have been little data on the conduction band of Te, because Te is always p type in the extrinsic range at low temperatures. In undoped Te crystals, however, the Hall coefficient changes sign at about 200 K, because the intrinsic carriers are thermally excited across the forbidden gap, reflecting a larger mobility of electrons than of holes. Accordingly, at temperatures above 200 K the cyclotron resonance of intrinsic electrons is observed in the conduction band if $\omega_c \tau$ is large enough under high magnetic fields. Figure 21 shows the experimental recordings of the cyclotron resonance absorption in Te observed near room temperature in megagauss fields (Miura *et al.*, 1979a). From the peak positions for $\mathbf{H} \parallel c$ and $\mathbf{H} \perp c$, the effective masses of electrons were obtained as listed in Table II. Figure 22 shows the photon-

TABLE II

EFFECTIVE MASS OF ELECTRONS IN THE CONDUCTION BAND OF TE

	Present experiment			Surface-inversion layer (von Ortenberg and Silberman, 1975)	Magnetoabsorption (Shinno *et al.*, 1973)
$\hbar\omega_c$	44.3 (meV)	73.4 (meV)	117 (meV)	3.68 (meV)	
$m_\perp^*$	0.167	0.186	—	0.117	0.104
$\sqrt{m_\perp^* m_\parallel^*}$	0.115	0.126	0.143	—	0.085
$m_\parallel^*$	0.079	0.085	—	—	0.070

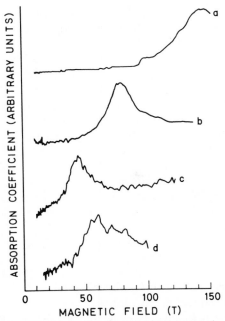

FIG. 21   Cyclotron-resonance absorption spectra in the conduction band of Te at various wavelengths: curve a, $H \perp c$, 10.6 μm; curve b, $H \perp c$, 16.9 μm, T = 270 K; curve c, $H \perp c$, 28.0 μm; curve d, $H \parallel c$, 28.0 μm.   The temperature was 300 K except for curve b, for which it was 270 K.   (From Miura *et al.,* 1979b.)

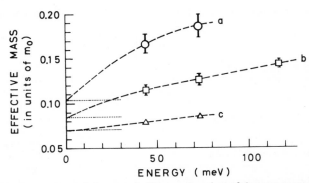

FIG. 22   Effective masses of electrons in Te as a function of the resonant photon energy: curve a, $m_{\perp}^{*}$; curve b, $\sqrt{m_{\perp}^{*}m_{\parallel}^{*}}$ curve c, $m_{\parallel}^{*}$.   The horizontal lines ($\cdots$) in the low field range indicate the values obtained by Shinno *et al.* (1973).

energy dependence of the effective masses. These values are considerably larger than the values obtained by Shinno et al. (1973) from the interband magnetoabsorption. Because the involved energy region of the cyclotron resonance is larger that that of the magnetoabsorption, the difference can be attributed to the nonparabolicity of the conduction band.

It is seen that there is a remarkably large anisotropy in the nonparabolicity. Namely, for $m_\perp^*$ the nonparabolicity is extremely large reflecting the small bandgap, but $m_\parallel^*$ does not depend so much on the energy. This anisotropic nonparabolicity can be elucidated by the anisotropic coupling of the conduction band and the valence band. According to Shinno et al. (1973), the effective Hamiltonian for the conduction band is very complicated because of the presence of the k-linear term. However, as a whole, the matrix element between the conduction band (H_1 band) and the nearest valence band (H_3 band) is nonzero only for P_x and P_y (Treusch and Sandrock, 1966). This means that $m_\perp^*$ is determined predominantly by the interaction with the uppermost valence band, whereas $m_\parallel^*$ is determined by the other remote bands. This leads to the large anisotropy of the energy dependence of the effective masses, as shown in Fig. 22.

C. QUANTUM CYCLOTRON RESONANCE

The Landau levels in the valence bands in Ge, Si, and III-V compounds show nonuniform spacings for low quantum numbers N because of the band degeneracy. When the two conditions $\omega_c \tau > 1$ and $\hbar\omega_c \gtrsim kT$ are satisfied, several split peaks are observed corresponding to different transitions between quantized levels. Such a quantum cyclotron resonance was first observed by Fletcher et al. (1955) in p-type Ge. Hensel and Suzuki (1974) reported a conclusive interpretation on the quantum cyclotron resonance in Ge. Detailed analysis of the quantum cyclotron resonance generally provides the most accurate information of the valence band structure, although the analysis is very complicated because of the k_H effect (Hensel and Suzuki, 1974).

In high magnetic fields, we can observe the quantum cyclotron resonance at not-too-low temperatures satisfying the condition $\hbar\omega_c \lesssim kT$. This is convenient for most crystals because the measurement can be performed at a high-temperature range where the carrier freeze-out does not occur. The quantum cyclotron resonance in p-type GaSb was measured by Suzuki and Miura (1975) at $\lambda = 119\,\mu$m in pulsed fields up to 33 T. The measurement was also done for Ge as a reference standard. The observed cyclotron resonance spectra for Ge and GaSb for the case of $\mathbf{H} \parallel \langle 100 \rangle$ are shown in Figs. 23 and 24, respectively. Detailed knowledge of the spectra of Ge (Hensel and Suzuki, 1974) was used to analyze the spectra of GaSb by a comparison between the two substances. It is necessary to include the

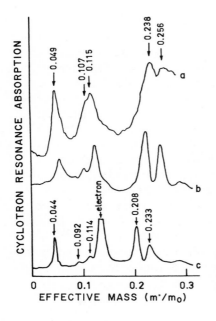

FIG. 23 Quantum cyclotron resonance spectra in p-type Ge, $\mathbf{H} \parallel \langle 100 \rangle$: curve a, $\lambda = 119 \ \mu$m, $\eta = 1.6$; curve b, $\lambda = 337 \ \mu$m, $\eta = 1.6$; curve c, $\lambda = 5.57$ mm, $\eta = 2.1$, where η is the quantum-effect parameter ($\eta = \hbar\omega_c/kT$). Curves b and c taken from Bradley *et al.* (1968) and Hensel (1962). The number attached to each peak represents the effective mass parameter. The horizontal axis, plotted with the effective mass, represents the magnetic field by the relation $m^*/m_0 = eH/m_0 c\omega$, where ω is the angular frequency of the radiation. (From Suzuki and Miura, 1975.)

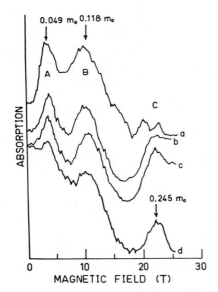

FIG. 24 Quantum cyclotron resonance absorption spectra in p-type GaSb, $\mathbf{H} \parallel \langle 100 \rangle$. The temperature of each sample was 59 K (curve a), 47 K (curve b), 40 K (curve c), and 29 K (curve d). (From Suzuki and Miura, 1975.)

TABLE III

COMPARISON OF THE VALUES OF THE BAND PARAMETERS FOR GaSb

Band parameter	Present experiment	Stress-modulated magnetoreflection[a]	Classical cyclotron resonance[b]	Theory[c]
m_l^*/m_0	0.046 ± 0.002	0.042 ± 0.02	0.052 ± 0.004	0.046
$m_{hh}^*/m_0 \langle 111 \rangle$	0.43 ± 0.12	0.040 ± 0.16	0.36 ± 0.03	0.529
$m_{hh}^*/m_0 \langle 100 \rangle$	0.29 ± 0.05	0.29 ± 0.09	0.26 ± 0.04	0.321
γ_1	12.4 ± 0.4	13.3 ± 0.4		11.80
γ_3	5.4 ± 0.2	5.7 ± 0.2		5.26
$\gamma_3 - \gamma_2$	1.5 ± 0.1	1.3 ± 0.2		1.23
κ	3.2 ± 0.1	3.5 ± 0.6		3.18

[a] From Reine *et al.* (1972).
[b] From Stradling (1966).
[c] From Lawaetz (1971).

split-off band and the conduction band to deal with the nonparabolicity simultaneously. To avoid the complexity arising from the k_H effect, resonance peaks caused by transitions at $k_H = 0$ are chosen for the analysis. The valence-band parameters for GaSb were obtained, as listed in Table III. These parameter values are in reasonably good agreement with the data of stress-modulated magnetoreflection (Reine *et al.*, 1972). The values of the classical effective masses calculated from these band parameters are also in good agreement with the results of the classical cyclotron resonance (Stradling, 1966). As for Ge, a shift of about 10% of the effective masses to larger values was observed at 119 μm in comparison with the data at a microwave frequency.

D. POLARON CYCLOTRON RESONANCE

In ionic crystals, conduction electrons strongly interact with the longitudinal-optical phonons (LO phonon) near the Γ point through the electrostatic potential of the polarization wave. The perturbation Hamiltonian of the electron–phonon interaction is expressed as

$$\mathcal{H}' = \sum_q i\hbar\omega_0/q[(\hbar/2m^*\omega_0)]^{1/4}[(4\pi\alpha/V)]^{1/2}$$
$$\times(a_q e^{iqr} + a_{-q}^+ e^{-iqr}), \tag{7}$$

where $\mathbf{q}$ is the wave vector of phonons, ω_0 the angular frequency of LO phonons, a_q and a_{-q}^+ the annihilation and creation operator for phonons, m^* the effective mass of electrons, V the volume of the system, and α the dimensionless coupling constant representing the strength of the electron–

phonon interaction:

$$\alpha = \frac{e^2}{2\hbar\omega_0} \left(\frac{2m^*\omega_0}{\hbar}\right)^{1/2} \left(\frac{1}{\epsilon_\infty} - \frac{1}{\epsilon_0}\right), \tag{8}$$

where ϵ_∞ and ϵ_0 are the high-frequency and static dielectric constants.

In polar crystals, the mass of polarons m_p^* is larger than the effective mass of bare electrons m_b^* because of the effect of the phonon cloud around the electrons. For $\alpha \ll 1$, m_p^* is given by

$$m_p^* = m_b^*(1 + \alpha/6). \tag{9}$$

At low magnetic fields where $\hbar\omega_c \ll \hbar\omega_0$, the effective mass obtained from the cyclotron resonance is m_p^*. However, at higher fields where $\hbar\omega_c > \hbar\omega_0$, the phonon cloud cannot follow the fast cyclotron orbital motion of electrons. Consequently, the mass we observe in the cyclotron resonance is m_b^*.

The Landau levels of polarons have been investigated by Larsen (1964) and Waldman *et al.* (1969) with variational methods and by Dickey *et al.* (1967) with a perturbational method. The energies of polarons in the $N = 0$ and $N = 1$ Landau levels are expressed by the Wigner–Brillouin perturbation theory, as follows (Dickey *et al.*, 1967):

$$\tilde{E}_0 = \frac{1}{2}\hbar\omega_c$$
$$- \frac{\alpha(\hbar\omega_0)^2}{2\pi^2} \int d^2k \sum_N \frac{|\mathcal{H}_{N0}(k)|^2}{(N+1/2)\hbar\omega_c + \hbar\omega_0 + \hbar^2k_z^2/2m_z^* - E_0}, \tag{10}$$

$$\tilde{E}_1 = \frac{3}{2}\hbar\omega_c$$
$$- \frac{\alpha(\hbar\omega_0)^2}{2\pi^2} P \int d^3k \sum_N \frac{|\mathcal{H}_{N1}(k)|^2}{E_0 + n\hbar\omega_c + \hbar\omega_0 + \hbar^2k_z^2/2m_z^* - E_1}, \tag{11}$$

where E_0 and E_1 are the energies of the unperturbed states of the $N = 0$ and $N = 1$ Landau levels. $\mathcal{H}_{Ni}(k)$ is the matrix element of the electron–phonon interaction between the Nth and ith Landau levels, which are given as follows (Dickey *et al.*, 1966):

$$\mathcal{H}_{N0}(k) = \frac{r_0^{1/2}}{(N!)^{1/2}} (ak_\perp)^n [\exp(-\tfrac{1}{2}a^2k_\perp^2)]/k, \tag{12}$$

$$\mathcal{H}_{N1}(k) = \frac{-r_0^{1/2}}{(N!)^{1/2}} \frac{N - a^2k^2}{(ak_\perp)^{1-n}} [\exp(-\tfrac{1}{2}a^2k_\perp^2)/k], \tag{13}$$

where

$$a = \left(\frac{\hbar c}{2eH}\right)^{1/2}, \qquad r_0 = \left(\frac{\hbar^2/2m^*}{\hbar\omega_0}\right)^{1/2}.$$

The energies $\tilde{E}_0$ and $\tilde{E}_1$ for a small α are shown in Fig. 25 as a function of the magnetic field. The magnetic field is scaled by $\hbar\omega_c$ with a unit of $\hbar\omega_0$. The energy of the $N = 1$ state splits in two and the effect is particularly large in the region where $\hbar\omega_c \simeq \hbar\omega_0$. This can be readily understood from Eq. (11). The break of the energy in the vicinity of $\hbar\omega_c \simeq \hbar\omega_0$ can be interpreted as the resonant interaction between the $N = 1$ with no-phonon state and the $N = 0$ with one-phonon state. This resonant interaction is called polaron-pinning effect. It was first observed in n-type InSb in which the effect can be observed at about 3.2 T owing to the small effective mass and small value of $\hbar\omega_0$ (Johnson and Larsen, 1966). However, the effect in InSb is not large, because of the small value of α (0.02).

The cyclotron resonance in n-type CdS and CdSe is of great interest because of their large electron–phonon interaction. The coupling con-

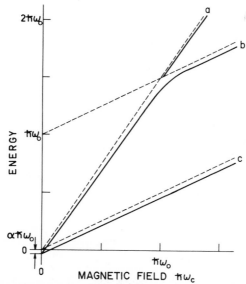

FIG. 25   The energy of the Landau levels for polarons as a function of magnetic field: curve a, $N = 1$, 0 phonon; curve b, $N = 0$, 1 phonon; curve c, $N = 0$, 0 phonon; ---, the unperturbed states, and ——, the perturbed states. The horizontal axis represents the magnetic field with a relation $\hbar\omega_c = \hbar eH/m_b^* c$, where m_b^* is the bare mass. Both the vertical and the horizontal axes are plotted with a unit of the LO phonon energy $\hbar\omega_0$.

stants are $\alpha = 0.6$ for CdS and $\alpha = 0.45$ for CdSe. In addition, the possibility of the existence of the piezoelectric polaron (Mahan and Hopfield, 1964) in CdS has been discussed because of the large piezoelectric interaction between electrons and acoustic phonons. Use of megagauss fields has enabled us to observe the cyclotron resonance in these materials over a wide frequency range across the Reststrahlung band. Experimental recordings of the cyclotron resonance spectra in CdS and CdSe are shown in Fig. 26 for various wavelengths and field directions (Miura *et al.*, 1979b). The temperature was varied in the range 100–300 K, but no significant temperature effect was observed. The apparent effective masses of electrons estimated from the peak positions are listed in Table IV for CdS and in Table V for CdSe, together with data obtained previously from low-field experiments. As for CdS, the effective masses at photon energies $\hbar\omega > \hbar\omega_0$ are considerably smaller than the values at lower fields. Particularly at $\hbar\omega = 44.3$ meV ($\lambda = 28.0$ μm), which is slightly larger than $\hbar\omega_0$, m_c^* is about 30% smaller

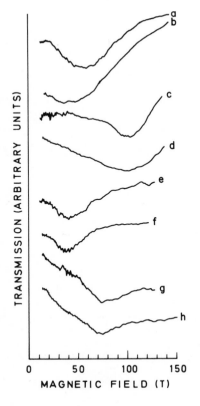

FIG. 26   Cyclotron-resonance absorption spectra in *n*-type CdS and CdSe. The wavelengths of the radiation and the direction of the field with respect to the *c* axis are as follows: Curve a: CdS, H ∥ *c*, 28 μm; curve b: CdS, H ⊥ *c*, 28 μm; curve c: CdS, H ∥ *c*, 16.9 μm; curve d: CdS, H ⊥ *c*, 16.9 μm; curve e: CdSe, H ∥ *c*, 28 μm; curve f: CdSe, H ∥ *c*, 28 μm; curve g: CdSe, H ∥ *c*, 16.9 μm; curve h: CdSe, H ⊥ *c*, 16.9 μm. The temperatures of the samples were 130–300 K. (From Miura *et al.*, 1979c.)

TABLE IV

VARIATION OF THE APPARENT EFFECTIVE MASS OF ELECTRONS
IN CdS WITH PHOTON ENERGY

$\hbar\omega$ (meV)	$\hbar\omega/\hbar\omega_0{}^a$	$T(k)$	$m_c^*/m_0(H \parallel c)$	$m_c^*/m_0(H \perp c)$	Reference
0.289	0.00765	1.3	0.171	0.162	Baer and Dexter (1964)
3.68	0.0974	3	0.16	—	Button et al. (1970)
6.36	0.168	3	0.17	—	Button et al. (1970)
4.96	0.131	~19–38	0.174	—	Nagasaka (1977)
10.5	0.278	35	0.188	—	Nagasaka et al. (1973)
44.3	1.172	~130–300	0.131	0.107	Miura et al. (1979c)
73.4	1.941	~130–300	0.164	0.158	Miura et al. (1979c)

[a] $\hbar\omega_0 = 37.8$ meV is the longitudinal optical (LO) phonon energy.

than the value at 10.5 meV ($\lambda = 119 \ \mu$m). A similar tendency is also seen for CdSe in Table V.

The polaron Landau levels $\tilde{E}_0$ and $\tilde{E}_1$ were calculated from Eqs. (10) and (11) (Miura et al., 1978a) by adjusting the bare masses m_b^* as fitting parameters. The resonant photon energy is calculated as a difference between E_1 and E_0 at each field. The calculated results of the resonant photon energy with the best fit values of m_b^* are shown in Fig. 27 for $H \parallel c$. The calculation and the experimental data for both CdS and CdSe are in excellent agreement. The break at the crossover point $\hbar\omega = \hbar\omega_0$ is very clear for both CdS and CdSe.

The best-fit values of the bare mass are $m_b^* = 0.165m_0$ for CdS and $m_b^* = 0.116m_0$ for CdSe. In this estimation, the effect of the nonparabolicity of the conduction band was taken into account based on a two-band model. If we estimate m_b^* only from the low-field data, $m_b^* = 0.155m_0$ is obtained with the aid of Eq. (9). That is to say, the value of m_b^* obtained from the

TABLE V

APPARENT EFFECTIVE MASS OF ELECTRONS IN CdSe

$\hbar\omega$ (meV)	$\hbar\omega/\hbar\omega_0{}^a$	$T(k)$	$m_c^*/m_0 (H \parallel c)$	$m_c^*/m_0 (H \perp c)$	Reference
3.68	0.140	—		0.120	Button and Lax (1970)
44.3	1.68	150 ~ 300	0.115	0.107	Miura et al. (1979b)
73.4	2.79	150 ~ 300	0.118	0.114	Miura et al. (1979b)

[a] $\hbar\omega_0 = 26.3$ meV is the LO phonon energy.

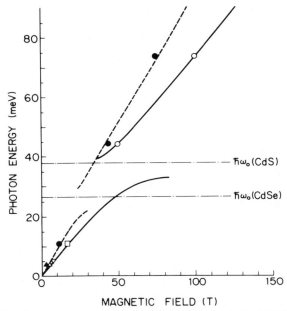

FIG. 27 Resonant photon energy as a function of the magnetic field for the cyclotron resonance in CdS (○) and CdSe (●). The magnetic field is in the direction of the c axis. Previous data by Nagasaka *et al.* (1973) (□), Nagasaka (1977) (▽), Button *et al.* (1970) (△), and Button and Lax (1970) (▲) are also shown, where the open symbols represent CdS and the closed symbols CdSe. $\hbar\omega_0$(CdS) and $\hbar\omega_0$(CdSe) represent the LO-phonon energies for CdS and CdSe. Theoretically calculated curves are shown for CdS (——) and CdSe (– – –). (From Miura *et al.*, 1978a.)

high-field experiments is considerably larger than that obtained from the low-field experiments. This difference can be ascribed to the piezoelectric-polaron effect. According to Miyake's theory (Miyake, 1968), the mass shift of the piezoelectric polaron is given by

$$-(\Delta m^*/m^*) \propto T^{2/3}H^{-1}. \tag{14}$$

Accordingly, the effective mass becomes smaller at lower fields, and this dependence seems to account for the discrepancy. If we take account of the temperature effect, however, the factor $T^{2/3}H^{-1}$ is not so different for the data listed in Table IV. To clarify this discrepancy, a unified theory of the polaron must include both the electron–LO-phonon interaction and the piezoelectric polaron interaction.

The discrepancy in m_b^* between low-field and high-field data is also found in CdSe. It should be noted that the absorption curve for CdSe has a long tail at the high-field side at $\lambda = 16.9\ \mu\mathrm{m}$, as shown in Fig. 26. Although the origin of this tail is not clear at present, it may be due to the phonon-assisted absorption.

E. Cyclotron Resonance Linewidth

In addition to the information on the energy-band parameters given by the peak position of the cyclotron resonance, measurement of the linewidth of the resonance offers us another useful tool to investigate the carrier-scattering mechanism. Extensive studies have been done on the various scattering mechanisms in Ge, Si, and InSb: impurity scattering (Apel et al., 1971; Shin et al., 1973), acoustic-phonon scattering (Fukai et al., 1964; Itoh et al., 1966), and carrier–carrier scattering (Kawamura et al., 1964). In the high-temperature region, the acoustic-phonon scattering is the dominant scattering mechanism. Bagguley et al. (1961) obtained from the cyclotron resonance linewidth the momentum relaxation time $\langle \tau \rangle$ of carriers that are scattered by acoustic phonons in Ge and Si at microwave frequencies. They showed that $\langle \tau \rangle$ obtained from the cyclotron resonance is almost the same as that obtained from the dc transport measurement. Fink and Braunstein (1974) found that the scattering time estimated by the cyclotron resonance linewidth at $\lambda = 337~\mu m$ is smaller by a ratio of 1.5 to 5 than the values at lower frequencies.

Meyer (1962) first pointed out that under the quantum limit condition $\hbar\omega_c \gg kT$ the cyclotron resonance linewidth cannot be described by the classical relaxation time. Meyer's argument is as follows: The classical transport theory breaks down when the radius of the cyclotron orbit becomes smaller than the wavelength of phonons that predominantly interact with electrons, determining their relaxation time. This condition is equivalent to the quantum condition $\hbar\omega_c > kT$. The relaxation time $\langle \tau \rangle$ is given as the inverse of the transition probability from the Landau level $N = 1$ to $N = 0$ by the deformation-potential interaction between electrons and phonons:

$$\frac{1}{\langle \tau \rangle} = \frac{1}{(2\pi)^2} \frac{E_1^2}{\rho v_s^2} \frac{kT}{\hbar} \int d^3q \frac{1}{2} r_c^2 q_t^2 \exp\left(-\frac{1}{2} r_c^2 q_t^2\right)$$

$$\times \left[\frac{\hbar v_s q / kT}{\exp(\hbar v_s q / kT) - 1} + \frac{1}{2} \frac{\hbar v_s q}{kT} \right]$$

$$\times \delta\left(\hbar\omega_c + \frac{\hbar^2 [k_z^2 - (q_z + k_z)^2]}{2m^*} \right), \tag{15}$$

where E_1 is the deformation-potential constant, $r_c = (\hbar c/eH)^{1/2}$, ρ the mass density, v_s the sound velocity, and q the wave vector of phonons. From Eq. (15) we can expect that $\langle \tau \rangle$ is decreased as the field is increased. This field dependence is contrary to the case of ionized-impurity scattering.

In very high magnetic fields, the quantum limit is achieved at relatively high temperatures so that the elastic scattering by acoustic phonons is predominant. The field dependence of the cyclotron resonance linewidth was studied for n-type Ge at room temperature in the high-field range using various laser lines (Miura and Kido, 1977). The experimental recordings of the cyclotron resonance in Ge are shown in Fig. 28. Part (a) of Fig. 29 shows the relaxation time $\langle \tau \rangle$ estimated from the linewidth of the cyclotron resonance corresponding to m_t^* for two samples with different impurity concentrations. In both the samples, for $\hbar\omega \geq 44$ meV, $\langle \tau \rangle$ was found to decrease with the increase in resonant photon energy almost in proportion to $\omega^{-1/2}$ and thus $H^{-1/2}$.

According to Meyer's theory, Eq. (15) is reduced to the form

$$\langle \tau \rangle \propto T^{-1}H^{-1/2} \tag{16}$$

when $\hbar\omega_c \gg kT \gg (\tfrac{1}{2}m^*v_s^2\hbar\omega_c)^{1/2}$ is satisfied. Therefore, as for the field dependence of $\langle \tau \rangle$, the experimental results are in agreement with the theory. However, no significant temperature dependence was observed in the experiment in the range 130–300 K, in contradiction to the theory.

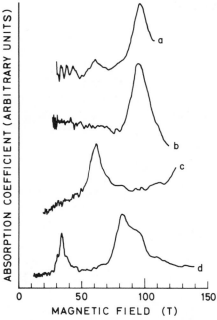

FIG. 28 Cyclotron-resonance absorption spectra in n-type Ge. The magnetic field B is applied in the $\langle 111 \rangle$ direction; $T = 130$ K. The wavelength of the radiation is 10.6 μm for curves a and b, 16.9 μm for curve c, and 28.0 μm for curve d. (From Miura et al., 1977.)

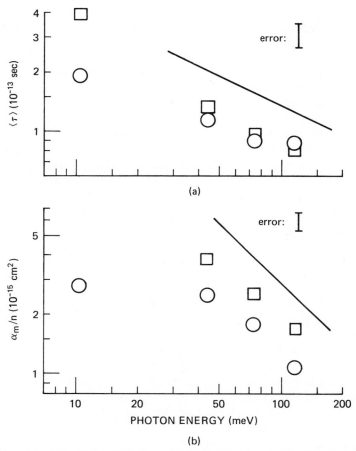

FIG. 29   Linewidth and absorption intensity in *n*-type Ge: (a) the resonant photon-energy dependence of the relaxation time of the carrier scattering, estimated from the cyclotron resonance linewidth; (b) the resonant photon-energy dependence of the absorption coefficient α_m at the peak of the cyclotron resonance divided by the carrier concentration *n*. ——— represents the dependence of the power law of the photon energy $\omega^{-1/2}$ in (a) and ω^{-1} in (b). Data are shown for two samples with different carrier concentrations: O, $n = 2.7 \times 10^{16}$ cm^{-3} for Ge(1); and □, $n = 2.2 \times 10^{15}$ cm^{-3} for Ge(2). (From Miura and Kido, 1977.)

In part (b) of Fig. 29, the intensity of the absorption line is plotted as a function of photon energy. Because the absorption is proportional to the carrier density, the ordinate is scaled with the quantity α_m/n, where α_m is the absorption coefficient at the peak and *n* the carrier density. It is seen that in both samples, the absorption was decreased with increasing magnetic field almost in proportion to H^{-1}. This dependence can be explained by the magnetic freeze-out effect and the transfer of electrons to other valleys that have a heavier effective mass. It is remarkable that the magnetic freeze-out effect is noticeable even at room temperature because of the very high magnetic fields.

One of the advantages of measuring $\langle \tau \rangle$ by the cyclotron resonance linewidth is the capability of obtaining "local" information in contrast to the dc transport measurements. In polycrystalline samples, for example, if the radius of the cyclotron motion becomes smaller than each grain size of the polycrystal, the scattering time $\langle \tau \rangle$ can be obtained independent of the scattering by the grain boundary. Takeda *et al.* (1983) measured the cyclotron resonance in p-type $Ge_{0.9}Si_{0.1}$ alloy at $\lambda = 337$ μm and at $\lambda = 119$ μm and found that the value of $\langle \tau \rangle$ estimated from the cyclotron resonance is a few times larger than that from the Hall effect. They ascribed the difference to the absence of the domain scattering in the former case.

F. Electron Spin Resonance in Ruby

Many paramagnetic substances show a nonlinear magnetic field dependence of the g factor in the very high-field range because of the presence of higher-order terms of the field (Motokawa et al., 1979). However, we can expect that the g factor in ruby is independent of magnetic field up to very high fields because of its energy-level scheme. The g factor of ruby was accurately determined in the microwave range as $g_{\parallel} = 1.9840 \pm 0.0006$ and $g_{\perp} = 1.9867 \pm 0.0006$ (Manenov and Prokhorov, 1955). Therefore, if the g factor is really constant up to the high fields, the electron spin resonance (ESR) in ruby would serve as a good means for calibrating very high field measurement by inductive probes.

The ESR for ruby was observed at $\lambda = 337$ μm and $\lambda = 119$ μm in the very high field range (Kido and Miura, 1982). In the absence of the field, the ground state of the Cr^{3+} ion is split in two. In the magnetic field, it is further split into four levels, $M_s = -\frac{3}{2}, -\frac{1}{2}, \frac{1}{2},$ and $\frac{3}{2}$, and three absorption lines are observed. Figure 30 shows the experimental recordings of the ESR in ruby. For 337 μm (Fig. 30b), the measurement was performed in the nondestructive field. The three absorption lines are well resolved. The central peak among the three, which should correspond to $g_{\parallel}\mu_B H$, was observed at $H = 32$ T, in excellent agreement with $H = 32.08$ T calculated from $H = \hbar\omega/g_{\parallel}\mu_B$ with $g_{\parallel} = 1.9840$. For 119 μm, the measurement was performed in megagauss fields. A single peak was observed at $H = 91$ T, which also agrees with the theoretical value of 91.0 T. Thus it was assured that the g factor in ruby is constant up to the megagauss range. The observed linewidth of 30 mT at $\lambda = 337$ μm was found to be caused by the homogeneity of the magnetic field. The observation of such sharp resonance peaks serves also for testing the response time of the detector system.

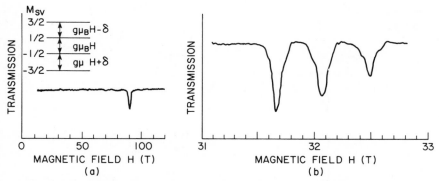

FIG. 30 Far-infrared spectra of electron-spin resonance in ruby: $H \parallel c$. (a) $\lambda = 119 \ \mu m$, $T = 300$ K; (b) $\lambda = 337 \ \mu m$, $T = 77$ K. (From Kido and Miura, 1982.)

IV. Far-Infrared Magnetooptical Spectroscopy in Semimetals

A. EFFECT OF FREE CARRIERS ON DIELECTRIC CONSTANTS

When the electromagnetic wave is incident on conductive substances such as semiconductors with high carrier concentration, semimetals, or metals, their dielectric response is mainly determined by the free-carrier effect. For fields varying as $\exp(i\omega t - i\mathbf{k}r)$, Maxwell's equation can be written as

$$-i\mathbf{k} \times \mathbf{E} = -(i\omega/c)\mu\mathbf{H}, \qquad (17)$$

$$-i\mathbf{k} \times \mathbf{H} = (1/c)(i\omega\epsilon_l + 4\pi\sigma)\mathbf{E}, \qquad (18)$$

where ϵ_l is the lattice part of the dielectric constant. Equations (17) and (18) are reduced to

$$-\mathbf{k} \times \mathbf{k} \times \mathbf{E} = [(\omega/c)^2\epsilon - (4\pi i\omega/c^2)\sigma]\mathbf{E}. \qquad (19)$$

Here we will deal only with the Faraday configuration, $\mathbf{k} \parallel \mathbf{H}$. Then, Eq. (19) leads to a dispersion equation given here:

$$-\mathbf{k}^2\mathbf{E} = (\omega/c)^2\epsilon_l \cdot \mathbf{E} - (4\pi i/\omega)\sigma \cdot \mathbf{E}$$

$$= (\omega/c)^2\epsilon^* \cdot \mathbf{E}, \qquad (20)$$

where we define the dielectric constant ϵ^*, which has a tensor form because the conductivity tensor σ has off-diagonal components in the magnetic field. In the classical Drude model, the conductivity tensor in the presence of a static magnetic field H can be obtained from the equation of motion of electrons, as follows:

$$m^* \, d\mathbf{v}/dt + m^*\mathbf{v}/\tau = -e[\mathbf{E}e^{i\omega t} + (\mathbf{v}/c) \times \mathbf{H}], \qquad (21)$$

where m^* is the effective mass, $\mathbf{v}$ the velocity, and τ the relaxation time of scattering of electrons. Putting $\mathbf{v} \propto e^{i\omega t}$ and recalling that the current $\mathbf{J}$ is given by

$$\mathbf{J} = \sigma \cdot \mathbf{E} = ne\mathbf{v}, \tag{22}$$

we obtain for the case of $\mathbf{H} \parallel z$ axis,

$$\sigma = 4\pi ne^2 \begin{bmatrix} \dfrac{-i(\omega - i/\tau)}{(\omega - i/\tau)^2 - \omega_c^2} & \dfrac{\omega_c}{(\omega - i/\tau)^2 - \omega_c^2} & 0 \\[3mm] \dfrac{-\omega_c}{(\omega - i/\tau)^2 - \omega_c^2} & \dfrac{-1(\omega - i/\tau)}{(\omega - i/\tau)^2 - \omega_c^2} & 0 \\[3mm] 0 & 0 & \dfrac{-i}{(\omega - i/\tau)} \end{bmatrix} \tag{23}$$

where n is the density of electrons. For circularly polarized radiation $E_x \pm iE_y$ we can obtain the effective dielectric constant at frequency ω from Eqs. (20) and (24) as

$$\epsilon_{\pm}^*(\omega) = \epsilon_l \left[1 - \frac{\omega_p^2}{\omega(\omega \mp \omega_c - i/\tau)} \right], \tag{24}$$

where ω_p is the plasma frequency represented by

$$\omega_p^2 = 4\pi ne^2/m^*\epsilon_l. \tag{25}$$

The wave vector component for each polarization is related to $\epsilon_{\pm}^*$ as follows:

$$k_{\pm} = (\omega/c)^2 \epsilon_{\pm}^*(\omega). \tag{26}$$

Equation (24) also holds for holes if we simply replace ω_c by $-\omega_c$. Therefore, for semimetals that have both electrons and holes, the dispersion equation is derived as

$$k_{\pm}^2 = \left(\frac{\omega}{c}\right)^2 \epsilon_l \left[1 - \frac{\omega_{ph}^2}{\omega(\omega \mp \omega_{ch} - i/\tau_h)} - \frac{\omega_{pe}^2}{\omega(\omega \pm \omega_{ce} - i/\tau_e)} \right], \tag{27}$$

where ω_{ce} and ω_{ch} are the cyclotron frequencies, ω_{pe} and ω_{ph} the plasma frequencies, and τ the relaxation time with subscripts e and h denoting electrons and holes, respectively.

The effective dielectric constant $\epsilon_{\pm}^*(\omega)$ is usually written as

$$\epsilon_{\pm}^*(\omega) = (N - i\kappa)_{\pm}^2, \tag{28}$$

where N is the refractive index and κ the extinction coefficient. From the dispersion relation (27), we can derive the reflectivity R by the relation

$$R = \frac{(N - 1)^2 + \kappa^2}{(N + 1)^2 + \kappa^2}. \tag{29}$$

On the other hand, the absorption of electromagnetic radiation is determined by κ.

Let us consider the case $\omega_p \gg \omega$. If $\omega_c < \omega$ and $\omega\tau > 1$ for both electrons and holes, $k_{\pm}$ becomes a pure imaginary and the wave cannot propagate in the media. However, when $\omega_c > \omega$ is satisfied for either electrons or holes in high magnetic fields, the propagation of the wave is allowed. For $\omega_{ce} \gg \omega$ and $\omega_{ch} \gg \omega$, Eq. (27) is expanded into the following form, neglecting the terms i/τ_e and i/τ_h, then

$$k_{\pm}^2 = (\omega/c)^2\epsilon_l[1 \pm \omega^{-1}(\omega_{ph}^2/\omega_{ch} - \omega_{pe}^2/\omega_{ce})$$
$$+ \omega_{ph}^2/\omega_{ch}^2 + \omega_{pe}^2/\omega_{ce}^2]. \tag{30}$$

In an ideal compensated semimetal in which the densities of electrons and holes are the same and the plasma mass is identical with the cyclotron mass, the second term in the square bracket in Eq. (30) is zero, and we obtain

$$k_{\pm}^2 = (\omega/c)^2\epsilon_l + (4\pi\omega^2\rho/H^2), \tag{31}$$

where ρ is the mass density,

$$\rho = n(m_e^* + m_h^*). \tag{32}$$

The electromagnetic wave that propagates with the dispersion relation (31) is the Alfvén wave. The Alfvén wave propagation has been extensively investigated for Bi by many authors (Faughnan, 1965; Isaacson and Williams, 1969; Kawamura et al., 1965; McLachlan, 1966; Takano and Kawamura, 1970; Williams and Smith, 1964).

In contrast to the classical treatment described here, the quantum-mechanical calculation gives the effective dielectric constant as follows:

$$\epsilon_{\pm}^*(\omega) = \epsilon_1 - \frac{2\pi}{\omega}\frac{e^2}{m_0^2}\frac{\hbar}{V}\sum_{\mu,\nu}$$
$$\times \left[\frac{f(E_\mu) - f(E_\nu)}{E_\mu - E_\nu}\frac{|\langle\mu|p^{\pm}|\nu\rangle|^2}{\hbar\omega - E_{\mu\nu} + (i\hbar/\tau_{\mu\nu})}\right], \tag{33}$$

where $\langle\mu|p^{\pm}|\nu\rangle$ is the momentum matrix element between the states μ and ν with energies E_μ and E_ν, $f(E)$ the Fermi distribution function, and $\tau_{\mu\nu}$ the relaxation time. When the quantum effect is significant under high magnetic fields, we have to rely on the quantum-mechanical expression [Eq. (33)] rather than the classical one. Expression (33) is applicable not only to the calculation of free-carrier contribution, but also to the contribution of the interband transition to the dielectric constant.

B. MAGNETOREFLECTION IN SEMIMETALS

In semimetals such as Bi and graphite, the plasma edge lies in the range $10-30\ \mu m$. Therefore the far-infrared radiation with longer wavelengths cannot propagate through the crystal under low magnetic fields for which $\omega_c < \omega$, as we saw in Section IV.A. In high magnetic fields that satisfy $\omega_c > \omega$, the crystal starts to permit the propagation, but the intensity of the radiation transmitted by the sample is usually very small because of the large free-carrier absorption and the large reflection at the surface. For investigating the cyclotron resonance and other inter-Landau-level transitions in such crystals, measurement of the magnetoreflection spectra provides the most direct information, because each optical transition of electrons is observed as a structure in the spectra, as we can expect from Eqs. (28), (29), and (33). Moreover, the magnetoreflection measurements are useful for studying the magnetoplasma modes.

We will first discuss the case of graphite. As is well known, graphite has a layer-type crystal structure, and its energy-band structure has an extremely large anisotropy. Within the layer the effective masses of electrons and holes are very small: $m_{ce} = 0.058m_0$, $m_{ch} = 0.039m_0$. On the other hand, in the direction perpendicular to the layer, the width of the E_3 band where both electrons and holes exist is only 38 meV, corresponding to the electron effective mass of $8.8m_0$. When we apply the magnetic fields perpendicularly to the layer, i.e., along the c axis, both electron and hole levels easily reach the quantum limit, whereas the k_H dispersion (k_H is the wave vector parallel to the field direction) is fairly flat. Landau levels in graphite were calculated by Nakao (1976) based on the Slonczewski–Weiss–McClure model (McClure, 1957; Slonczewski and Weiss, 1958). Figure 31 shows the Landau levels of graphite when the magnetic field of 25 T is applied parallel to the c axis. The abscissa is scaled by the quantity $\xi = c_0/2\pi \cdot k_H$, where c_0 is the lattice constant along the c axis. Only four levels, denoted by $-1^{\pm}$ and $0^{\pm}$, remain below the Fermi level, including the spin degeneracy. Other

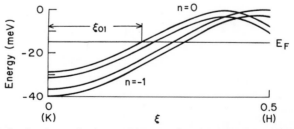

FIG. 31   Landau-level energies in graphite as a function of $\xi = k_z c_0/2\pi$ (c is the lattice constant along the c axis), when the magnetic field of 25 T is applied along the c axis of the crystal; $2\xi_{01}$ the wave vector at which the charge-density wave is considered to appear.

Landau levels are far apart at this field. Electrons and holes are located in these levels, centered at the K point and the H point, respectively. This peculiar energy diagram resembles the one-dimensional energy band.

Recently, a striking anomaly was observed in the magnetoresistance in graphite under high magnetic fields at low temperatures (Iye et al., 1982; Nakamura et al., 1983a; Tanuma et al., 1981). The magnetoresistance abruptly jumps up at a critical magnetic field. The critical field increases as the temperature is raised. Yoshioka and Fukuyama (1981) ascribed this anomaly to the charge-density-wave (CDW) transition that is characteristic of the quasi–one-dimensional energy-level scheme shown in Fig. 31. According to their theory, electrons in the $N = 0$ spin-up level become unstable against the formation of the charge-density wave with a wave vector of $2\zeta_{01}$. Such a phase transition may also show up in the magnetoreflection spectra in graphite under high magnetic fields.

Nakamura et al. (1983b, 1984) measured the magnetoreflection spectra in graphite under high pulsed magnetic fields up to 45 T at 337, 119, 28.0 and 10.6 μm. Instead of the anomaly mentioned previously, they found a new structure that cannot be interpreted by any electronic transition between Landau levels. Figure 32 shows the magnetoreflection spectra observed for

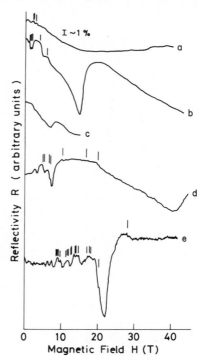

FIG. 32 Magnetoreflection spectra in graphite (HOPG) at various wavelengths. Curve a: hole-active, 119 μm; curve b: electron-active, 119 μm; curve c: nonpolarized, 337 μm; curve d: nonpolarized, 28 μm; curve e: nonpolarized, 10.6 μm. $\mathbf{H} \parallel c$, $T \simeq 14$ K. The vertical bars represent the field positions at which the transition between Landau levels should occur. (From Nakamura et al., 1983b.)

highly oriented pyrolitic graphite (HOPG) by Nakamura *et al.* (1983b). At relatively low fields, many structures are observed corresponding to the transitions between Landau levels at the K point ($k_z = 0$), as shown by vertical bars. The overall line shapes, including the positions of the structures, are in good agreement with the ones calculated on the basis of Nakao's computer program (1976), if one uses the parameter values given by Mendez *et al.* (1980) and $\gamma_3 = 0.30$ eV (Doezema *et al.*, 1979; Schroeder *et al.*, 1971). The experimentally observed line shape below 8 T for $\lambda = 119 \ \mu$m is consistent with previous data reported by Doezema *et al.* The last transition between the Landau levels $N = 0$ and $N = 1$ at the K point appears as a sharp shoulder at about 6 T for $\lambda = 119 \ \mu$m and at about 20 T for $\lambda = 28 \ \mu$m. Above the field corresponding to this last transition, a very large structure is observed at 7.5 T for $\lambda = 337 \ \mu$m, at 14.5 T for $\lambda = 119 \ \mu$m, and at 40 T for $\lambda = 28 \ \mu$m. These extra structures cannot be assigned as transitions between Landau levels. In the data for $\lambda = 119 \ \mu$m, the extra structure is observed only for the electron-active polarization.

The relation between the photon energy and the magnetic field where the extra structure appears is shown in Fig. 33. The experimental data are shown by triangles. The energy of the extra structure starts from zero and increases faster than a linear function with increasing magnetic field. The positions of the last inter-Landau-level transition (0, 1) are shown by solid circles. These points fit well to the line drawn for the classical cyclotron resonance of electrons using the effective mass $m_e = 0.058 \ m_0$ (Suematzu and Tanuma, 1972).

The extra structure can be interpreted in terms of the electron–hole coupled-plasma mode. A large structure appears in the magnetoreflection spectra when the real part of the dielectric constant crosses zero. For the system that has both electrons and holes, the condition $\epsilon^{\pm} = 0$ leads to

$$1 - \frac{\omega_{\text{ph}}^2}{\omega(\omega \mp \omega_{\text{ch}})} - \frac{\omega_{\text{pe}}^2}{\omega(\omega \pm \omega_{\text{ce}})} = 0 \qquad (34)$$

from Eq. (27), neglecting the terms i/τ_{h} and i/τ_{e}. Equation (34) can be regarded as a cubic equation for frequency ω:

$$\omega^3 \pm (\omega_{\text{ch}} - \omega_{\text{ce}})\omega^2 - [(\omega_{\text{pe}}^2 + \omega_{\text{ph}}^2) + \omega_{\text{ce}}\omega_{\text{ch}}]\omega \pm K = 0, \qquad (35)$$

where

$$K = \omega_{\text{ph}}^2 \omega_{\text{ce}} - \omega_{\text{pe}}^2 \omega_{\text{ch}}. \qquad (36)$$

For an ideal compensated semimetal with parabolic energy bands, $n_e = n_h$, $m_{\text{ce}} = m_{\text{pe}}$, and $m_{\text{ch}} = m_{\text{ph}}$, so that the last term of Eq. (35), K, vanishes. In this case, the positive roots of Eq. (35) give the magnetoplasma modes for electron-active and hole-active polarization, respectively. These modes are

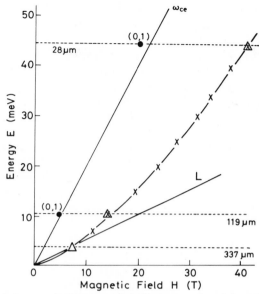

FIG. 33   The relation required between the photon energy and the magnetic field for the extra new structure to appear in the magnetoreflection spectra in graphite: $\triangle$, the experimental data; L, the theoretical line based on the classical model for the electron–hole coupled-plasma mode; X, the theoretical points calculated by the quantum-mechanical formulation of the electron–hole coupled-plasma mode.   The points denoted by $(0, 1)$ show the experimental data for the transition between the Landau levels $N = 0$ and $N = 1$; ω_{ce} shows the theoretical line for this transition.   (From Nakamura *et al.*, 1983c.)

the usual type of magnetoplasma for electron-active polarization, respectively.   The frequency of these modes tends to

$$\omega_p = \sqrt{\omega_{pe}^2 + \omega_{ph}^2} \tag{37}$$

as the magnetic field becomes zero.

In graphite, however, K is not necessarily zero, and the presence of this term gives rise to another mode with ω, that starts from zero at $H = 0$ and increases with increasing field. The third mode was found to correspond to the extra structure of the present discussion. Depending on the sign of K, this mode should be either electron-active or hole-active. In the case of graphite, the former is the case, being consistent with the experiment of Nakamura *et al.* (1983b).   They calculated the line shape of the structure in the magneto-reflection using Eq. (29) and obtained a reasonably good agreement with the experiment by assuming $m_{ce} = 0.058\, m_0$, $m_{ch} = 0.039\, m_0$, $n_e = n_h = 3 \times 10^{18}$, $\hbar\omega_p = 0.44$ eV, $\epsilon_1 = 5.0$, and adjusting $A = m_{ph}/m_{pc}$ for each wavelength.   The line L in Fig. 33 was obtained by taking a value of $A = m_{ph}/m_{pc}$

that is in agreement with the experimental data for $\lambda = 337 \, \mu$m.   As can be seen in Fig. 33, the classical approach described here fails to explain the superlinear dependence of ω of the extra structure on the field.   For such an extreme quantum limit of the highly nonparabolic bands as in the present case, a quantum mechanical treatment of the dielectric constant is required to theoretically reproduce the structures in the magntoreflection spectra, starting with Eq. (33).   The Landau-level energies and the matrix elements between them should be calculated by numerically solving determinants with infinite dimension.   Nakamura *et al.* (1984) carried out such a quantum-mechanical calculation, based on the Slonczewski–Weiss–McClure model of the energy bands in graphite (McClure, 1957; Slonczewski and Weiss, 1958).   The matrix elements in Eq. (33) necessary for calculating the dielectric constant are obtained from the following expression for $p^{\pm}$ in a matrix form as follows (Dresselhaus and Dresselhaus, 1965):

$$\frac{h}{2m} p^{-} = \begin{bmatrix} \dfrac{\hbar^2}{2m} k^{-} & 0 & \pi_{13} & 0 \\[2ex] 0 & \dfrac{\hbar^2}{2m} k^{-} & -\pi_{23} & 0 \\[2ex] 0 & 0 & \dfrac{\hbar^2}{2m} k^{-} & \pi_{33} \\[2ex] \pi_{13} & \pi_{23} & 0 & \dfrac{\hbar^2}{2m} k^{-} \end{bmatrix} \tag{38}$$

where $p^{+} = [p^{-}]^{\dagger}$, $\pi_{13} = (1/\mu)(-\gamma_0 + 2\gamma_4 \cos \pi\xi)$, $\pi_{23} = (1/\mu)(-\gamma_0 - 2\gamma_4 \cos \pi\xi)$, $\pi_{33} = (1/\mu)2\sqrt{2}\gamma_3 \cos \pi\xi$, $\xi = (c_0/2\pi)k_z$, and $k^{\pm} = k_x \pm ik_y$. Here γ_i's are the band parameters.

By means of the expansion with the harmonic-oscillator functions, the matrix (38) is transformed into an indefinite-dimensional matrix. Employing this matrix and the eigenvector of the Landau levels, the matrix elements $\langle \mu | p^{\pm} | \nu \rangle$ can be calculated.   As a final result, Nakamura *et al.* (1984) succeeded in reproducing the line shape of the extra structure as well as the structures arising from the transitions between the Landau levels.   The theoretical points of the extra structure obtained from the quantum-mechanical calculation are shown in Fig. 33.   The agreement between theory and experiment is excellent.   Nakamura *et al.* (1984) pointed out that the transition between the Landau levels $N = -1$ and $N = 0$ gives a slight contribution to the position of the extra structure.

The magnetoreflection spectra in Bi and Bi–Sb alloys were extensively investigated (Hiruma *et al.*, 1981) as well as the magnetoabsorption spectra (Hiruma *et al.*, 1983) and the transport phenomena (Hiruma and Miura,

1983). Bi is a semimetal that has a very small bandgap (15.3 meV) between the conduction band and the valence band at the L points. As a result, some of the effective-mass tensor components are extremely small. Moreover, because the energy-band overlap between the conduction band at the L points and the valence band at the T point where holes exist (hole band) is also small (37.5 meV), the application of high magnetic fields yields a large change in the carrier density. When we apply the fields parallel to the binary axis, one of the three electron pockets at the three L points has a relatively heavy cyclotron mass and the other two have a light cyclotron mass. Accordingly, the electrons populating in the two kinds of pockets are called heavy electrons and light electrons, respectively. For $H \parallel$ binary axis, the energy of the lowest Landau level of the light electrons decreases with increasing field because of the large spin–orbit interaction. Similarly, the lowest Landau level of the valence band at the L points increases in energy with increasing field, so that the bandgap at the L points first decreases when the field is increased, as shown in Fig. 34. However, as the interaction between the two levels repels them from each other, the bandgap starts to increase from the crossing point H_c (Baraff, 1965). At a high-enough field H_T, the overlap between the conduction band and the hole band diminishes and the magnetic-field-induced semimetal–semiconductor transition takes place. Meanwhile, the Fermi level and carrier concentration exhibit large

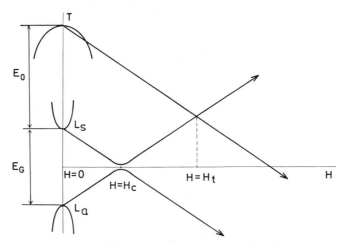

FIG. 34   The shift of the band edges (the lowest Landau levels) in Bi when the high magnetic fields are applied in the direction of the binary axis: L_s, L_a, and T represent the conduction band at the L point, the valence band at the L point, and the hole band at the T point, respectively; E_o and E_g represent the band overlap and the bandgap energies, respectively. At the field $H = H'_T$ the magnetic-field-induced semimetal-to-semiconductor transition takes place. (From Hiruma and Miura, 1983.)

and complicated dependences on the magnetic field, which show up in the magnetoreflection and the magnetoabsorption spectra.

Figure 35 shows the magnetoreflection spectra in Bi at $\lambda = 119 \ \mu$m (Hiruma *et al.*, 1981). The cyclotron resonances of light electrons, heavy electrons, and holes are observed as structures, in this order, from the lower to the higher fields. Theoretical curves calculated by Eqs. (27) and (29) are also shown. Because the experiment was done with an unpolarized radiation, the experimental data should be compared with the theoretical curve that shows the average of the electron-active ($H > 0$ in Fig. 35) and the hole-active ($H < 0$) radiations. By fitting the theoretical curve to the experimental data, the cyclotron effective masses are obtained. The effective masses for Bi, $Bi_{99.5}Sb_{0.5}$, and $Bi_{95.6}Sb_{4.4}$ at various photon energies are listed in Table VI.

It should be noted that for Bi and $Bi_{99.5}Sb_{0.5}$ the effective mass of heavy electrons decreases with increasing photon energy, contrary to the usual energy dependence of the effective mass in a nonparabolic band. The reason for the decrease is ascribed to the rapid decrease of the Fermi level with increasing field, namely, as the field becomes higher, the electronic transitions take place between the Landau levels with lower energies. It can be seen in Table VI that the electron effective masses decrease as the Sb content is increased, although the hole mass does not change significantly.

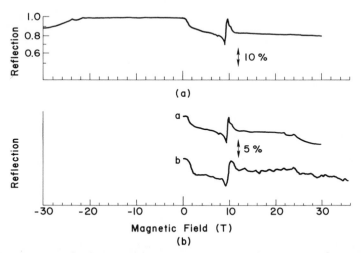

FIG. 35   Magnetoreflection spectra in Bi when the magnetic field is applied along the binary axis for $\lambda = 119 \ \mu$m, Faraday configuration. (a) Theoretical curve: Spectra for holes and electrons are shown in the positive and the negative sides of the magnetic fields, respectively; (b) theoretical curve a, the average of the spectra in (a) for both electrons and holes, is shown with experimental curve b. (From Hiruma *et al.*, 1981.)

TABLE VI

EFFECTIVE MASSES IN BI AND BI–SB ALLOYS
IN UNITS OF m_0 FOR H ∥ BINARY AXIS

Sample	Photon energy (meV)		
	3.68	3.99	10.45
Bi			
Electrons	—	—	0.009 ± 0.001
	0.115 ± 0.002	0.115 ± 0.005	0.109 ± 0.004
Holes	0.23 ± 0.01	0.23 ± 0.01	0.261 ± 0.005
$Bi_{99.5}Sb_{0.5}$			
Electrons	—	—	0.007 ± 0.002
	0.110 ± 0.002	0.113 ± 0.005	0.102 ± 0.003
Holes	0.22 ± 0.01	0.23 ± 0.01	0.266 ± 0.005
$Bi_{95.6}Sb_{4.4}$			
Electrons	—	—	0.0045 ± 0.002
	0.076 ± 0.007	0.079 ± 0.02	—
Holes	0.232 ± 0.02	0.22 ± 0.03	0.22 ± 0.02

C. MAGNETOTRANSMISSION AND ALFVÉN-WAVE PROPAGATION

As was described in Section IV.A, the electromagnetic wave can propagate through compensated semimetals as an Alfvén wave when the magnetic field is applied along the direction of the wave propagation and the condition $\omega_c > \omega$ is fulfilled. As is evident from Eq. (31), the real part of the dielectric constant or the wave vector of the Alfvén wave is dependent on the magnetic field. Therefore, when radiation is transmitted by a sample with a thickness d, the Fabry–Perot interferogram appears as the magnetic field is varied. The condition for the occurrence of transmission maxima of the first-order interference is given by

$$k_r d = \pi N, \qquad N = 1, 2, 3, \ldots, \tag{39}$$

where k_r is the real part of the wave vector (k_+) and d the thickness of the sample. The condition for the second-order interference is given by

$$k_r d = (\pi/2)M, \qquad M = 1, 2, 3, \ldots. \tag{40}$$

The intensity of the higher-order interference is weaker than the fundamental one if the damping of the Alfvén wave is large. By knowing the field position H_N of each interference maximum, the mass density ρ is obtained from Eqs. (31) and (39) as follows:

$$\rho_{ij} = (H_N^2/4\pi)[(N\pi/\omega d)^2 - c^{-2}\epsilon_{jj}], \tag{41}$$

where ϵ_{jj} is the lattice dielectric constant represented as ϵ_l in Eq. (31). The mass density ρ and ϵ_l should be treated as tensors for anisotropic crystals such as most of semimetals. In Eq. (41), the subscripts i and j denote the direction of the magnetic field and the direction of the electric vector of the wave, respectively.

Many studies have been done on the Alfvén wave propagation in Bi in the microwave range (Takano and Kawamura, 1970; Williams and Smith, 1964). In such high magnetic fields as the pulsed-field region, the Alfvén wave propagation can be observed in the far-infrared range. Hiruma *et al.* (1983) investigated the Alfvén wave propagation at 337-μm wavelength in high magnetic fields up to 40 T and found an anomaly in the dielectric constant. Figure 36 shows the experimentally observed magnetotransmission spectra in Bi when the magnetic field is applied in the direction of the binary axis of the crystal. The electric-field vector of the radiation is in the direction of the bisectrix. The Fabry–Perot interference is clearly observed; the peak positions of the interference are shown by solid lines. In addition to the Fabry–Perot interference, several transmission minima are observed on the envelope of the interferogram. It should be noted that while the positions of the interference peaks depend on the thickness of the sample, the minima on the envelope appear at identical field positions in samples with different thicknesses. Those minima are caused by the optical Shubnikov–

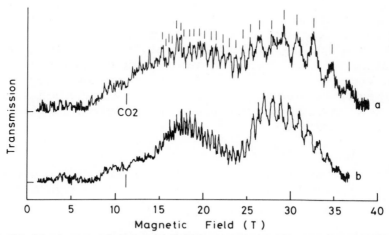

FIG. 36   Magnetotransmission spectra in Bi for the radiation with a wavelength of 337 μm, $T = 6.8$ K. The experimental data are shown for two samples with different thicknesses: curve a, 0.999 $\pm$ 0.002 mm; curve b, 1.608 $\pm$ 0.002 mm. The magnetic fields are applied along the binary axis; the electric vector of the radiation is in the direction of the bisectrix axis; CO2 indicates the position of the combined resonance; and the vertical bars represent the Fabry–Perot maxima. (From Hiruma *et al.*, 1983.)

de Haas effect, as will be mentioned later, except for the minimum designated by CO2, which is caused by the combined resonance of holes.

For **H** ∥ binary axis, the mass density ρ_{ij} is expressed as follows (Takano and Kawamura, 1970):

$$\rho_{xy} = N_A m_3 + (N_B + N_C)\frac{(m_1 + 3m_2)m_3 - m_4^2}{m_1 + 3m_2} + N_h M_3, \qquad (42)$$

$$\rho_{xz} = N_A m_2 + (N_B + N_C)\frac{4m_1 m_2}{m_1 + 3m_2} + N_h M_1, \qquad (43)$$

where x, y, and z denote the binary, bisectrix, and trigonal axes of the crystal; N_A, N_B, N_C, and N_h the carrier densities of electrons in the A, B, and C pockets and holes, respectively; M_1 and M_3 the effective masses of holes; and m_1, m_2, m_3, and m_4 the effective masses of electrons. Equations (42) and (43) show that the Fabry–Perot interferograms due to the Alfvén-wave transmission should be quite different if the direction of the electric-field vector **E** of the radiation is different, even if the magnetic field is in the same direction. In fact, for **H** ∥ binary, **E** ∥ trigonal, an interferogram entirely different from Fig. 36 is observed even in the same samples. On the other hand, the transmission minima on the envelope appear at the same fields as in the case of **E** ∥ bisectrix, because the magnetic-field direction is the same.

To obtain the mass density ρ_{ij} from the experimentally observed interferogram, proper values should be determined for the fringe indices N for each sample, for each peak, and for the lattice dielectric constant ϵ_{ij}. Hiruma et al. (1983) determined these values and obtained ρ_{xy} and ρ_{xz}. Figure 37 shows the magnetic-field dependence of ρ_{xz} thus obtained for three different samples. This figure also shows the theoretical line calculated from Eq. (43)

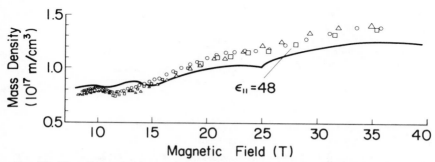

FIG. 37 Plot of mass density ρ_{xz} as a function of the magnetic field for Bi: **H** ∥ binary axis, **E** ∥ trigonal axis. Sample thicknesses were 0.999 ± 0.002 mm (□), 1.209 ± 0.002 mm (△), and 1.608 ± 0.003 mm (○). The experimental points are obtained by assuming $\epsilon_\parallel = 48$. The theoretical curve (——) was calculated from Eq. (43). (From Hiruma et al., 1983.)

for comparison. The agreement between experiment and theory is excellent when the value $\epsilon_\parallel (=\epsilon_{zz}) = 48$ is chosen.

Figure 38 shows the magnetic field dependence of ρ_{xy} for the case **E** ∥ bisectrix. The theoretical line is also plotted for comparison. For this direction of E, there is a considerable discrepancy between experiment and theory if a constant dielectric constant $\epsilon_\perp (=\epsilon_{yy}) = 100$ is chosen. The discrepancy increases as the magnetic field is increased. The experimental line cannot coincide with the theoretical one when the constant $\epsilon_\perp$ is assumed. Hiruma et al. (1983) concluded that the lattice dielectric constant $\epsilon_\perp$ has a magnetic field dependence. If a larger value of $\epsilon_\perp$ is assumed for higher fields, the experimental data come down and can be fitted to the theoretical line, as shown in Fig. 38. The dielectric constant $\epsilon_\perp$ was obtained in this way and plotted in Fig. 39 as a function of magnetic field. A very remarkable feature is that, although $\epsilon_\perp$ stays almost constant up to 14 T, it increases considerably as the magnetic field is increased. This magnetic field dependence forms a sharp contrast to $\epsilon_\parallel$, which is almost constant up to the highest field.

The anomalously large magnetic-field dependence is explained by Hiruma et al. (1983) in terms of the interband transition of electrons. Namely, the lattice dielectric constant, exclusive of the contribution of free carriers, is expressed as a sum of the contribution of the core electrons and the interband transitions, as given here:

$$\epsilon_l = \epsilon_c + \epsilon_{int}. \tag{44}$$

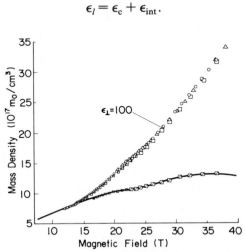

FIG. 38 Mass density ρ_{xy} versus the magnetic field for Bi: **H** ∥ binary axis, **E** ∥ bisectrix. Sample thicknesses were 0.999 ± 0.002 mm (□), 1.209 ± 0.002 mm (△), and 1.608 ± 0.003 mm (○). The experimental points (□, △, ○) were obtained by assuming $\epsilon_\perp = 100$. The theoretical curve (——) was calculated from Eq. (42); the experimental points (□) were obtained by assuming increase in $\epsilon_\perp$ with increasing magnetic field. (From Hiruma et al., 1983.)

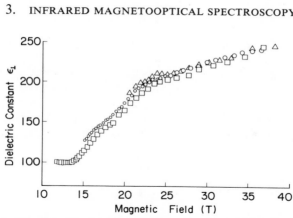

FIG. 39   Plot of the dielectric constant $\epsilon_\perp$ in Bi as a function of the magnetic field.   (From Hiruma *et al.*, 1983.)

The term ϵ_{int} is expressed in the same form as the second term in Eq. (33), where μ and ν should be taken in different bands.   Although ϵ_c rarely depends on the magnetic field, ϵ_{int} sometimes depends on it if the Landau levels are greatly altered by the field.

In the case of Bi, when the magnetic field is applied along the binary axis the transition between the lowest Landau levels in the light-electron pockets should contribute to a large magnetic-field dependence of ϵ_l because of the smallness of the energy gap and the large field dependence of the gap, as shown in Fig. 34.   Hiruma *et al.* (1983) found that this transition has a significant contribution to ϵ_l only when the electric vector is parallel to the bisectrix axis because of a relatively large interband matrix element.   Thus they explained an anisotropy of the anomalous magnetic-field dependence of the lattice dielectric constant.

In the magnetotransmission spectrum in semimetals, the oscillation of the envelope with a lower frequency than the interference, as seen in Fig. 36, gives important information about the Landau levels.   This oscillation corresponds to the optically detected Shubnikov–de Haas effect.   It is related to the imaginary part of the wave vector k_i, because k_i determines the absorption of the radiation in the sample.   The origin of k_i is the free-carrier absorption.   When each Landau level crosses the Fermi level, the scattering rate of electrons is increased, thus the free-carrier absorption of light is also increased, causing the Shubnikov–de Haas oscillation in the optical absorption.   Hiruma and Miura (1983) investigated the Shubnikov–de Haas effect in Bi by means of the dc magnetoresistance.   The resonance fields in the optically detected Shubnikov–de Haas effect are in good agreement with those in the dc magnetoresistance.   By plotting the field positions of the Shubnikov–de Haas peaks on the lines of the Landau levels, the Fermi level can be estimated as a function of magnetic field, as shown in Fig. 40. The peaks of the Shubnikov–de Haas oscillation are of holes in fields higher than 15 T, because all the Landau levels in the heavy-electron pocket are above the Fermi level. For the light-electron pockets, only the lowest Landau level exists below the Fermi level and exhibits a rather complicated magnetic-field dependence, as shown in Fig. 40. The energy dispersion of the lowest Landau level was derived by Vecchi *et al.* (1976) in the following form:

$$E^{\pm}_{j=0}(\xi) = \pm \left[[\epsilon\sqrt{1+\xi^2} - G\beta^*(H/\sqrt{1+\xi^2}) \right.$$
$$- \tfrac{1}{2}|L_{\perp}|^2\beta^*H(1+\xi^2) - (|L_{\perp}|^2/\epsilon)b_1^2(\beta^*H)^2]^2$$
$$+ \left[i\xi\frac{1}{\sqrt{1+\xi^2}}G\beta^*H + Q\beta^*H \right.$$
$$\left. + i\xi\frac{|L_{\perp}|^2b_1}{2\epsilon}(\beta^*H)^2 \right]^2 \Big]^{1/2}, \tag{45}$$

where $\epsilon = \tfrac{1}{2}E_g$ (E_g is the bandgap), $\xi^2 = 2\hbar^2k_H^2/E_gm_H$ (m_H is the effective mass in the applied field direction), $\beta^* = e\hbar/m_c^*c$ (m_c^* is the cyclotron effective mass), and G, Q, $L_{\perp}$, b_1 are the band parameters. The values of the parameters were determined by Vecchi *et al.* (1976) from the magnetooptical data below 15 T. If their parameters are used, we can predict that the magnetic-field-induced semimetal–semiconductor transition should occur at about 40 T. This field corresponds to H_T in Fig. 34. The field H_T also

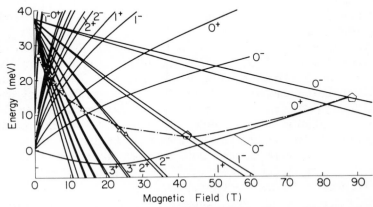

FIG. 40　Landau-level energies (———) and the Fermi E_F level (—·—) in Bi as a function of magnetic field: **H** ∥ binary; □, light electron; ○, heavy electron; △, hole. The experimental points for the Fermi level were obtained from the Shubnikov–de Haas effect and the far-infrared magnetotransmission. (From Hiruma and Miura, 1983.)

corresponds to the position of the last peak of the Shubnikov–de Haas oscillation, because the lowest Landau levels of holes and electrons cross the Fermi level at H_T.

The magnetic-field-induced semimetal–semiconductor transition was explored in megagauss fields by Miura *et al.* (1981). As the dc magnetoresistance is difficult to measure in megagauss fields, the far-infrared magnetotransmission was measured at 337 μm to monitor the magnetoresistance. Figure 41 shows the experimental result. At 88 T, the transmission of the far-infrared suddenly increases, corresponding to the semimetal–semiconductor transition. So the transition field H_T can be determined as 88 T. In thinner samples, the Shubnikov–de Haas peak for the next-lowest Landau level ($N = 1^\pm$) was observed at 43 T. From these field values, together with the Shubnikov–de Haas peak positions observed in the dc magnetoresistance at lower fields, the Landau-level diagram was constructed as shown in Fig. 40. To obtain agreement with the experiment, the values of the band parameters had to be modified from those determined by Vecchi *et al.* (Hiruma and Miura, 1983).

At the field of the semimetal–semiconductor transition, the zero-gap state is realized. In the vicinity of the zero-gap state, various models of the

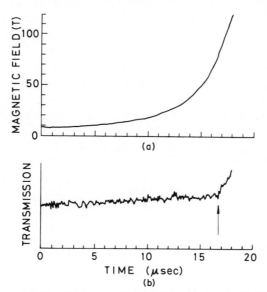

FIG. 41   (a) Magnetic field and (b) transmission for an experimental recording of the far-infrared magnetotransmission in Bi for wavelength 337 μm in megagauss fields: $\mathbf{H}$ ∥ binary axis; $T = 4.5$ K. Thickness of the sample was 0.62 mm. The magnetic-field-induced semimetal-to-semiconductor transition takes place at the point indicated by the arrow. (From Miura *et al.*, 1982.)

electronic phase transitions have been proposed, such as the excitonic phase transition (Fukuyama and Nagai, 1971) or the gas–liquid-type phase transition (Nakajima and Yoshioka, 1976). So far, no sign of such electronic phase transitions has been observed near H_T. According to Yoshioka (1978), various types of phase transition may take place at low temperatures ($T \lesssim 0.1$ K) in a magnetic field of about 10 T, where the next-lowest Landau levels of $N = 1$ play an important role. In the vicinity of the semimetal–semiconductor transition where the lowest levels of $N = 0$ are involved, we can expect the possibility of the phase transitions at a reasonably high temperature. The dimensionless parameter γ for excitons representing the magnitude of the magnetic fields relative to the binding energy of excitons R^* is given by

$$\gamma = \hbar\omega_c/2R^* = [H(\text{T})/2.35 \times 10^5][(m_0/m^*)\epsilon_1]^2, \tag{46}$$

where ϵ_1 is the dielectric constant and m^* the reduced mass of excitons. At $H_T = 88$ T, we obtain $\gamma = 2890$, and the binding energy of excitons in Bi is then estimated to be 4.6 meV, according to the YKA theory (Yafet *et al.*, 1956). The electronic phase transitions mentioned previously remains for future exploration.

The magnetoplasma-wave propagation was investigated in graphite by Nakamura and Miura (1982). Figure 42 shows the experimental recordings of the magnetotransmission at 337- and 311-μm wavelengths on the samples of HOPG. Because the absorption is large in graphite, very thin samples were measured, so that the period of the Fabry–Perot interference is much longer than in the case of Bi. In terms of the Alfvén wave, the mass density obtained from the observed interferogram increases significantly as the magnetic field is increased. This increase is much larger than the theoretically expected increase (Nakamura and Miura, 1982). The discrepancy between experiment and theory is because the magnetoplasma wave cannot be regarded as a pure Alfvén wave. As is discussed in Section IV.B, K in Eq. (36)

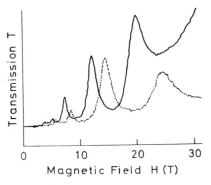

FIG. 42 Far-infrared magnetotransmission spectra in HOPG 611 graphite for wavelengths of 337 μm (——) and 311 μm (· · ·) showing the Fabry–Perot interferogram: $\mathbf{H} \parallel c$ axis; $T = 6.8$ K. (From Nakamura and Miura, 1982.)

plays a significant role in the magnetoreflection spectra. Accordingly, the second term in the square bracket in Eq. (30) does not diminish completely in the case of graphite, because of the difference between the plasma mass and the cyclotron effective mass. As a result, the wave vector cannot be represented simply by Eq. (31), using the mass density. If we calculate the line shape of the Fabry–Perot interferogram using the dielectric constant obtained in Section IV.B, we can reproduce theoretically the interferogram that is in good agreement with the experimental results (Nakamura, 1983). The Fabry–Perot interferogram resulting from the magnetoplasma-wave transmission in graphite would exhibit some anomalies if the CDW transition is brought about at low temperatures. This is worth investigating in future.

V. Faraday Rotation and Spin–Flip Transition in Magnetic Substances

A. FARADAY ROTATION IN MAGNETIC SUBSTANCES

So far, we have mainly discussed the behavior of the conduction electrons in high magnetic fields. In this section, we will consider the phenomena related to the electron spins. In recent years, much attention has been focused on magnetic substances optically transparent in the infrared region. The magnetooptical properties caused by spontaneous magnetization in ferromagnetic or ferrimagnetic substances are of practical importance for application to various optical devices. Also, the mechanism of the magnetooptical phenomena and particularly their relation with magnetization have been studied by many authors as a fundamental problem in magnetism. Usually, the Faraday rotation in magnetic substances involves components proportional to the magnetization. Consequently, the measurement of the Faraday rotation is a useful means for obtaining information about the magnetization of the substances. Especially in the megagauss-field range, where the measurement of the magnetization is extremely difficult, the Faraday rotation serves as an alternative experimental technique to investigate the magnetic properties.

Let us first consider the Faraday rotation in magnetic substances. When the magnetization is parallel to the z direction, which is one of the principal axes of the crystal, the dielectric constant can be expressed as

$$\epsilon = \begin{bmatrix} \epsilon_\perp & \epsilon' & 0 \\ -\epsilon' & \epsilon_\perp & 0 \\ 0 & 0 & \epsilon_\parallel \end{bmatrix}. \tag{47}$$

The expression (47) holds also for the case when the magnetic field is applied along the z direction in paramagnetic substances. When linearly polarized light is incident on a sample in the z direction, the angle of the Faraday rotation is given by

$$\theta = (\omega d/2c)(n_- - n_+), \tag{48}$$

where d is the thickness of the sample, and n_+ and n_- are the refractive index for the left-handed and the right-handed circularly polarized light, respectively, as already defined by Eq. (28). If $n_\pm^2 \gg \kappa^2$, Eq. (48) can be transformed into

$$\theta = (\omega d/2cn) \, \text{Im} \, \epsilon', \tag{49}$$

where $n = \frac{1}{2}(n_+ + n_-)$. The Kerr effect, the rotation of the polarization in the reflection of the light, is given similarly by

$$\theta_k = -n^{-1} \, \text{Re}(\epsilon'/\epsilon_\perp - 1). \tag{50}$$

Equation (49) shows that the Faraday rotation is calculated from the off-diagonal component of the dielectric constant. It is represented as

$$\epsilon'(\omega) = \frac{4\pi n}{\hbar} \sum_{m,n} [f(E_m) - f(E_n)] \frac{\omega}{\omega_{nm}} \frac{\omega - i\Gamma}{(\omega_{mn}^2 - \omega^2 + \Gamma^2) + 2i\omega\Gamma}$$
$$\times (|p_{mn}^+|^2 - |p_{mn}^-|^2), \tag{51}$$

where $p^\pm = p_x \pm p_y$ are the electric dipole moment for the left-handed and the right-handed circularly polarized light, $\hbar\omega_{mn}$ energy difference between the states m and n, $f(E)$ the distribution function, and Γ the damping factor. The Faraday rotation occurs if $|p_{mn}^+| \neq |p_{mn}^-|$, or if $f(E)$ is different among different states. In magnetic substances in which the populations of electrons among different spin states $f(E_m)$ in magnetic ions are different, there is a large contribution of the spin states to the Faraday rotation, as is evident from Eq. (51). The difference in the population among different spin states determines the magnetization of the substance. Therefore, a large part of the Faraday rotation is proportional to the magnetization of the substance (Shen and Bloembergen, 1964).

For ions in an exchange field, Crossley et al. (1969) demonstrated that the dominant contribution to the Faraday rotation is a term expressed as

$$\theta = K\omega^2/(\omega_0^2 - \omega^2)M, \tag{52}$$

where M is the magnetic moment of the ion, $\hbar\omega_0$ the characteristic energy separation for the involved transition, and K a constant. For ferrimagnetic substances that have several sublattice moments, the Faraday rotation is given as the sum of the contributions from each sublattice, as shown here:

$$\theta = \sum_i A_i(\omega)M_i, \tag{53}$$

where M_i is the magnetic moment of the sublattice and $A_i(\omega)$ is the coefficient.

In ferrimagnetic substances such as iron garnets, besides the electric-dipole transition, the magnetic-dipole transition resulting from the ferromagnetic resonances brings about a Faraday rotation:

$$\theta_M = (2\pi nd/c)\gamma_{eff}M, \qquad (54)$$

for ω much larger than the resonance frequency, where γ_{eff} is the effective gyromagnetic ratio (Wangness, 1954). In YIG, this component of the Faraday rotation is $61°/cm$ in the near-infrared range ($\lambda = 5 - 8 \mu m$), where the contribution from the electric-dipole transition [Eq. (53)] is very small. For shorter wavelengths, Eq. (53) gives a more significant contribution to the Faraday rotation than the magnetic term (54).

As can be seen in Eq. (52), the Faraday rotation shows resonance structures when the photon energy is swept across the energy of some electron transitions involved (Shen and Bloembergen, 1964). In the infrared range, when $\omega < \omega_0$ the Faraday rotation usually increases as the photon energy is increased.

B. PARAMAGNETIC FARADAY ROTATION IN MAGNETIC SEMICONDUCTORS

The relation between the Faraday rotation and magnetization can be investigated in optically transparent magnetic substances. Europium chalcogenides are magnetic semiconductors that are reasonably transparent in the infrared range. They are well known to show large magnetooptical effects caused by the electronic transitions in Eu^{2+} ions. Many studies have been done on the Faraday rotation and the magnetic dichroism in Eu chalcogenides in connection with magnetism (Suits and Argyle, 1965). In EuO, singularities were observed in the Faraday rotation under high magnetic fields up to 150 T (Druzhinin et al., 1976). Druzhinin et al. attributed the singularities to the formation of some giant ferromagnetic molecules called "ferron."

Suekane et al. (1983) measured the Faraday rotation in EuS and EuSe in high magnetic fields up to 170 T, at various temperatures. An He – Ne laser was employed as a source of linearly polarized infrared radiation with a wavelength of $1.15 \mu m$. Figure 43 shows the observed signal of the Faraday rotation in EuS as a function of the magnetic field. It can be seen that the period of the oscillation of the signal resulting from the Faraday rotation is prolonged as the field is increased. In addition, the amplitude of the oscillation is decreased with increasing field because of the magnetoabsorption. In Fig. 44, the Faraday rotation angle is plotted as a function of the magnetic field for two different samples. The Brillouin function for $J = 7/2, g = 2$ is plotted in the same figure for comparison. The experimental data can be

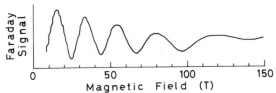

FIG. 43   The Faraday rotation signal as a function of the magnetic field for EuS at wave-length 1.15 μm: $\mathbf{H} \parallel \langle 100 \rangle$; $T = 300$ K.   The thickness of the sample was 0.40 mm.   (From Suekane *et al.*, 1983.)

well fitted to the Brillouin function representing the magnetization of Eu^{2+} ions.   To the knowledge of the author, this is the first experiment that has distinctly demonstrated the bending of the Brillouin function at room temperature, owing to the very high magnetic fields.

Figure 45 shows the magnitude of the Faraday rotation for EuS at various temperatures between 58 and 300 K.   The solid lines indicate the theoretically calculated magnetization curve for Eu^{2+} ions, taking into account the exchange interaction among them.   With exception of the data at $T = 133$ K, the experimental points fit the magnetization curve.   The reason for the discrepancy at $T = 133$ K is not known at present.   It is quite evident that the Faraday rotation is almost proportional to the magnetization.   For EuSe, it was found that the Faraday rotation is proportional to the magnetization.

The temperature dependence of the Faraday rotation was investigated in

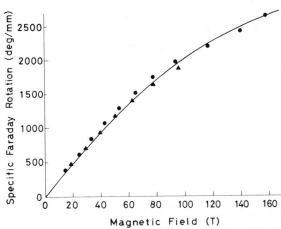

FIG. 44   The Faraday rotation angle as a function of the magnetic field for EuS: $T = 300$ K; ●, sample 1; ▲, sample 2; ——, the Brillouin function for $J = 7/2$ and $g = 2$.   (From Suekane *et al.*, 1983.)

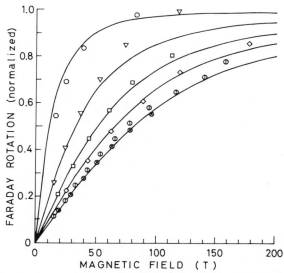

FIG. 45 The Faraday rotation angle in EuS as a function of the magnetic field at different temperatures: O, 58 K; ▽, 133 K; □, 196 K; ◇, 244 K; ⊗ and ⊕, 300 K; ——, theoretical curves for the magnetization at each temperature; **H** ‖ ⟨100⟩. (From Suekane *et al.*, 1983.)

more detail in nondestructive pulsed fields up to 30 T (Suekane *et al.*, 1983). In this range, a more accurate measurement is possible. In low-enough fields, the Faraday rotation is almost proportional to the field, and thus the Verdet constant V can be defined at each temperature. It was found that a linear relationship holds between $1/V$ and the magnetic field. For EuS, the extrapolation of the line representing this relation to $1/V = 0$ crosses the temperature axis at $T = 17$ K. Considering that the Curie temperature of EuS is $T = 16.5$ K, V is proportional to the susceptibility that obeys the Curie–Weiss law. For EuSe, the intercept on the temperature axis of the $1/V$ versus the T line is $T = 8.1$ K. However, in the case of EuSe, because of the complex magnetic structure at low temperature, this temperature cannot be compared directly with a particular ordering temperature.

C. SPIN–FLIP TRANSITION IN FERRIMAGNETIC IRON GARNETS

Since the discovery of large Faraday rotation in YIG ($Y_3Fe_5O_{12}$) (Dillon, 1958), the magnetooptical properties in ferrimagnetic iron garnets have attracted much attention. Iron garnets constitute a large class of mixed oxides with a rather complicated cubic structure. They have a transparent "window" in the wavelength region from 1 to 5 μm. The measurement of the Faraday rotation in this window region is a useful tool for the investigation of the spin structure.

YIG is the simplest member of this series, because it consists of only two magnetic sublattices, formed with Fe^{3+} ions. The tetrahedral and the octahedral sites for Fe^{3+} ions are usually called the d and the a site, respectively. In each unit formula $Y_3Fe_5O_{12}$, there are three Fe^{3+} ions in the d site and two Fe^{3+} ions in the a site. In other rare-earth–iron garnets (RIG) there is in addition the c site for the rare-earth ions R^{3+}, so that they have three magnetic sublattices.

The superexchange interaction between magnetic ions in each sublattice is known to be all negative. Because of the large negative-exchange interaction between the d and the a sites, the magnetization of these sublattices is ordered in the antiparallel direction, forming a ferrimagnetic configuration below the ferrimagnetic Curie temperature. In RIG, the third sublattice, the c site is parallel to the a site. When we apply a high-enough field in comparison to the molecular field caused by the exchange interaction, spins in all the sublattices flip to the direction of the external field. In this process of the spin–flip transition, there occurs the spin–cant phase where the sublattice magnetizations cant with each other in the intermediate-field range. These changes of the spin configuration in the external applied fields can be conveniently investigated by means of the Faraday rotation in the infrared range.

Let us first consider the case of a two-sublattice substance such as YIG. Figure 46 illustrates the change of the spin configuration in applied magnetic fields. When the applied field H_0 is lower than a critical field H_{c1}, the magnetic moments M_a and M_d of the two sublattices are antiparallel, forming a ferrimagnetic configuration (Stage I). When H_0 exceeds H_{c1}, the magnetic moment M_a that is antiparallel to the applied field starts to tilt

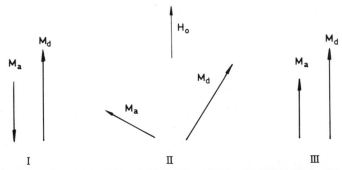

FIG. 46   The configuration of the sublattice magnetizations in YIG in high magnetic fields: H_0 is the external magnetic field, M_a and M_d the sublattice magnetizations. Region I is the ferrimagnetic phase, with $H_0 < J_{c1}$, region II the spin–cant phase, with $H_{c1} < H_0 < H_{c2}$, and region III the ferromagnetic phase. (From Miura *et al.*, 1978b.)

toward the direction of the applied field. Then the magnetic moment M_d also starts to tilt, because of the exchange interaction with the a site (Stage II). This is the spin–cant phase. When H_0 exceeds the second critical field H_{c2}, all the spins become parallel to the direction of the applied field (Stage III).

The transitions from one magnetic phase to the other can be analyzed on the basis of the molecular field model (Miura *et al.*, 1978b). The molecular fields acting on the a and the d sites are represented by

$$H_a = H_0 + n_{aa}M_a + n_{ad}M_d, \tag{55}$$

$$H_d = H_0 + n_{ad}M_a + n_{dd}M_d, \tag{56}$$

where the coefficients n_{ij} are proportional to the exchange coefficients between the sites i and j by the following relation:

$$n_{ad} = 2Z_{ad}J_{ad}/\mu N(g\mu_B)^2 = 2Z_{da}J_{ad}/\lambda N(g\mu_B)^2, \tag{57}$$

$$n_{aa} = 2Z_{aa}J_{aa}/\lambda N(g\mu_B)^2, \tag{58}$$

$$n_{dd} = 2Z_{dd}J_{dd}/\mu N(g\mu_B)^2. \tag{59}$$

Here μ_B is the Bohr magneton, Z_{ij} the numbers of the nearest neighbor j-site ions surrounding an i-site ion, N the number of Fe^{3+} ions per unit volume, and λ and μ the fractions of the Fe ions in the a and the d sites, respectively. The sublattice magnetizations M_a and M_d are determined by the molecular fields H_a and H_d, respectively. When H_0 becomes large in comparison with the other molecular field terms in Eqs. (55) and (56), the ferrimagnetic phase becomes unstable, undergoing a transition to the spin–cant phase. For the spin–cant phase to be realized, the lower and the upper critical fields are given by

$$H_{c1} = -n_{ad}(M_d - M_a), \tag{60}$$

$$H_{c2} = -n_{ad}(M_a + M_d). \tag{61}$$

Because both M_a and M_d are dependent on temperature, H_{c1} and H_{c2} depend on temperature.

Miura *et al.* (1978b) calculated the magnetic-phase diagram for YIG using the various sets of exchange coefficients previously reported. Figure 47 shows the magnetic-phase diagram for YIG with parameter values given by Anderson (1964). It can be seen that the spin–cant phase should exist only below room temperature. It should be pointed out that the extrapolation of the boundary between the ferrimagnetic phase and the ferromagnetic phase to $H_0 = 0$ gives the temperature T_0, which is much higher than the ferrimagnetic Curie temperature $\theta_f = 559$ K. When we sweep the magnetic field at constant temperature below 300 K, phase transitions should be observed at

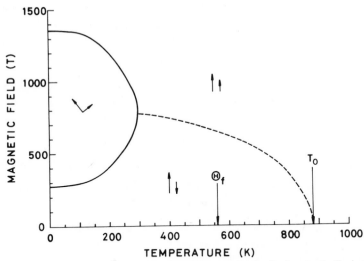

FIG. 47   A magnetic phase diagram for YIG: θ_f stands for the ferrimagnetic Curie temperature, with $H_0 > H_{c2}$. (From Miura *et al.*, 1978b.)

H_{c1} and H_{c2}. The magnetic-phase diagram is very sensitive to the choice of the exchange parameters J_{ad}, J_{aa}, and J_{dd}. So far, the exchange coefficients J_{ij} have been determined by the measurements of the temperature dependence of the spontaneous magnetization, the dispersion of the magnons, and so on. However, there have been considerable discrepancies among the data because of the uncertainty involved in the determination by means of these methods. On the other hand, the spin–flip transition is the most direct way to determine J_{ij}. Therefore, the accurate values of the exchange parameters should be determined by observing the spin–flip transitions.

Figure 48 shows the experimental recordings of the Faraday rotation observed for YIG, Ga-doped YIG, and GdIG in megagauss magnetic fields (Miura *et al.*, 1977). In YIG, the rotation angle decreased with increasing field and changed its sign at about 60 T. Because the Faraday rotation reflects the sublattice magnetizations, as discussed in the previous sections, any abrupt change in the sublattice magnetizations caused by the spin–flip transition would show up in the Faraday rotation. The magnetic-field dependence of the rotation angle is almost linear up to 170 T, and no sign of the spin–flip transition is observed in this field range. This result is reasonable if we rely on the phase diagram of Fig. 47.

When we reduce the large exchange fields in YIG by substitution of nonmagnetic ions for Fe^{3+}, the critical field for the transition can be greatly lowered. In gallium-substituted YIG, $Y_3Ga_xFe_{5-x}O_{12}$, Ga^{3+} is known to

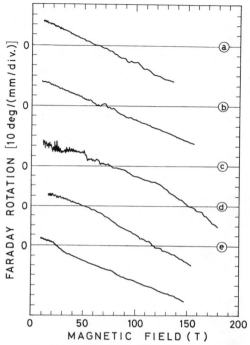

FIG. 48   The Faraday rotation angle as a function of the magnetic field for various iron garnets.   Curve a: YIG, 300 K; curve b: YIG, 177 K; curve c: $Y_3Ga_{0.5}Fe_{4.5}O_{12}$, 136 K; curve d: $Y_3Ga_1Fe_4O_{12}$, 267 K; curve e: GdIG, 298 K.   The zeros in the ordinate are shown for each curve. (From Miura $et\ al.$, 1977).

substitute preferentially on the d site.   For small x, the net magnetization of $Y_3Ga_xGe_{5-x}O_{12}$ decreases almost linearly with x, and it crosses zero at about $x = 1.25$.   Depending on x, the spontaneous magnetization exhibits various types of temperature dependence of ferrimagnets in Neel's theory, such as R, Q, or N types (Miura $et\ al.$, 1978b).   Figure 48 shows the experimental results for the crystals with $x = 0.5$ and $x = 1.0$.   In these samples, the spin–flip transition is clearly observed, i.e., the slope of the Faraday rotation changes at the fields indicated by arrows.   These points correspond to H_{c1}, the critical field for the transition from the ferrimagnetic phase to the spin–cant phase.   The transition is also observed in GdIG, which has a compensation point near room temperature.

The critical fields for the spin–flip transition in $Y_3Ga_xFe_{5-x}O_{12}$ was analyzed by Miura $et\ al.$ (1978b).   It was found that the experimental value of H_{c1} is much lower than the theoretical value both for $x = 0.5$ and $x = 1.0$, if we rely on the exchange coefficients given by Anderson (1964).   The discrepancy is caused by the inadequacy of the employed molecular-field model in the case of the diluted alloys. In this model, the local fields acting on a magnetic ion owing to neighboring ions are averaged and regarded as uniform. The exchange coefficients J_{ij} are assumed to be constant even in the alloys. However, the substitution of Ga may alter the environment of the magnetic ion in a different fashion.

Therefore, the exchange coefficients $|J_{ij}|$ themselves may decrease with increasing x. If we permit the decrease of $|J_{ij}|$ with increasing x, the experimental results can be interpreted within the framework of the molecular-field model.

Another interesting feature in the data shown in Fig. 48 is the large decrease of the rotation angle in YIG with increasing field. The rate of the decrease is 0.5 deg/T · mm at $T = 300$ K for $\lambda = 1.06$ μm. The change of the sublattice magnetization is not large enough to explain this large change. This implies the existence of a term proportional to the external field in addition to the terms included in Eq. (53). The mechanism of this term is not clear at the moment (Dillon *et al.*, 1979).

Other than YIG, rare-earth–iron garnets (RIG) have another magnetic sublattice, the c site of the rare-earth ions R^{3+} (R = Sm, Eu, Gd, Tb, Dy, Ho, Er, Tm, Yb, Lu), besides the a and the d sites for Fe^{3+} ions. So the magnetic structure of RIG should be treated as a three-sublattice system. Because the exchange coefficient between the a and the d sites $|J_{ad}|$ is much larger than the other coefficients $|J_{ac}|$ or $|J_{dc}|$, the coupling between the a and the d sites for Fe^{3+} ions is very stable in comparison to the couplings between the Fe^{3+} sites and the R^{3+} site. Therefore, assuming the a and d sites to form one sublattice, RIG can be treated approximately as a two-sublattice system (Clark and Callen, 1968) at relatively low magnetic fields. This approximation of RIG can explain some of the magnetic properties of RIG, such as the temperature dependence of the spontaneous magnetization. However, it is not an adequate approximation for analyzing the spin–flip transition (Tanaka *et al.*, 1983).

The sublattice magnetization of the c site M_c is in the direction opposite to the sum of those of the d and the a sites $M_d + M_a$. The temperature dependence of M_c is larger than that of $M_d + M_a$, and the relative magnitudes of $|M_c|$ and $|M_d + M_a|$ change with varying temperature. Consequently, in several crystals of RIG (R = Gd, Tb, Dy, Ho, Er), the total magnetization $M = M_c + M_d + M_a$ reverses its direction crossing zero at a compensation temperature T_c. This is called the N-type ferrimagnet. In the vicinity of T_c, the magnetic properties of RIG and the antiferromagnets are similar. If we neglect the magnetic anisotropy energy, the critical field H_{c1} becomes zero at $T = T_c$. The magnetic-phase diagram of the three-sublattice system was extensively investigated by Féron *et al.* (1973), Kharchenko *et al.* (1976), and

Akihiro (1980). In the extreme case $|J_{ad}| \gg |J_{ac}|, |J_{dc}|$, the spin–flip transition takes place successively in two stages at low temperatures $T < T_c$. First the sum $M_d + M_a$ flips toward the direction of M_c that is parallel to the field direction, and then M_a flips toward the field direction. So the spin–cant phase is divided into two regions corresponding to each stage. However, in usual cases, these two regions are mingled, and the spin–cant phase in the phase diagram takes a complicated form.

Because $H = H_{c1}$ is realized at low fields near T_c, the spin–flip transition has been investigated for GdIG in the temperature range close to T_c (Kharchenko et al., 1976). GdIG provides the simplest case of the spin–flip transition because the Gd ion is in the S state and the anisotropy of the ions is small. Nevertheless, the magnetic-phase diagram in the vicinity of T_c seems quite complicated because of the existence of metastable states (Kharchenko et al., 1976).

The spin–flip transition was investigated in GdIG, TbIG, DyIG, and ErIG at ISSP by the Faraday rotation in high magnetic fields up to the megagauss range. An example of the results for GdIG is shown in Fig. 48. Figure 49 shows the Faraday rotation in TbIG at $T = 80$ K (Yang et al., 1980). Reflecting the change of the sublattice magnetization in the c site, the rotation is a nonlinear function of the magnetic field below H_{c1}, indicated by the arrow. To explain this nonlinear magnetic field dependence, it is necessary to introduce a term $C|M_c|^2 H_0$ in addition to the right side of Eq. (53). Figure 50 shows the critical fields H_{c1} for the transition from the ferrimagnetic phase to the spin–cant phase in TbIG, as a function of temperature for $\mathbf{H} \parallel \langle 100 \rangle$ and $\mathbf{H} \parallel \langle 111 \rangle$. At the compensation point $T_c = 245$ K, H_{c1} becomes almost zero. In the temperature range below T_c, there is a large anisotropy in H_{c1} depending on the field direction with respect to the crystal-

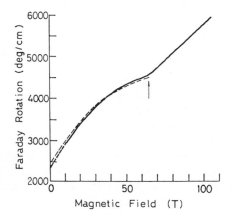

FIG. 49 The Faraday rotation angle as a function of the magnetic field for TbIG: $\mathbf{H} \parallel \langle 111 \rangle$; $T = 80$ K; $\lambda = 1.06$ μm; ——, experimental data; – – –, the theoretical curve. The arrow indicates the point of transition from the ferrimagnetic phase to the spin–cant phase. (From Yang et al., 1980.)

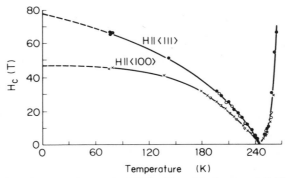

FIG. 50 A magnetic phase diagram for TbIG. The critical magnetic field H_{c1} is shown for **H** ∥ ⟨100⟩ and **H** ∥ ⟨111⟩. Experimental points ● are from Faraday rotation measurement (**H** ∥ ⟨111⟩), ○ from magnetic measurement (**H** ∥ ⟨111⟩), ▽ from magnetic measurement (**H** ∥ ⟨100⟩), and × from Faraday rotation measurement (**H** ∥ ⟨100⟩). The solid lines are drawn smoothly connecting the experimental points, and the broken lines are the extrapolation of the solid lines to the lower temperature. (From Yang *et al.*, 1980.)

lographic axes. It is reasonable that the anisotropy in H_{c1} is large, considering the large anisotropy of the orbital magnetization in Tb^{3+} ions. Although it is difficult to analyze the anisotropy of Tb^{3+} because of its complexity at low temperatures, the anisotropy in H_{c1} at $T = 0$ K was estimated to be 32 T.

In RIG, six exchange coefficients are involved in the molecular-field model. The determination of all of these coefficients requires a large amount of experimental data. It is considered to be a good approximation to assume J_{ad}, J_{aa}, and J_{dd} are the same in all RIG as in YIG. With accurate values of these coefficients for YIG, the determination of the other coefficients becomes much simpler. In this respect, it is strongly desirable to construct the whole magnetic-phase diagram for YIG, as in Fig. 47, by the experiments in very high magnetic fields.

VI. Summary

Owing to the development of techniques for the generation of high pulsed-magnetic fields and for data acquisition in recent years, much progress has been achieved in infrared and far-infrared magnetooptical spectroscopy in semiconductors, semimetals, and magnetic substances. Employing molecular-gas lasers in the far-infrared and infrared ranges and a YAG laser and an He–Ne laser in the near-infrared range, various new phenomena have been found characteristic of the extremely high magnetic fields. Especially in very high fields in the megagauss range, many new possibilities have been opened up, some examples of which are discussed in this chapter.

In the cyclotron resonance in semiconductors, the resonant photon energy becomes so high that it can become comparable to or even exceed various excitation energies, such as the energy bandgap, the height of the hump of the camel's-back structure, or the optical-phonon energy. Hence different types of nonparabolicities and the polaron problem can be studied extensively. The samples with low mobility or large-mass carriers can be studied by means of the cyclotron resonance in extremely high fields. Besides the resonance fields, the linewidth provides information on the carrier-scattering mechanism in the ultraquantum limit.

The far-infrared magnetooptical spectroscopy in semimetals is a vast area of research which has benefitted from the application of very high magnetic fields is very useful. The dielectric function that determines the response to far-infrared radiation is sensitively dependent on the magnetic field in semimetals. A variety of new aspects in magnetoreflection and magnetotransmission have been observed in Bi and graphite. In particular, in these crystals electronic-phase transitions have been theoretically predicted in very high magnetic fields, so that magnetooptical spectroscopy will be a powerful means for exploring these phase transitions in the future.

In magnetic substances, magnetization can be investigated by means of the Faraday rotation. In ferrimagnetic substances such as iron-garnet crystals, the spin–flip transitions can be studied by the Faraday rotation in megagauss fields.

At ISSP we have done such investigations as mentioned here, using very high magnetic fields in the megagauss range and nondestructive pulsed fields up to 45 T. In these pulsed fields, optical measurements in the visible range, transport measurements, and magnetization measurements are possible in addition to the infrared spectroscopy. Among various experiments, infrared spectroscopy using lasers will continue to be employed as one of the most convenient tools to investigate solid-state physics in high pulsed magnetic fields.

Rapid growth in techniques for generating even higher fields is continuing. We are conducting a new project at ISSP to generate very high fields up to 1000 T by installing very large condenser banks. Therefore, new results will be obtained in higher fields in the near future.

ACKNOWLEDGMENTS

The author is indebted to his co-workers, Professor S. Chikazumi, G. Kido, K. Nakao, N. Miyajima, I. Oguro, K. Kawauchi, K. Suzuki, K. Hiruma, M. Akihiro, H. Katayama, K. Nakamura, T. Itakura, M. Suekane, Y. Iwasa, T. Tanaka, T. Osada, T. Kikuchi, Dr. F. M. Yang, Dr. J. F. Dillon, Jr., Dr. L. G. van Uitert and many other collaborators for their contribution to the research discussed in this chapter. The author is thankful to Professor F. Herlach for critically reading the manuscript.

REFERENCES

Akihiro, M. (1980). "Spin Phase Transition in Ferrimagnetic Oxides under Megagauss Fields." Ph. D. Thesis, Department of Physics, University of Tokyo, Tokyo, Japan.

Anderson, E. E. (1964). *Phys. Rev. A* **134**, 1581–1585.

Apel, J. R., Pohler, T. O., Westgate, C. R., and Joseph, R. I. (1971). Phys. Rev. B **4**, 436–451.

Baer, W. W., and Dexter, R. N. (1964). *Phys. Rev.* **135**, 1338–1393.

Bagguley, D. M. S., Stradling, R. A., and Whiting, J. S. S. (1961). *Proc. R. Soc. A* **262**, 340–379.

Baraff, G. A. (1965). *Phys. Rev. A* **137**, 842–853.

Bimberg, D., Skolnick, M. S., and Sanders, L. M. (1979). *Phys. Rev. B* **19**, 2231–2245.

Bradley, C. C., Button, K. J., Lax, B., and Rubin, L. G. (1968). *IEEE J. Quantum Electron.* **QE-4**, 733–737.

Button, K. J., and Lax, B. (1970). *Bull. Amer. Phys. Soc.* **15**, 365.

Button, K. J., Gebbie, H. A., and Lax, B. (1966). *IEEE J. Quantum Electron.* **QE-2**, 202–207.

Button, K. J., Lax, B., Cohn, D. R., and Dreybrodt, W. (1970). *Proc. 10th Int. Conf. Phys. Semicond.*, Cambridge, Massachusetts (S. P. Keller, J. C. Hensel, and F. Stern, eds.), pp. 154–157. Atomic Energy Commission, Washington, D.C.

Carter, A. C., Dean, P. J., Skolnick, M. S., and Stradling, R. A. (1977). *J. Phys. C* **10**, 5111.

Clark, A.E., and Callen, E. (1968). *J. Appl. Phys.* **39**, 5972–5982.

Crossley, W. A., Cooper, R. W., and Page, J. L. (1969). *Phys. Rev.* **181**, 896–904.

Dean, P. J., and Thomas, D. G. (1966). *Phys. Rev.* **150**, 690–703.

Dickey, D. H., Johnson, E. J., and Larsen, D. M. (1967). *Phys. Rev. Lett.* **18**, 559–602.

Dillon, J. F., Jr. (1958). *J. Appl. Phys.* **29**, 539–541.

Dillon, J. F., Jr., Chen, E. Yi, Blank, S., Bonner, W. A., and van Uitert, L. G. (1979). J. Magn. Magn. Mat. **11**, 152–156.

Doezema, R. E., Datars, W. R., Shaber, H., and van Schyndel, A. (1979). *Phys. Rev.* **B19**, 4224–4230.

Dresselhaus, G., and Dresselhaus, M. S. (1965). *Phys. Rev.* **140**, A401–412.

Dresselhaus, G., Kip, A. F., and Kittel, C. (1955). *Phys. Rev.* **98**, 368–384.

Druzhinin, V. V., Pavlovskii, A. I., Samokhvalov, A. A., and Tatsenko, O. M. (1976). *JETP Lett.* **23**, 233–236.

Faughnan, B. W. (1965). *J. Phys. Soc. Japan* **20**, 574–591.

Féron, J. L., Fillion, G., and Hug, G. (1973). *J. Phys.* **34**, 247–256.

Fink, D., and Braunstein, R. (1974). *Solid State Commun.* **15**, 1627–1631.

Fletcher, R. C., Yaeger, W. A., and Merritt, F. R. (1955). *Phys. Rev.* **100**, 747–748.

Foner, S., and Kolm, H. H. (1957). *Rev. Sci. Instrum.* **28**, 799–807.

Fowler, C. M., Garn, W. B., and Caird, R. S. (1960). *J. Appl. Phys.* **31**, 588–594.

Fowler, C. M., Caird, R. S., Garn, W. B., and Erickson, D. J. (1976). *IEEE Trans. Magn.* **12**, 1018–1023.

Fukai, M., Kawamura, H., Sekido, K., and Imai, I. (1964). *J. Phys. Soc. Japan* **19**, 30–39.

Fukuyama, H., and Nagai, T. (1971). *J. Phys. Soc. Japan* **31**, 812–822.

Hensel, J. C. (1962). *Proc. 6th Int. Conf. Phys. Semicond.*, Exeter, England, pp. 281.

Hensel, J. C., and Suzuki, K. (1974). *Phys. Rev. B* **9**, 4219–4257.

Herlach, F., Davis, J., Schmidt, R., and Spector, H. (1974). *Phys. Rev. B* **10**, 682–687.

Hiruma, K., and Miura, N. (1983). *J. Phys. Soc. Japan* **52**, 2118.

Hiruma, K., Kido, G., and Miura, N. (1977). *Solid State Commun.* **31**, 1019–1022.

Hiruma, K., Kido, G., and Miura, N. (1981). *Solid State Commun.* **38**, 859–863.

Hiruma, K., Kido, G., and Miura, N. (1983). *J. Phys. Soc. Japan* **52**, 2550.

Humphreys, R. G., Rossler, U., and Cardona, M. (1978). *Phys. Rev. B* **18**, 5590–5605.

Isaacson, R. T., and Williams, G. A. (1969). *Phys. Rev.* **177**, 738–746.

Itoh, R., Fukai, M., and Imai, I. (1966). *J. Phys. Soc. Japan (Suppl.)* 21, 357–361.

Iye, Y., *et al.* (1982). *Phys. Rev. B* **25**, 5478–5485.

Johnson, E. J., and Larsen, D. M. (1966). *Phys. Rev. Lett.* **16**, 655–659.

Kapitza, P. L. (1924). *Proc. Roy. Soc. A* **105**, 691–710.

Kawamura, H., Saji, H., Fukai, M., Sekido, K., and Imai, I. (1964). *J. Phys. Soc. Japan* **19**, 288–297.

Kawamura, H., Nagata, S., Nakama, T., and Takano, S. (1965). *Phys. Lett.* **15**, 111–112.

Kharchenko, N. F., Eremenko, V. V., Gnatchenko, S. L., Belyi, L. I., and Kabanova, E. M. (1976). *Sov. Phys. JETP* **41**, 531–539.

Kido, G., and Miura, N. (1978). *Appl. Phys. Lett.* **33**, 321–322.

Kido, G., and Miura, N. (1982). *Appl. Phys. Lett.* **41**, 569–571.

Kido, G., Miura, N., Kawauchi, K., Oguro, I., and Chikazumi, S. (1976). *J. Phys. E. Sci. Instrum.* **9**, 587–592.

Kido, G., Miura, N., Akihiro, M., Katayama, H., and Chikazumi, S. (1981). *In* "Physics in High Magnetic Fields" (S. Chikazumi and N. Miura, eds.), pp. 72–81. Springer-Verlag, Berlin and New York.

Kido, G., *et al.* (1983). *In* "High Field Magnetism" (M. Date, ed.), pp. 339–318. North-Holland Publ., Amsterdam.

Kurti, N. (1975). *In* "Physique sous Champs Magnetiques Intenses," pp. 15–20. Centre Nacional de la Reserche Scientifique, Paris, France.

Landwehr, G. (1981). *In* "Physics in High Magnetic Fields" (S. Chikazumi and N. Miura, eds.), pp. 2–11. Springer-Verlag, Berlin and New York.

Larsen, D. M. (1964). *Phys. Rev. A* **135**, 419–426.

Lawaetz, P. (1971). *Phys. Rev. B* **4**, 3460–3467.

Lawaetz, P. (1975). *Solid State Commun.* **16**, 65–67.

Lax, B., Zeiger, H. J., and Dexter, R. N. (1954). *Physica* **20**, 818–828.

Lax, B., Mavroid, J. G., Zeiger, H. J., and Kiys, R. J. (1961). *Phys. Rev.* **122**, 31–35.

Leotin, J., *et al.* (1975). *Solid State Commun.* **16**, 363–366.

Mahan, G. D., and Hopfield, J. J. (1964). *Phys. Rev. Lett.* **12**, 241–243.

Manenov, A. A., and Prokhorov, A. M. (1955). *Sov. Phys. JETP* **1**, 611.

McClure, J. W. (1957). *Phys. Rev.* **108**, 612–618.

McLachlan, D. S. (1966). **147**, 368–375.

Mendez, E., Misu, A., and Dresselhaus, M. S. (1980). *Phys. Rev. B* **21**, 827–836.

Meyer, H. J. G. (1962). *Phys. Lett.* **2**, 259–260.

Miura, N., and Chikazumi, S. (1979). *Japan J. Appl. Phys.* **18**, 553–564.

Miura, N., and Kido, G. (1977). *Proc. 13th Int. Conf. Phys. Semicond.* (F. G. Fumi, ed.), pp. 1149–1152. Tipografia Marves, Rome, Italy.

Miura, N., and Tanaka, S. (1970). *Phys. Status Solidi* **42**, 257–266.

Miura, N., Kido, G., and Chikazumi, S. (1976b). *Solid State Commun.* **18**, 885–888.

Miura, N., Kido, G., Suzuki, K., and Chikazumi, S. (1976a). *Proc. 3rd Int. Conf. Appl. High Magnetic Fields in Semicond. Phys.* (G. Landwehr, ed.), pp. 441–476. University of Würzburg, Würzburg, Federal Republic of Germany.

Miura, N., *et al.* (1977). *Physica B* **86-88**, 1219–1220.

Miura, N., Kido, G., and Chikazumi, S. (1978a). *Proc. 4th Int. Conf. Appl. High Magnetic Fields in Semicond. Phys.* (J. F. Ryan, ed.), pp. 233–252. University of Oxford, Oxford, England.

Miura, N., Oguro, I., and Chikazumi, S. (1978b). *J. Phys. Soc. Japan* **45**, 1534–1541.

Miura, N., Kido, G., Akihiro, M., and Chikazumi, S. (1979a). *J. Magn. Magn. Mat.* **11**, 275–283.

Miura, N., Kido, G., and Chikazumi, S. (1979b). *In* "The Physics of Selenium and Tellurium" (E. Gerlach and P. Grosse, eds.), pp. 110–112. Springer-Verlag, Berlin and New York.

Miura, N., Kido, G., and Chikazumi, S. (1979c). *Proc. Int. Conf. Phys. Semicond.,* 14th Edinburgh (B. L. H. Wilson, ed.) 1109–1112. The Institute of Physics, Bristol and London.

Miura, N., Kido, G., Miyajima, H., Nakao, K., and Chikazumi, S. (1981). *In* "Physics in High Magnetic Fields" (S. Chikazumi and N. Miura, eds.), pp. 64–70. Springer-Verlag, Berlin and New York.

Miura, N., Hiruma, K., Kido, G., and Chikazumi, S. (1982). *Phys. Rev. Lett.* **49**, 1339–1342.

Miura, N., Kido, G., Suekane, M., and Chikazumi, S. (1983a). *Physica B,C* **117–118**, 66–68.

Miura, N., Kido, G., Suekane, M., and Chikazumi, S. (1983b). *J. Phys. Soc. Japan* **52**, 2838.

Miyake, S. J. (1968). *Phys. Rev.* **170**, 726–732.

Mooradian, A., and Fan, H. Y. (1966). *Phys. Rev.* **148**, 873–885.

Motokawa, M., Kuroda, S., and Date, M. (1979). *J. Appl. Phys.* **50**, 7762–7767.

Nagasaka, K. (1977). *Phys. Rev. B* **15**, 2273–2277.

Nagasaka, K., Kido, G., and Narita, S. (1973). *Phys. Lett. A* **45**, 485–486.

Nakajima, S., and Yoshioka, D. (1976). *J. Phys. Soc. Japan* **40**, 328–333.

Nakamura, K. (1983). Ph. D. Thesis, Department of Applied Physics, University of Tokyo, Tokyo, Japan.

Nakamura, K., and Miura, N. (1982). *Solid State Commun.* **42**, 119–122.

Nakamura, K., Kido, G., and Miura, N. (1983b). *Solid State Commun.* **47**, 349.

Nakamura, K., Osada, T., Kido, G., Miura, N., and Tanuma, S. (1983a). *J. Phys. Soc. Japan* **52**, 2875.

Nakamura, K., Kido, G., Nakao, K., and Miura, N. (1984). *J. Phys. Soc. Japan* **53**, 1164.

Nakao, K. (1976). *J. Phys. Soc. Japan* **40**, 761–768.

Nakao, K. Doi, T., and Kamimura, H. (1971). *J. Phys. Soc. Japan.* **30**, 1400–1413.

Narita, S., Kido, G., and Nagasaka, K. (1972). *J. Phys. Soc. Japan* **33**, 1488.

Pidgeon, C. R., and Brown, R. N. (1966). *Phys. Rev.* **146**, 575–583.

Pidgeon, C. R., Groves, S. H., and Feinleib, J. (1967). *Solid State Commun.* **5**, 677–680.

Reine, M., Aggarwal, R. L., and Lax, B. (1972). *Phys. Rev. B* **5**, 3033–3049.

Roth, L. M., Lax, B., and Zwerdling, S. (1959). *Phys. Rev.* **114**, 90–104.

Shen, Y. R., and Bloembergen, N. (1964). *Phys. Rev. A* **133**, 515–665.

Shin, E. E. H., Argyres, P. N., and Lax, B. (1973). *Phys. Rev.* **137**, 3572–3579.

Shinno, H., Yoshizaki, R., Tanaka, S., Doi, T., and Kamimura, H. (1973). *J. Phys. Soc. Japan* **35**, 525–533.

Shroeder, P. R., Presselhaus, M. S., and Javan, A. (1971). *J. Phys. Chem. Solids* **32**(Suppl. 139).

Slonczewski, J. C., and Weiss, P. R. (1958). *Phys. Rev.* **109**, 272–279.

Stradling, R. A. (1966). *Phys. Lett.* **20**, 127–128.

Suekane, M., Kido, G., Miura, N., and Chikazumi, S. (1983). *J. Magn. Magn. Mat.* **31-34**, 589–590.

Suematsu, H., and Tanuma, S. (1972). *J. Phys. Soc. Japan* **33**, 1619–1628.

Suits, J. C., and Argyle, B. E. (1965). *J. Appl. Phys.* **36**, 1251–1252.

Suzuki, K., and Miura, N. (1975). *J. Phys. Soc. Japan* **39**, 148–154.

Suzuki, K., and Miura, N. (1976). *Solid State Commun.* **18**, 233–236.

Suzuki, K., Miura, N., Uchida, S., and Tanaka, S. (1976). *Phys. Status Solidi B* **76**, 787–796.

Tachikawa, K. (1981). *In* "Physics in High Magnetic Fields" (S. Chikazumi and N. Miura, eds.), pp. 12–23. Springer-Verlag, Berlin and New York.

Takano, S., and Kawamura, H. (1970). *J. Phys. Soc. Japan* **28**, 348–359.

Takeda, K., *et al.* (1983). *Phys. Status Solidi B* **115**, 369–379.

Tanaka, T., Nakao, K., Kido, G., Miura, N., and Chikazumi, S. (1983), *J. Magn. Magn. Mat.* **31-34**, 773–774.

Tanuma, S., Inada, R., Furukawa, A., Takashi, O., and Iye, Y. (1981). *In* "Physics in High Magnetic Fields" (S. Chikazumi and N. Miura, eds.), pp. 316–319. Springer-Verlag, Berlin and New York.

Treusch, J., and Sandrock, R. (1966). *Phys. Status Solidi* **16**, 487–497.
Vecchi, M. P., Pereira, J. R., and Dresselhaus, M. S. (1976). *Phys. Rev. B* **14**, 298–317.
von Ortenberg, M., and Silberman, R. (1975). *Solid State Commun.* **17**, 617–620.
Waldman, J., *et al.* (1969). *Phys. Rev. Lett.* **23**, 1033–1037.
Wangness, R. K. (1954). *Phys. Rev.* **95**, 339–345.
Williams, G. A., and Smith, G. E. (1964). *IBM J. Res. Dev.* **8**, 276–283.
Yafet, Y., Keyes, R. W., and Adams, E. N. (1956). *J. Phys. Chem. Solids.* **1**, 137–142.
Yang, F. M., Miura, N., Kido, G., and Chikazumi, S. (1980). *J. Phys. Soc. Japan* **48**, 71–76.
Yoshioka, D. (1978). *J. Phys. Soc. Japan* **45**, 1165–1173.
Yoshioka, D., and Fukuyama, H. (1981). *J. Phys. Soc. Japan* **50**, 725–726.

CHAPTER 4

Spectral Thermal Infrared Emission of the Terrestrial Atmosphere

Gert Finger and Fritz K. Kneubühl

Physics Department, ETH
Zürich, Switzerland

I. Introduction

The infrared properties of the terrestrial atmosphere are of interest for many scientific and technical reasons. Most important are the two atmospheric spectral windows of high transmittance, from $\lambda = 3$ to $5 \ \mu$m and from $\lambda = 8$ to $14 \ \mu$m. The relevant absorption process in these two regions is the water-vapor continuum absorption, also called anomalous water-vapor absorption or excess absorption. The continuum absorption still eludes a complete theoretical explanation. Therefore, various semiempirical models (Kneizys et al., 1980; LaRocca, 1975; McClatchey et al., 1972; Roberts et al., 1976) meet the immediate need for a reliable prediction of the infrared propagation in the two atmospheric windows. These models have been applied previously to many problems related to the terrestrial atmosphere.

We mention first the problem of radiative energy transfer within the atmosphere. The energy balance of the earth and its atmosphere is widely determined by infrared absorption and emission of atmospheric gases and water in clouds (Houghton, 1977). Part of the incident solar energy is reemitted into space by thermal infrared radiation. The water-vapor continuum absorption reduces the infrared flux emitted from the surface of the earth, similar to the greenhouse effect of CO_2. Recently, Hummel (1982) estimated that the anomalous water-vapor absorption causes an increase in the ground temperature up to 2 K.

A related problem was studied by the authors (Zürcher et al., 1982) in past years. In the two infrared windows the radiative loss of the ground is not counterbalanced by the thermal emission of the atmosphere under clear-weather conditions. During clear winter nights this results in a radiative energy deficit of the ground up to 90 W/m². Because a black surface at ambient air temperature shows its intensity maximum near $\lambda = 10\ \mu m$, the atmospheric window from $\lambda = 8$ to 14 μm is relevant to the energy transfer. Consequently, the authors demonstrated that, by application of mirror coatings to building envelopes in this spectral region, the energy consumption of houses can be reduced by up to 20% in central Europe during winter.

The two atmospheric infrared windows are also of importance to infrared astronomy, remote sensing, range finding, communication, air pollution monitoring, retrieval of temperature and moisture soundings from satellites (Curtis et al., 1974; Ellis et al., 1973) and related topics. For the solution of all these problems, system designers and users of electrooptical systems are bound to know transmission, absorption, and emission in the windows under various environmental conditions.

The continuum absorption in the atmospheric windows is not only of technical significance but also of considerable scientific interest. At present it is not known to what extent water-vapor continuum absorption can be attributed to absorption in distant wings of water-vapor monomer lines (Clough et al., 1980; Elsasser, 1938; Thomas and Nordstrom, 1982) or to absorption by clusters of water molecules, e.g., dimers, trimers, etc.† The

† Bohlander (1979), Suck et al. (1979, 1982), Viktorova and Zhevakin (1966, 1975), and Varanasi et al. (1968).

formation of water clusters is the first step in the nucleation process and therefore the key to the development of a microscopic theory on homogeneous nucleation (Daee et al., 1972) in the atmosphere.

Hitherto, different experimental methods have been applied to the study of the properties of the two atmospheric infrared windows. Absorption measurements of the water-vapor continuum both in the laboratory and in the field have been reviewed for the $\lambda = 8 - 14\ \mu m$ spectral region by Burch and Gryvnak (1980), Dianov-Klokov et al. (1981), and Adiks et al. (1975).

Ground-based spectral thermal-emission measurements have been performed for many years. Oetjen et al. (1960) measured sky-radiance spectra from $\lambda = 1$ to 20 μm and their dependence on the angle of elevation from a mobile laboratory at different locations in the United States. Harrison (1976) compared measured emission spectra with radiative transfer calculations. Grassl (1973) analyzed sky-emission data in terms of water-vapor continuum absorption. Ben-Shalom et al. (1980) compared sky-emission data measured from the ground with LOWTRAN predictions (Selby et al., 1980). Long-time measurements of thermal sky radiance on a routine basis were carried out later by Berdahl to evaluate the potential of radiative cooling (Berdahl and Fromberg, 1982).

Ground-based sky-emission measurements are complicated because the real atmosphere represents an inhomogeneous optical path. The contributions of the different atmospheric layers to the total sky radiance observed from the ground can be determined by the direct measurement of the height profile of the atmospheric thermal radiance. Hence, Lee (1973) measured the continuum absorption in the $\lambda = 8 - 14\ \mu m$ atmospheric window with a tethered-balloon-borne radiometer in 1972. In a more elaborate experiment, Coffey (1977) installed the "selective chopper radiometer (SCR)" developed for the *Nimbus* 5 satellite in an aircraft to study height profiles of radiance in the $\lambda = 10 - 12\ \mu m$ spectral region. Atmospheric-window studies from an aircraft were also reported by Valero et al. (1982) and Platt (1971). Huppi et al. (1974) developed a radiometer with a liquid-nitrogen-cooled chopper for aircraft observations of the infrared emission of the atmosphere in the spectral region $\lambda = 3.5 - 14.8\ \mu m$. A liquid-helium-cooled circular-variable-filter radiometer (Wyatt, 1975) and a Michelson interferometer (Stair et al., 1983) were flown on rockets for observing atmospheric emission spectra at altitudes above 45 km. Murcray et al. (1978) designed a liquid-helium-cooled grating spectrometer for the measurement of emission spectra of subarctic stratospheric constituents from a balloon platform. They measured high-quality spectra during the ascent of the balloon in the $\lambda = 8 - 14\ \mu m$ atmospheric window, but they did not analyze their data with respect to water-vapor continuum emission. Their instrument was equipped with a warm window, which probably limited the accuracy of absolute calibration.

For our own experiment, we installed a Barnes Model 12-550 MKII radiometer in an aircraft, directed toward the zenith, to observe the height profile of the atmospheric thermal-emission spectra from the ground to the top of the troposphere during ascent and descent. We did not use cooled optics, yet we took special care to perform an absolute in-flight calibration of the instrument for the retrieval of continuum emission in the $\lambda = 8-14\ \mu m$ atmospheric window.

In Section II we present an outline of band-model techniques and a description of the LOWTRAN computer model that we used for a comparison of our measured spectra with those calculated from radiosonde data. A definition of the water-vapor continuum absorption as well as a description of the semiempirical model for the continuum absorption that is incorporated in our computer programs follows in Section III. Details of measurement techniques, aircraft, radiometers, and calibration procedures are given in Section IV. Our experimental results from broadband, continuous, and discrete spectral thermal-emission measurements from heights between 425 and 8500 m above sea level are reviewed and discussed in Section V. In Section VI the various interpretations of the water-vapor continuum are compared with the experimental results. Both the line shape and the aggregate formalism are taken into consideration.

II.   The LOWTRAN Model

A.   BAND MODELS

The molecular absorption of infrared radiation in the atmosphere is complicated because of the complex structure of vibrational energy levels of the infrared-active atmospheric gases. From the spectroscopic parameters of all absorption lines, the monochromatic transmission $\tau(v)$ can be evaluated by a line-by-line calculation.

For comparison with experimental transmission spectra of moderate spectral resolution, the monochromatic transmission function $\tau(v)$ must be averaged over the spectral resolution interval Δv, resulting in the low-resolution transmission function $\bar{\tau}(v, \Delta v)$. The spectral resolution of our radiometer was $\Delta v = 15$ cm^{-1} at $v = 1000$ cm^{-1}. Hence, many atmospheric absorption lines of typical halfwidth $\alpha = 0.05$ cm^{-1} are contained in the spectral resolution interval Δv. Because of the complexity of molecular line structure, it is extremely time-consuming to calculate $\bar{\tau}(v, \Delta v)$ by a line-by-line calculation, especially for inhomogeneous atmospheric paths from the bottom to the top of the atmosphere. For this reason band models were used.

Band models assume an array of lines that have chosen line strengths S, halfwidths α, and line spacings δ that are adjusted to represent the line structure of a real absorption band in the frequency interval ranging from v to $v + \Delta v$. In Goody's (1952) band model, lines are considered to be spaced randomly and to have a distribution of line strengths following Poisson statistics, with a mean line strength σ. For the Lorentz line shape the resulting average transmission $\bar{\tau}$ is

$$\bar{\tau} = \exp\left[-\frac{\rho L \sigma \alpha}{\delta(\alpha^2 + \rho L \sigma \alpha/\pi)^{1/2}}\right], \tag{1}$$

where ρ is the density of the absorber gas and L the length of the absorption path. Two limiting cases are of interest:

(1) In the weak-line approximation defined by $\rho L \sigma/\alpha \ll 1$, absorption in the line centers is small. The argument of the exponential function in Eq. (1) reduces to $\rho L \sigma/\delta$ and does not depend on the halfwidth α.

(2) In the strong-line approximation defined by $\rho L \sigma/\alpha \gg 1$, line centers are completely absorbing and the argument of the exponential function in Eq. (1) becomes $(\rho L \sigma \alpha \pi)^{1/2}/\delta$. The halfwidth α remains in the argument. Because the line center of an atmospheric absorption line of typical line strength $S = 10^4$ cm^{-1}/g·cm^{-2} is completely opaque for absorption paths $L > 20$ cm, the strong-line approximation is valid (Houghton, 1977). Most of the radiation is absorbed by the wings of the lines. Therefore, the transmission is less than exponential, as can be seen from the $(\rho L)^{1/2}$ dependence of the argument in the strong-line limit.

The LOWTRAN transmission functions $\bar{\tau}(v)$ that we used for our calculations are a generalization of Goody's random-band model (McClatchey et al., 1972). The low-resolution transmission $\bar{\tau}$ is represented by a single parameter function f:

$$\bar{\tau} \quad (v, \Delta v = 20 \text{ cm}^{-1}) = f[C(v) \cdot \rho_m \cdot L], \tag{2}$$

where $C(v)$ represents a frequency-dependent coefficient and ρ_m an equivalent absorber density described by

$$\rho_m = \rho[(p/p_0)\sqrt{T_0/T}]^n, \tag{3}$$

where $p_0 = 1$ atm, $T_0 = 273$ K, and n is an empirical constant. For $n = 0$ or $n = 1$, Eq. (2) reverts to the weak-line or strong-line approximation, respectively, because the halfwidth α of a pressure-broadened line is proportional to $pT^{-1/2}$. The forms of the transmission function f and the parameter n were determined empirically from laboratory measurements and line-by-line calculations. To illustrate the similarity of the empirical LOWTRAN transmission functions and the Goody model, we show an analytical fit to the LOWTRAN transmission function of water vapor derived by Gruenzel

(1978) in 1978:

$$\bar{\tau}(v, \Delta v = 20 \text{ cm}^{-1}) = \exp\{-[C(v)(p/\sqrt{T})^{0.9}\rho L]^{0.55}\}. \tag{4}$$

Equation (4) is nearly identical with the strong-line approximation of the Goody model.

The LOWTRAN transmission functions f and parameters n are tabulated for water vapor, ozone, nitric acid, and uniformly mixed gases. The uniformly mixed gases comprise CO_2, N_2O, CH_4, CO, and O_2 and are described by only one separate transmission function. Mean values of n are 0.9 for H_2O, 0.75 for the uniformly mixed gases, and 0.4 for ozone. The spectral resolution of the LOWTRAN transmission functions is $\Delta v = 20 \text{ cm}^{-1}$. They cover the spectral range from $\lambda = 0.25$ to 28.5 μm.

The LOWTRAN transmission functions were first published in 1972 by McClatchey et al. in the AFCRL report "Optical Properties of the Atmosphere." They are represented in the form of prediction charts scaled in such a manner that spectral transmission averaged over 20 cm^{-1} can easily be predicted once the equivalent absorber amount $\rho_m L$ is known for any chosen atmospheric path. Furthermore, aerosol extinction, molecular scattering, and H_2O continuum absorption in the $\lambda = 8-14$ μm atmospheric windows are incorporated. Six model atmospheres containing height profiles of relevant meteorological data are also included. In 1973 this report was made available in the form of the well-documented computer code LOWTRAN 2, written as a FORTRAN program (Selby and McClatchey, 1972). A revised version, the computer code LOWTRAN 3, was published in 1975 (Selby and McClatchey, 1975). In 1976 LOWTRAN 3B was published (Selby et al., 1976). Major additions were the inclusion of the $\lambda = 3.5-4.2$ μm water-vapor continuum attenuation, the strong temperature dependence of H_2O continuum absorption, and a reduction of the foreign-broadening term for continuum absorption in the $\lambda = 8-14$ μm region. Aerosol models were extended and improved. The computer code LOWTRAN 4, published in 1978 (Selby et al., 1978), contained an additional transmission function for nitric acid and a routine that calculates atmospheric thermal emission from LOWTRAN transmission functions. An option permits replacement of the model atmospheres by measured radiosonde data. LOWTRAN 5, published in 1980 (Kneizys et al., 1980), contains new altitude-dependent and relative-humidity-dependent aerosol models and new fog models.

With minor modifications we use the computer code LOWTRAN 4 for our calculations. The water-vapor continuum absorption, as it is incorporated in the LOWTRAN programs, has not been changed since the version LOWTRAN 3B. Because we neglect aerosol extinction in our experiment, changes from LOWTRAN 4 to LOWTRAN 5 are not relevant for this study.

B. Transmission Calculation

In Fig. 1 we present the result of a LOWTRAN transmission calculation for the important $\lambda = 8 - 14\ \mu$m atmospheric window. For this purpose the computer code LOWTRAN 3B was employed. It is designed for the calculation of the spectral transmittance of the atmosphere for different atmospheric paths specified by the user. The transmission is calculated for a slant path from ground to space. The base height is 600 m above sea level. The zenith angle is 45°. The selected model atmosphere is "midlatitude summer" (McClatchey *et al.*, 1972).

The total transmission is the product of all transmission functions describing the different absorption processes. In addition to the transmission

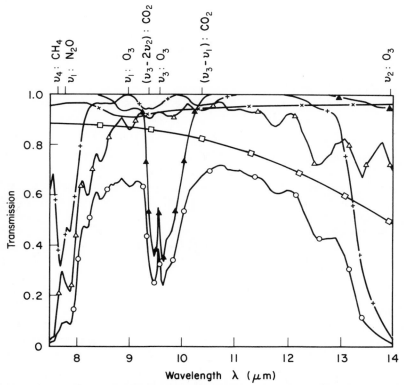

FIG. 1   The contribution of atmospheric gases to the total transmission of the atmosphere for an absorption path from ground to space in the 8–14 μm region: base height 600 m above sea level; elevation angle 45°; O, total transmission; △, water vapor; +, uniformly mixed gases CO_2, N_2O, CH_4; ✕, aerosol scattering; □, water-vapor continuum; ▲, ozone. For water-vapor continuum see Roberts *et al.* (1976).

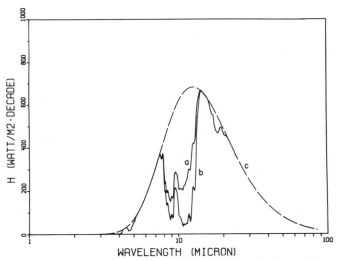

FIG. 2   Atmospheric thermal irradiance with (curve a) and without (curve b) contribution of water-vapor continuum emission: height, 547 m; temperature, 17.3°C; — —, a Planck curve at ambient air temperature. (Zurich, August 14, 1979.)

functions for molecular absorption discussed in Section II.A, effects of continuum absorption, molecular scattering, and aerosol extinction are also included. In Fig. 1 the contribution of each absorption process to the total absorption is shown separately. The wing of the v_2 vibration of water vapor limits transmission of the window region near $\lambda = 7\ \mu m$. At $\lambda = 14\ \mu m$ the wing of the strong $v_2 - CO_2$ vibration band terminates the atmospheric window. In the center of the window at $\lambda = 9.6\ \mu m$, the characteristic v_3 vibration of ozone in the stratosphere can be seen. For clear atmospheres, scattering can usually be neglected for thermal radiation (Berdahl and Fromberg, 1982). And, most important, in the transparent region between the absorption bands an excess absorption is observed, which cannot be explained by monomolecular water-vapor absorption when assuming standard line profiles. The physical origin of the water-vapor continuum absorption is still a matter of discussion.[†] However, it is the limiting factor for the transparency of the $\lambda = 8 - 14\ \mu m$ spectral window. The semiempirical absorption used in our LOWTRAN calculations will be discussed in Section III.C. Figure 1 also shows the total spectral transmission of the atmosphere.

[†] Bohlander (1979), Carlon (1982), Clough et al. (1980), Elsasser (1938), Gebbie (1982), Suck et al. (1979, 1982), Thomas and Nordstrom (1982), Varanasi et al. (1968), and Viktorova and Zhevakin (1966, 1975).

C.   Emission Calculation

On the basis of the low-resolution transmission functions discussed previously, we developed an algorithm for the calculation of the atmospheric thermal emission. For this purpose the atmosphere was divided into 33 layers. If we assume local thermodynamic equilibrium, the calculated monochromatic-radiance contribution I_n of each atmospheric layer n to the total atmospheric thermal radiance I arriving at the earth surface is

$$I = \sum_{n=1}^{33} P(v, T_n)(1 - \tau_{n, n+1}) \tau_{1,n} = \sum_{n=1}^{33} I_n, \tag{5}$$

where $P(v, T_n)$ is the Planck function ($W/cm^2 \cdot sr \cdot cm^{-1}$), T_n the temperature of layer n in Kelvin, $\tau_{n, n+1}$ monochromatic transmission from the bottom of layer $n + 1$ to the bottom of layer n, $\tau_{1, n}$ the monochromatic transmission from the bottom of layer n to the earth surface, and $(1 - \tau_{n, n+1}) = \epsilon_n$ the monochromatic emissivity of layer n.

If the resolution of monochromatic radiance I in Eq. (5) is degraded to $\Delta v = 20$ cm^{-1}, care must be taken when averaging the product of the two monochromatic τ's. Using the identity $\tau_{1, n} \cdot \tau_{n, n+1} = \tau_{1, n+1}$ and folding Eq. (5) with the apparatus function of spectral slit width Δv, Eq. (5) can be written as follows (Berdahl and Fromberg, 1982; Selby *et al.,* 1978):

$$\bar{I} = \sum_{n=1}^{33} P(v, T_n)(\bar{\tau}_{1, n} - \bar{\tau}_{1, n+1}). \tag{6}$$

Because the Planck function represents a smoothly varying function of v, it is regarded as a constant and taken outside the folding integral. The transmission functions $\bar{\tau}_{1, n}$ and $\bar{\tau}_{1, n+1}$ are the low-resolution transmission functions of the LOWTRAN model discussed in Section II.A. The computer codes LOWTRAN 4 and 5 contain a routine that calculates the atmospheric thermal radiance on the basis of Eq. (6).

The algorithm of Eq. (6) is the analog of the equation of radiative transfer for nonscattering atmospheres with zero background (Chandrasekhar, 1950):

$$I(v) = \int_{\tau_\infty}^{1} P[v, T(z)] \, d\tau(v, z), \tag{7}$$

where z is the height in meters, $T(z)$ the temperature at height z in Kelvin, and τ_∞ the transmission through the whole atmosphere.

As an illustration the atmospheric thermal irradiance computed with the aid of computer code LOWTRAN 4 is shown in Fig. 2 for the entire spectral range of atmospheric thermal emission. The computed atmospheric thermal-radiance spectra from all possible directions have been averaged according to a Lambertian horizontally oriented surface. The irradiance is important for radiative-energy-balance considerations of building surfaces (Zürcher *et al.*, 1982).

Curve a in Fig. 2 includes the contribution of water-vapor continuum emission as it is currently incorporated in the LOWTRAN program. In curve b the contribution of continuum emission is not taken into account. Also shown is a Planck curve at ambient air temperature. Meteorological input data of model atmosphere "midlatitude summer" were used for the calculation.

Figure 2 demonstrates that the major contribution to the atmospheric thermal irradiance in the spectral window region from $\lambda = 8$ to 14 μm results from water-vapor continuum emission, which is of utmost importance not only for the atmospheric propagation but also for the global energy balance.

III.   Water-Vapor Continuum Absorption

A.   DEFINITION OF EXCESS ABSORPTION

The excess absorption, which is also called continuum absorption, cannot be measured directly. It represents the difference between observed and predicted absorption and is calculated using line-by-line methods based on specific line-shape formulas. Standard line-shape formulas for the collision broadening are summarized here (Gross, 1955; Lorentz, 1906; Van Vleck and Weisskopf, 1945):

Simple Lorentz (Lorentz, 1906),

$$K(v) = \frac{S}{\pi} \frac{1}{(v - v_0)^2 + \alpha^2};$$

Full Lorentz (Van Vleck and Weisskopf, 1945),

$$K(v) = \frac{S}{\pi} \frac{v}{v_0} \left[\frac{\alpha}{(v + v_0)^2 + \alpha^2} - \frac{\alpha}{(v - v_0)^2 + \alpha^2} \right];$$

Van Vleck – Weisskopf (Van Vleck and Weisskopf, 1945),

$$K(v) = \frac{S}{\pi} \left(\frac{v}{v_0} \right)^2 \left[\frac{\alpha}{(v + v_0)^2 + \alpha^2} + \frac{\alpha}{(v - v_0)^2 + \alpha^2} \right];$$

Gross (Gross, 1955),

$$K(v) = \frac{S}{\pi} \frac{4v^2\alpha}{(v^2 - v_0^2)^2 + 4v^2\alpha^2};$$

where S is the line strength [cm^{-1}/(molecules·cm^{-2})], v_0 the resonant frequency (cm^{-1}), and α the halfwidth (cm^{-1}).

The basic assumption for all these line shapes is instantaneous collision among molecules. This implies that the line shapes are calculated exclusively by impact theory, where the duration of transient effects of collisions is negligible. Thus, all line shapes are appropriate near the line centers, where $|v - v_0| \ll v_0$, but none correctly describes the absorption in the extreme wings. Yet exactly this is required to determine whether the excess window absorption results from the influence of neighboring strong bands.

Most line-by-line calculations use a tabulation of resonant frequencies v_0, line strengths S, line widths α, and energies E of the lower states of all observed atmospheric absorption lines, which is compiled on a magnetic tape by the Air Force Geophysics Laboratories (McClatchey *et al.*, 1973). For these calculations, only lines within a certain specified spectral interval of, typically, 20 cm^{-1} centered at the frequency v are assumed to contribute to the absorption at that frequency. In absorption spectra calculated on the basis of any line shape mentioned here the absorption in the window regions is underestimated by the amount of excess absorption. If all spectral lines from $v = 0$ cm^{-1} to 5000 cm^{-1} and their respective contributions to absorption in the $\lambda = 8 - 14$ μm window are taken into account, it can be seen from Fig. 3 of Thomas and Nordstrom (1982) that for "simple Lorentz" and "full Lorentz" line shapes the predicted absorption is also less than the experimentally observed absorption. However, for the van Vleck–Weisskopf line shape the prediction exceeds measured absorption. Yet, by the impact approximation the validity range of these line shapes is restricted to $|v - v_0| < 3$ cm^{-1} (Goody, 1964) and should not be extended from $v = 0$ to 5000 cm^{-1}.

B. EXPERIMENTAL FACTS ON EXCESS ABSORPTION

Although the mechanism of excess absorption in the infrared windows is still a matter of discussion, there are three generally accepted characteristics.

(1) The excess absorption represents a continuum. It changes smoothly with wavelength. This means that there are no spectral lines that could be assigned directly to a rotational or vibrational transition of some absorbing species occurring in the atmosphere.

(2) The excess absorption depends on the square of the partial pressure of the water vapor. This implies that at least two water molecules are involved in the absorption process.

(3) The excess absorption shows a strong negative temperature dependence, which can be explained by an activation energy of some bound state.

The first two characteristics permit us to label the excess absorption as water-vapor continuum absorption. In spite of the fact that the underlying physics is still poorly understood, the necessity for describing absorption in the relevant infrared windows led to the development of a semiempirical formula, which is also incorporated in the LOWTRAN program. The problem of negative temperature dependence will be discussed in Sections III.C, VI.B, and VI.C.

C. SEMIEMPIRICAL MODEL FOR THE WATER-VAPOR CONTINUUM ABSORPTION

From the line-shape formulas stated in Section III.A, we find that the absorption coefficient K is proportional to the halfwidth α in the extreme wings where $|v - v_0| \gg \alpha$. The halfwidth α^0 of a pressure-broadened line of a mixture of water-vapor and nitrogen at a fixed temperature T_0 can be represented by a sum of the contribution owing to self broadening α_s^0 and foreign broadening α_f^0, where the superscript 0 refers to the temperature $T_0 = 296$ K (Burch and Gryvnak, 1980), as

$$\alpha^0 = \alpha_s^0 p_{H_2O} + \alpha_f^0 p_{N_2} = \alpha_s^0 [p_{H_2O} + \gamma(p - p_{H_2O})], \tag{8}$$

where p_{H_2O} is the partial pressure of water vapor in atmospheres, p_{N_2} the partial pressure of foreign gas in atmospheres, $p = p_{H_2O} + p_{N_2}$ the total pressure in atmospheres, and $\gamma = \alpha_f^0/\alpha_s^0$ the ratio of foreign broadening to self broadening.

Laboratory measurements (Burch, 1971; McCoy et al., 1969) in long-path absorption cells show that collisions between two water molecules are much more efficient in broadening water-vapor lines than collisions between a water molecule and a molecule of a foreign carrier gas, e.g., N_2 or O_2. Thus the foreign-to-self-broadening ratio γ is much smaller than unity. Assuming that the absorption coefficient of the water-vapor continuum $K_{H_2O\ cont}^0$ is proportional to the halfwidth α^0 of the water-vapor lines neighboring the atmospheric windows, Roberts et al. (1976) postulates the following formula:

$$K_{H_2O\ cont}^0(v) = C(v)(p_{H_2O} + \gamma[p - p_{H_2O}]). \tag{9}$$

In the long-wavelength atmospheric window, which ranges from $\gamma = 8$ to 14 μm, the LOWTRAN models 4 and 5 (Kneizys et al., 1980; Selby et al., 1978) use $\gamma = 0.002$. In this spectral region the coefficient $C(v)$ in Eq. (9) decreases with increasing wave number v. A fit to experimental data yields (Roberts et al., 1976)

$$C(v) = a + b \exp(-\beta v), \tag{10}$$

where $a = 1.25 \times 10^{-22}$/molecules (cm^2/atm); $b = 1.67 \times 10^{-19}$/molecules (cm^2/atm); $\beta = 7.87 \times 10^{-3}$ cm; and v is the wave number (cm^{-1}). The wavelength dependence can be interpreted as a wing of a broad water-vapor-dimer librational-absorption mode at $\lambda = 16$ μm (Wolynes and Roberts, 1978), which will be discussed in Section VI.B.

The strong negative temperature dependence of the water-vapor continuum absorption involves a bound state between two water molecules, which can be described by an activation energy E_2, required for the dissociation of two hydrogen-bonded water molecules (Varanasi *et al.*, 1968). Consequently, the temperature dependence of the absorption coefficient $K_{H_2O\ cont}(v, T)$ is represented by an exponential factor $\exp(-E_2/kT)$, or synonymously

$$K_{H_2O\ cont}(v, T) = K^0_{H_2O\ cont}(v) \exp[T_B(1/T - 1/T_0)], \qquad (11)$$

where $K^0_{H_2O\ cont}$ is the absorption coefficient at $T_0 = 296$ K and T_B is estimated to be 1800 K (Roberts *et al.*, 1976). This corresponds to a binding energy $-E_2 = kT_B = 0.155$ eV $= 3.58$ kcal/mol, which is in reasonable agreement with *ab-initio* calculations of the binding energy of a linear water-vapor-dimer molecule (Curtiss and Pople, 1975). Therefore, the transmission function τ for water-vapor continuum absorption in the $\lambda = 8 - 14\ \mu m$ window as incorporated in our LOWTRAN calculations is given by the following equations (Kneizys *et al.*, 1980):

$$\tau = \exp(-\sigma_{H_2O\ cont} L)$$

and

$$\sigma_{H_2O\ cont} = K_{H_2O\ cont} \cdot W_{H_2O} \qquad (12)$$
$$= C(v)\{p_{H_2O} \exp[T_B(1/T - 1/T_0)] + \gamma(p - p_{H_2O})\}W_{H_2O},$$

where $\sigma_{H_2O\ cont}$ is the extinction coefficient (cm^{-1}), L the length of absorption path in centimeters, $W_{H_2O} = p_{H_2O}/kT$ the water-vapor concentration in molecules per cubic centimeter, $C(v)$ the wave-number-dependent coefficient (molecules$^{-1} \cdot$ cm^2/atm), p_{H_2O} the partial pressure of water vapor in atmospheres, p the total pressure in atmospheres, γ the foreign- to self-broadening coefficient, and T the temperature in Kelvin.

In the short-wavelength atmospheric window at $\gamma = 3.5 - 4.2\ \mu m$ the LOWTRAN model for water-vapor continuum absorption is based on the laboratory measurements of Burch *et al.* (1971) and White *et al.* (1975). The extinction coefficient used by the LOWTRAN model is represented by the following formula:

$$\sigma'_{H_2O\ cont} = C'(v)[p_{H_2O} + \gamma'(p - p_{H_2O})] \exp[T'_B(1/T - 1/T_0)]. \qquad (13)$$

The prime (') indicates the short-wavelength window. The attenuation coefficient $C'(v)$ for the temperature $T_0 = 296$ K was evaluated by an extrapolation of the measurements reported by Burch *et al.* (1971). From limited experimental data the temperature T'_B is estimated to be 1350 K (Kneizys *et al.*, 1980). The exponential temperature dependence appears to be weaker in the $\lambda = 8 - 14\ \mu m$ window. The foreign- to self-broadening coefficient is

$\gamma' = 0.12$. Hence, nitrogen broadening in the short-wavelength window is more significant than in the $\lambda = 8 - 14 \, \mu m$ window.

The semiempirical formulas (12) and (13) for the extinction coefficient of water-vapor continuum absorption $\sigma_{H_2O\,cont}$ meet conditions (2) and (3) of Section III.B, which are imposed by experimental observations. Because the foreign- to self-broadening ratio γ is small, and because the water-vapor concentration W_{H_2O} is proportional to p_{H_2O}, the extinction coefficient is proportional to the square of the partial pressure p_{H_2O} of water vapor, according to condition (2). The strong negative temperature dependence of condition (3) is taken into account by the exponential factor $\exp(-E_2/kT)$.

IV. Experimental Setup

A. Techniques of Measurement

Because the absorption in the atmospheric infrared windows is very weak, absorption or emission measurements in these spectral regions require long optical paths. In the laboratory, White-type multipass cells† are widely used. Other laboratory measurements are based on the sensitive technique of photoacoustics.* However, all laboratory measurements on water vapor close to saturation are handicapped by heat-pipe and related effects. Also, it is difficult to simulate atmospheric conditions at temperatures as low as $-50°C$ in an absorption cell. On the other hand, there is an uncertainty about the difference between the laboratory atmosphere and the real atmosphere, because little is known about how ions produced by ultraviolet radiation and cosmic rays in the stratosphere influence the infrared continuum absorption. For these reasons it is necessary to supplement laboratory measurements with reliable field experiments.

The real atmosphere above a ground-based observer offers the longest optical path for emission measurements. If a radiometer is pointed upward, the thermal emission I incident from the zenith, which originates from all atmospheric layers above the observer, can be measured against the cold space. The thermal emission of the space in the spectral region $\lambda = 2.5 - 40 \, \mu m$ is negligible because the cosmic-background temperature is $T_{cb} = 2.9 \, K$ (Woody and Richards, 1981). When measurements of the emission from the zenith are continuously monitored during ascent and descent of the aircraft, the dependence of atmospheric thermal radiance $I(H)$ on height H

---

† Bignell (1970), Burch *et al.* (1971), Eng and Mantz (1980), McCoy *et al.* (1969), Montgomery (1978), Watkins *et al.* (1979), and White *et al.* (1975).

* Bean and White (1981), Hinderling *et al.* (1982), Peterson (1978), and Shumate *et al.* (1976).

in the atmospheric windows is described by the formulas

$$I(H) = \int_H^\infty P(z)\tau(H, z)\sigma(z)\,dz,$$

$$\tau(H, z) = \exp\left[-\int_H^z K(z)\rho(z)\,dz\right] = \exp\left[-\int_H^z \sigma(z)\,dz\right],$$

(14)

where $P(z)$ is the Planck function at height z (W/cm$^2 \cdot$ sr $\cdot \mu$m), $\tau(H, z)$ the transmission from height z to height H, $\sigma(z)$ the extinction coefficient at height z (m^{-1}), $K(z)$ the absorption coefficient (molecules$^{-1} \cdot$ m^2), $\rho(z)$ the density in molecules per cubic meter, and z the height in meters.

The term $\tau(H, z)$ is exponential even for moderate spectral resolution, because in this case it describes the transmission of water-vapor continuum. Equation (14) is in accordance with Eq. (7) for a continuum emission. Differentiating Eq. (14) with respect to H, the absorption coefficient $K(H)$ of the atmospheric layer at height H can be evaluated from the gradient of the measured radiance profile $I(H)$, as can be seen in Eq. (15):

$$K_{H_2O\,cont}(H) = \frac{dI(H)/dH}{\rho(H)[P(H) - I(H)]}.$$

(15)

Within the troposphere, the physical parameters of the emitting atmospheric layers vary over a large range that is hardly accessible to laboratory experiments. These parameters can be related to continuum absorption by use of Eq. (15).

B. AIRCRAFT

For our measurements we used a Pilatus Porter aircraft having a Pratt and Whitney turboprop engine. It carried two passengers and a payload of 500 kg. This allowed operation of a radiometer up to a height of 9000 m above sea level. The cabin was not pressurized; therefore, the operating staff used oxygen masks. The window in the cabin roof between the wings of the aircraft was removed and replaced by a metal plate 75 × 120 cm. This plate was used as the experimental platform. A conical dome was mounted on the back end of this platform. It was 26 cm high and had an opening at the top 21 cm in diameter. Through this opening the atmospheric radiation entered the radiometer. The dome prevented contamination of the radiometer optics by the exhaust gases of the engine.

The radiometer for broadband measurements was mounted directly inside the dome in a cardan suspension holding the optical axis of the radiometer in a vertical position. The radiometer for spectral measurements was mounted horizontally under the cabin roof in a sealed Plexiglas housing that prevented moist cabin air from condensing on optical components of the radiometer. The housing was purged with dry nitrogen gas.

Radiometer and control electronics were mounted in a transportable rack that could be fixed to the floor of the aircraft cabin. Electrical power was provided by two 40-Ah/24-V–NiCd batteries. The voltage was transformed to 220 V. The dissipated heat of the transformer was used to warm up the dry-nitrogen purging gas. The batteries enabled us to perform measurements when the engine was turned off and the aircraft was sailing. Tests showed that the exhaust gases of the engine did not disturb the measurements.

PT 100 thermistors were mounted on the cabin roof and under the wing of the plane for measurement of the outside air temperature. The result was in excellent agreement with the radiosonde data. We also tried to measure the dew point with a Peltier-cooled mirror. Unfortunately, we could not use the data, probably because of iced conveyance pipes. The height was calculated from a pressure sensor mounted in the electronics rack. Global solar radiation was measured by a solarimeter mounted on the cabin roof.

C. RADIOMETERS

For broadband measurements we used a commercial Heimann KT4 radiometer. The detector of the radiometer was an uncooled Bi–metal-film bolometer that is sensitive from $\lambda = 0.6$ to $40\ \mu m$. The film has a square resistance of $R = 188.5\ \Omega$, where 50% of incident radiation is absorbed by the metal film independent of wavelength (Carli, 1977). The detectivity of the detector was D^* (500 K, 12.5 Hz, 1 Hz) $= 10^8\ cm \cdot Hz^{1/2}/W$. The radiometer was equipped with a Cassegrain telescope with an entrance aperture of $D = 80$ mm. The radiation was chopped with 30 Hz against an internal reference blackbody. The blackbody temperature was not stabilized but was measured. This permitted electronic correction of the chopped detector signal for changes of blackbody temperature. During the flights the contribution of diffuse solar radiation to the detector signal was measured with a simple glass filter and subtracted from the total detector signal measured without a filter.

For spectral measurements we used a Barnes Spectral Master Model 12-550 MKII radiometer. The photodetectors of the radiometer were cooled with liquid nitrogen. An InSb detector for the wavelength range between $\lambda = 2.57$ and $5.50\ \mu m$ and an HgCdTe detector sensitive from $\lambda = 5.50$ to $14.51\ \mu m$ were integrated into a single sandwich detector. The peak detectivity of the photovoltaic InSb detector at the chopping frequency of 1000 Hz and at the wavelength $\lambda_{pk} = 5.5\ \mu m$ was D^* (λ_{pk}, 1000 Hz, 1 Hz) $= 1.3 \times 10^{11}\ cm \cdot Hz^{1/2}/W$, and that of the intrinsic photoconductive

HgCdTe detector at a wavelength of $\lambda_{pk} = 13\ \mu$m was D^* $(\lambda_{pk}, 1000$ Hz, 1 Hz$) = 9.2 \times 10^9$ cm $\cdot$ Hz$^{1/2}$W.

In the optical layout of the radiometer system (Fig. 3), care was taken that the atmospheric thermal radiation was only reflected by mirrors and did not pass any window before it was chopped. Thus, problems arising from window emission, which at high altitudes can easily reach levels comparable with atmospheric radiation, were avoided.

The atmospheric thermal radiation incident through the dome in the cabin roof enters the radiometer telescope after reflection by a plane mirror. The Cassegrain telescope has an aperture of $D = 100$ mm. Radiation is chopped with a frequency of 1000 Hz against an on-axis reference blackbody stabilized at 56°C.

The filter assembly of the radiometer consists of three circular-variable-filter (CVF) wheel segments. The thickness of the narrow-band interference filters increases continuously along a circular segment by a factor of two. This allows us to scan the wavelength continuously from $\lambda = 2.57$ to 4.57 μm, from $\lambda = 4.41$ to 8.12 μm, and from $\lambda = 8.06$ to 14.51 μm for each filter segment. The spectral resolution of the CVFs is $\Delta\lambda/\lambda = 1.5\%$ with a peak transmission of 70%.

Laboratory tests of the HgCdTe detector proved that the detector responsivity is lowered by 26% when the ambient air pressure is lowered to $p_{amb} =$

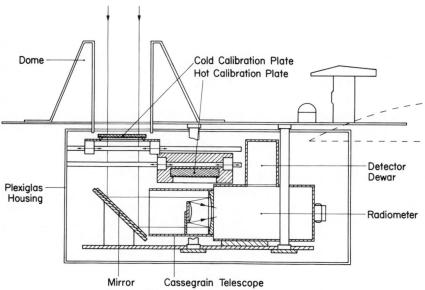

FIG. 3 Schematic arrangement of radiometer and calibration system in the aircraft.

300 mb. According to the vapor-pressure equation of liquid nitrogen, variations of ambient pressure causes changes of detector temperature and responsivity. Delicate two-stage, needle-valve regulators are not suited for detector stabilization because of mechanical vibration and shock under flight conditions. Therefore, we designed a simple and rugged pressure-regulation system. With the aid of a rubber membrane, the pressure in the detector Dewar is compared with the pressure of a temperature-stabilized reference volume. Thus we achieved a pressure control of $\Delta p_{reg}/p_{reg} = 3 \times 10^{-3}$ in the detector Dewar for ambient air pressures varying between 1000 mb and 100 mb. The remaining pressure fluctuations correspond to radiance fluctuations of 1 μW/cm$^2 \cdot$ sr $\cdot \mu$m. This was tolerable for our measurements.

Because the nonzero reflectivity and emissivity of the optical elements mounted in front of the chopper contribute to the detector signal, we measured systematic deviations between the atmospheric thermal radiance taken during the ascent and during the descent of the aircraft. The radiance taken during the ascent appeared to be larger than during the descent, which can be explained by the cooling of the optical elements in flight. The introduction of a temperature stabilization of all critical optical components reduced the systematic deviations to 30 μW/cm$^2 \cdot$ sr $\cdot \mu$m. This was still too large for the measurement of the weak-continuum emission. Further improvement could only be achieved by the in-flight calibration of the radiometer system to be discussed in Section IV.E.

D. ELECTRONICS

The complete experimental system is controlled on-board by a CBM 3032 microcomputer. Figure 4 is a schematic diagram of the electronics. User port 1 is multiplexed to exchange control commands and data with the radiometer-control unit. An autonomous microprocessor of the radiometer-control unit controls the analog processing of the detector signal by the locked-in-amplifier, the position of the circular-variable filter, and the stabilization of the chopper frequency and the reference-cavity temperature. Programs can be loaded via user port 2 from mini-DCR cassettes. With the aid of the high-resolution graphics (HRG), measured spectra that have been calibrated in flight can be plotted on the terminal display of the on-line CBM. A digital multimeter and scanner for analog sensor readings, a magnetic tape for data storage, an $X - Y$ recorder, a printer, and the position control of the calibration plates are connected to the computer via the IEEE bus.

E. CALIBRATION

Because we made an absolute measurement of the thermal emission of the atmosphere, we had to calibrate our radiometers. The calibration proce-

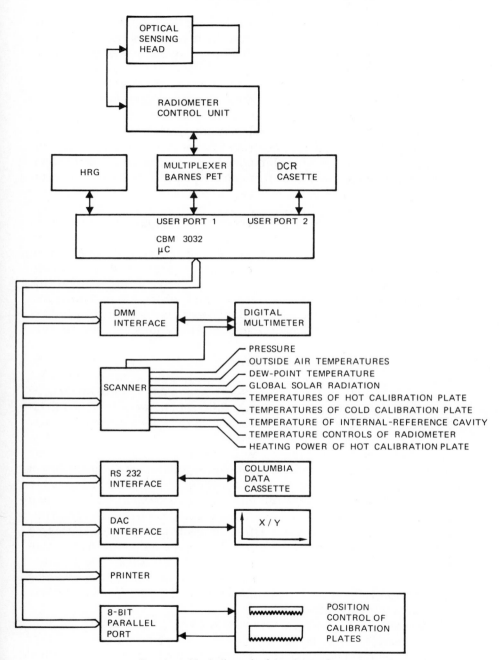

FIG. 4   A block diagram of the electronic system.

dure was performed in the laboratory before the flights for both the broad-band and the spectral radiometers. For the reduction of the atmospheric emission measured during flights, laboratory calibration was used for the broadband measurements and in-flight calibration for spectral measurements. For the latter, the laboratory calibration served for simulation of flight conditions and for testing the pressure stabilization of the detector.

For the laboratory calibration, we used a blackbody that can be cooled down gradually from 373 K to the liquid-nitrogen temperature of 77 K. This blackbody is imaged into the detector plane at the bottom of the LN_2 Dewar of the radiometer by a collimating telescope mounted in front of the radiometer telescope and by the radiometer optics. The complete system, including calibration and radiometer optics, can be evacuated to a pressure lower than 10^{-3} Torr.

To eliminate any drift of the overall radiometer system, it was necessary to perform an in-flight calibration immediately before and after the measurement of every atmospheric spectrum. The best way to account for all elements of the optical system and to avoid errors caused by aberrations and diffraction losses is to use precisely the same optical arrangement for the calibration blackbody as for the measurement of the atmospheric spectrum. This necessitated the construction of two large-aperture blackbodies of 15-cm diameter with a temperature distribution homogeneous and controllable within $0.1°C$.

The two calibration blackbodies were mounted directly above the reflecting mirror below the dome of the cabin roof and could be moved in and out of the optical path (Fig. 3). The accuracy of the calibration was found to be improved by increasing the temperature difference between the two calibration plates. One calibration plate must be cold and show a thermal emission comparable with the emission of the cold atmosphere.

To avoid condensation, the lowest temperature of the calibration plate must be above the dew-point. The plate was exposed to the turbulent flow of the outside air, which acted as cooling medium. The temperature of the cold plate followed the temperature changes of outside air during ascent and descent of the aircraft. Therefore, its heat capacity was reduced by decreasing the plate thickness to 6 mm. The warm-calibration plate was stabilized at 49°C. An aluminium plate 27 mm thick was used to achieve the optimum temperature homogeneity at the emitting front side. The plate was heated uniformly over the entire area of the back side by a meandering Nikrothal foil 100 μm thick. The side wall and the back of the calibration plate were thermally insulated.

To increase the emissivity, wedge-shaped grooves were cut into the emitting front surface with a wedge angle of 45°. The temperature needed for data reduction and heating control was measured by vapor-deposited PT 100 thermistors and Cu–Co thermoelements of minimum heat capacity. The temperature sensors were attached directly on the wedge-shaped aluminum grooves under the emitting infrared-black coating.

The measurement and calibration procedure consisted of three steps: The temperatures and the emission spectra of the hot and the cold calibration plates were measured for the calculation of the spectral responsivity of the radiometer system. Then, an emission spectrum of the atmospheric layers situated above the aircraft was taken after both calibration plates had been removed from the optical path. Additional data recorded with the atmospheric spectrum were the meteorological data from the analog sensors and temperatures of critical components of the radiometer. In the last step, both calibration plates were moved into the optical path to find a new stationary thermal state. During this time, the central computer CBM 3032 stores measured data on the magnetic tape, calculates the atmospheric thermal emission obtained by absolute in-flight calibration, and displays the result on the CRT screen.

V. Experimental Results

A. AIRPLANE AND RADIOSONDE-BALLOON FLIGHTS

The flights of our radiometers were performed with a Pilatus Porter aircraft over the Swiss Plateau. Takeoff and landing took place at the air base in Emmen, Lucerne, whereas the maximum height of more than 8000 m was reached above the station of the radiosonde balloons at Payerne, Vaud, which is located about 110 km from Emmen. Exceptionally clear days with visual ranges of more than 30 km were selected for the flights. Thus, aerosol scattering could be neglected in the wavelength range from $\lambda = 2.5$ to 25 μm, which simplified emission calculations. The days of flight were chosen according to meteorological conditions that guaranteed that the water-vapor continuum emission dominated over cloud, fog, or smog emission in the atmospheric windows.

The flight for the broadband measurement began on August 21, 1980 at 1424 h and ended at 1623 h. The aircraft climbed continuously to a height of 8400 m and then descended, while broadband radiance values were recorded every 7 sec. Average climbing and descending velocity of the aircraft was about 5 m/sec.

The flight for spectral measurements began on April 16, 1982 at 1345 h and ended at 1540 h. The measurement program consisted of two interchanging phases. During phase one the aircraft stayed at the following constant heights: during ascent successively at 394, 1445, 3070, 5445, 7324, and 8208 m and during descent, successively at 8208, 5654, 4270, 2785, and 968 m.

Continuous spectra of atmospheric thermal emission in the wavelength

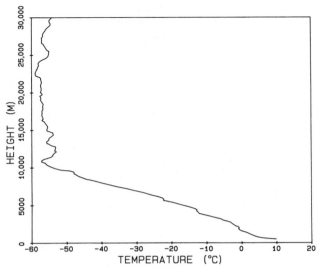

Fig. 5 Temperature as a function of height as recorded by the radiosonde in Payerne on April 16, 1982.

range from $\lambda = 8.2$ to $14.5\ \mu m$ were recorded and calibrated absolutely, calculated, and displayed on line. The whole procedure took 260 sec. During phase two, when the aircraft was climbing or descending to the next height level, atmospheric radiance was measured at ten selected wavelength channels and calibrated with two calibration plates, repeatedly. Each measurement cycle, including calibration, took 140 sec. By this procedure the height dependence of the atmospheric radiance was examined.

Because more measurement time per wavelength channel was available for recording the radiance at discrete wavelengths instead of taking continuous spectra, the time constant for the lock-in electronics was increased by a factor of 10 ($\tau_{el} = 1.5$ sec), thus improving the signal-to-noise ratio. Besides, height resolution of the radiance profiles could be refined because less time was needed for each measurement cycle.

Meteorological radiosonde balloons, which are routinely launched at the "station aérologique" in Payerne, measured height profiles of standard meteorological parameters simultaneously with our Pilatus Porter flights. The balloon that measured the partial pressure of ozone and temperature began at 1200 h (Dütsch, 1984). The other balloon, for pressure, temperature, and relative humidity measurements, began at 1300 h (Ricker, 1980–1982). The balloons reached a height of 30 km after 90 min.

Height profiles of temperature, relative humidity, and ozone concentration taken on April 16, 1982 are shown in Figs. 5, 6, and 7. The relative

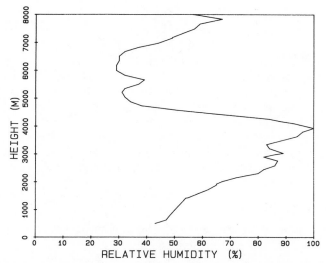

FIG. 6 Relative humidity as a function of height as recorded by the radiosonde in Payerne on April 16, 1982.

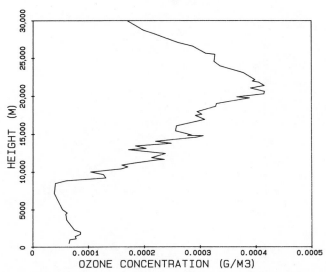

FIG. 7 Ozone concentration as a function of height as recorded by the radiosonde in Payerne on April 16, 1982.

humidity was measured with a hygroscopic skin. It represents our most
critical meteorological parameter. The accuracy of the humidity data are
± 15% at a height of 7000 m (Phillips *et al.,* 1982). At heights above 8000 m
the humidity profile measured by the radiosonde is useless. Instead, profiles
of water-vapor concentration were taken from the standard atmosphere
"midlatitude summer" and "midlatitude winter" (McClatchey *et al.,* 1972)
for humidity profiles for August 21, 1980 and April 16, 1982, respectively.
Although the measured relative humidity in Fig. 6 reached 100% in a thin
layer at a height of 4000 m, the atmosphere was optically clear. This can be
explained by the fact that the Swiss radiosonde tends to overestimate relative
humidity compared with other sondes (Phillips *et al.,* 1981). Deviations
increased with height.

B. Broadband Measurement

Figure 8 shows the height profile of the integrated atmospheric thermal
emission

$$I(H) = \int_{\lambda=2.7\,\mu m}^{\infty} I(\lambda, H)\,d\lambda \tag{16}$$

incident from the zenith measured during the flight of August 21, 1980. The
upper curve corresponds to the ascent, the lower curve to the descent.

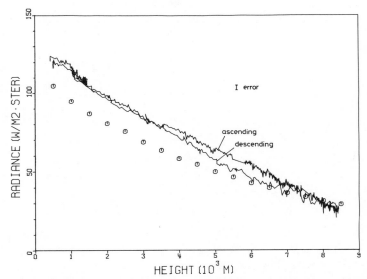

Fig. 8   Broadband atmospheric thermal radiance as a function of height: spectral bandpass
from $\lambda = 2.7$ to 40 μm: ——, measured; $\oplus$, calculated using radiosonde data.

Radiance data taken during ascent are 5 W/m² higher than those taken during descent at heights between 5000 and 6500 m, yet they agree within the experimental error at all other heights. The experimental error is estimated to be 3 W/m². During the flight, checks were performed at height intervals of 500 m to determine whether the detector signal resulted from solar radiance at wavelengths shorter than $\lambda = 2.7\,\mu m$. At these wavelengths the detector sensitivity was low but not zero. For the clear-weather conditions with little diffuse solar radiance, corrections for heights larger than 4000 m were less than 1%.

Results of LOWTRAN calculations using radiosonde data are represented in Fig. 8 by circles. For the calculations, the atmosphere was assumed to be opaque at wavelengths $\lambda > 25\,\mu m$. Using the theoretical far-infrared atmospheric transmission spectra of Traub and Stier (1976), the error introduced by the above assumption was estimated to be less than 5 W/m² at a height of 8000 m.

Within the lower troposphere, the calculated thermal emission is 12 W/m² lower than the experimentally determined emission. This difference indicates that LOWTRAN predictions underestimate water-vapor continuum emission, which is the most important absorption mechanism in the lower troposphere within the atmospheric windows.

C. CONTINUOUS SPECTRA OF ATMOSPHERIC THERMAL EMISSION

Spectra of the thermal emission incident from the zenith are presented in Figs. 9 and 10. The spectra were measured successively at the heights indicated. Figures 9 and 10 correspond to the ascent and to the descent, respectively. Because the thickness of the atmospheric layers above the observing platform decreases with increasing height, the total intensity of the atmospheric thermal radiance emitted by these layers is decreasing.

The spectral features of the emission spectra include the long wavelength wing of the ν_2 bending mode of water vapor near $\lambda = 8\,\mu m$, the ν_3 vibration of ozone centered at $\lambda = 9.6\,\mu m$, and the short-wavelength wing of the strong ν_2 deformation–vibration of carbon dioxide at $\lambda = 14\,\mu m$. Furthermore, a less intense water-vapor vibration–emission can also be observed at $\lambda = 12.6\,\mu m$. The residual emission between the monomolecular vibration bands in the wavelength region of highest atmospheric transparency results from the water-vapor continuum emission. In contrast to the intensities of the H_2O and CO_2 emission bands, the ozone emission band is not essentially diminished up to an observation height of 8200 m. This is because the maximum concentration of the ozone layer is at a level of 20,000 m, well above the maximum height reached by the Pilatus Porter aircraft.

The arrows in Figs. 9 and 10 mark the wavelength channels where height

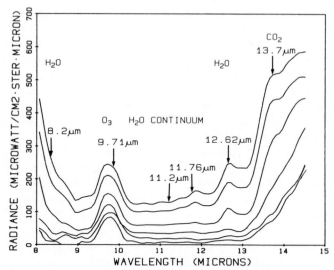

FIG. 9   Atmospheric thermal emission spectra recorded at various heights during ascent. Arrows mark discrete wavelength channels (see Section V.D). Measurements were taken successively at heights of 394 (top curve), 1445, 3070, 5445, 7324, and 8208 m (bottom curve).

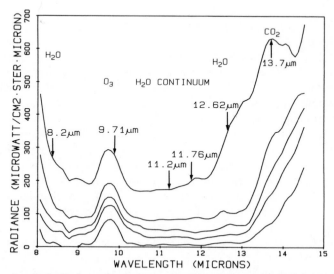

FIG. 10   Atmospheric thermal emission spectra recorded at various heights during descent. Arrows mark discrete wavelength channels (see Section V.D). Measurements were taken successively at heights of 8208 (bottom curve), 5654, 4270, 2785, and 968 m (top curve).

profiles of the atmospheric thermal emission were measured (see Section V.D).

The spectrum measured before the takeoff of the aircraft at a height of 394 m (Fig. 9) shows less intensive thermal emission of the atmosphere than the spectrum taken at the end of the flight at a height of 968 m (Fig. 10). This is not surprising, because the places where the spectra were recorded are separated by a distance of 100 km. Local inhomogeneities near the ground are responsible for this inconsistency. On the other hand, the local inhomogeneities of the atmosphere are much less pronounced at heights above 3000 m.

In Fig. 11 two spectra measured at similar heights are compared, one measured during the ascent at 5654 m, the other during the descent at 6554 m. The spectra are in good agreement within the overall error of the absolute calibration, which is estimated to be better than 10 W/cm$^2 \cdot$ sr$\cdot\mu$m at the wavelength of $\lambda = 11.2\ \mu$m. This demonstrates that the in-flight calibration successfully compensates the drift of the sensitivity of the complete system, i.e., drift of detector responsivity and temperature drifts of nonideal optical components.

The two measured spectra in Fig. 11 are also compared with the result of a LOWTRAN calculation discussed in Section II, which is plotted as a dotted line. Radiosonde data were used for this calculation. The center of the

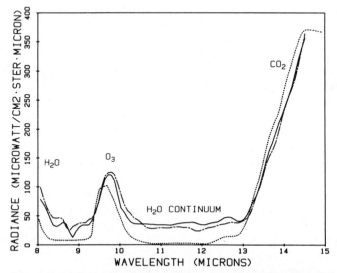

FIG. 11 Comparison of measured and calculated atmospheric thermal-emission spectra: ——, measured during ascent at 5445 m; — · —, measured during descent at 5654 m; · · ·, LOWTRAN calculation using radiosonde data from Payerne. Flight 23, April 16, 1982.

experimentally measured ozone vibration seems to be shifted about $\Delta\lambda =$ 0.15 μm toward longer wavelengths when compared with the calculated spectrum. The shift is of the same order as the spectral resolution of the CVF and is caused by a calibration error in the wavelength scale of the filters provided by the manufacturer of the radiometer.

Apart from this inconsistency, both band intensities and spectral shapes of measured and calculated spectra agree well within the regions of water vapor, ozone, and carbon-dioxide emission bands. At the short-wavelength end between $\lambda = 8$ μm and $\lambda = 9$ μm, and at the long-wavelength end between $\lambda = 13.5$ μm and $\lambda = 14.5$ μm, the errors of the experimental spectra increase because the detector sensitivity and the transmission of the continuous-filter-wheel segments are reduced by one order of magnitude.

There is one remarkable difference between theoretically predicted and experimentally observed spectra. In the spectral region of high atmospheric transparency between $\lambda = 8.5$ μm and $\lambda = 12.5$ μm, the model of water-vapor continuum emission used for the LOWTRAN calculation (Roberts *et al.,* 1976) predicts a radiance of 2.72 μW/cm^2 · sr · μm, whereas the measured radiance amounts to values as high as 30 μW/cm^2 · sr · μm. In this wavelength region, infrared-active monomolecular gases do not contribute substantially to the atmospheric thermal emission, except for ozone.

The following arguments exclude experimental artifacts as possible causes for this discrepancy. First, when our system climbed to heights of 8000 m the registered radiance fell some microwatts and rose when the system descended. Second, the radiances measured at the same heights during ascent and descent agree within the experimental errors, as demonstrated in Fig. 11. This proves that the calibration procedure lacks systematic errors. At the heights considered, local and temporal inhomogeneities of the atmosphere are not relevant. Third, the ozone emission band at $\lambda = 9.7$ μm obviously sits on a background of water-vapor continuum emission. If the calculated spectrum shown in Fig. 11 (dotted curve) is increased by 30 μW/cm^2 · sr · μm, which corresponds to the difference between observed and calculated water-vapor continuum emission, measured and calculated ozone-emission-band intensities are in good agreement. The integrated ozone-emission-band intensity of the calculated spectrum, corresponding to the area under the dotted curve of Fig. 11 in the wavelength region of the O$_3$ emission, is 54.6 μW/cm^2 · sr. The integrated ozone-emission-band intensity of the measured spectra, after subtraction of the observed contribution of continuum emission, amounts to 52.6 μW/cm^2 · sr. From these considerations we conclude that the observed excess emission is a real atmospheric effect of the water-vapor continuum emission. Hence, the modeling of water-vapor continuum emission by the LOWTRAN program is insufficient.

D. HEIGHT PROFILES OF ATMOSPHERIC THERMAL EMISSION AT DISCRETE WAVELENGTHS

Nine separate wavelength channels were chosen to record the height dependence of atmospheric thermal emission. These channels with wavelengths in the range $\lambda = 8-15$ μm are marked with arrows in Figs. 9 and 10. Height profiles of the atmospheric radiance incident from the zenith are represented in Figs. 12–20. Figures 12 and 13 show the height dependence of the emission in the water-vapor bands in the channels at $\lambda = 12.62$ μm and $\lambda = 8.20$ μm, whereas Figs. 14–17 exhibit the height profiles of the controversial water-vapor continuum emission in the four channels at $\lambda = 11.76, 11.20, 4.68,$ and 3.81 μm. For comparison we also present in Fig. 18 the ozone band in the channel at $\lambda = 9.71$ μm and in Figs. 19 and 20 the carbon-dioxide bands in the channels at $\lambda = 13.70$ and 4.30 μm.

In Figs. 12–20 the LOWTRAN model predictions on the basis of *in-situ* measurements of all relevant meteorological parameters by radiosondes are represented by solid lines. Emission measurements recorded during ascent of the aircraft are indicated by asterisks (∗) and those recorded during descent by circles (O). As before, it should be noticed that the data taken both during ascent and during descent are fairly consistent. This demonstrates that neither local inhomogeneities of the atmosphere nor any drift of the radiometer seriously affected the measurement.

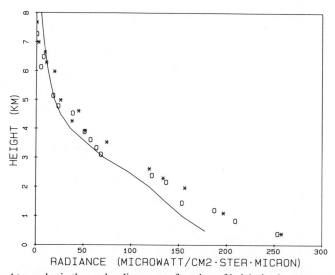

FIG. 12 Atmospheric thermal radiance as a function of height in the water-vapor band at $\lambda = 12.62$ μm: ∗, O, measured radiance during ascent and descent, respectively; ——, LOWTRAN calculation using radiosonde data from Payerne. Flight 23, April 16, 1982.

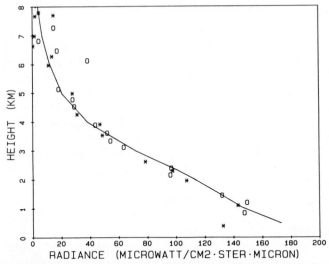

FIG. 13 Atmospheric thermal radiance as a function of height in the wing of v_2 water-vapor band at $\lambda = 8.20 \, \mu m$: *, O, measured radiance during ascent and descent, respectively; ———, LOWTRAN calculation using radiosonde data from Payerne. Flight 23, April 16, 1982.

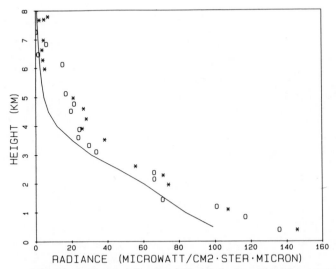

FIG. 14 Atmospheric radiance as a function of height in the region of water-vapor continuum emission at $\lambda = 11.76 \, \mu m$: *, O, measured radiance during ascent and descent, respectively; ———, LOWTRAN calculation using radiosonde data from Payerne. Flight 23, April 16, 1982.

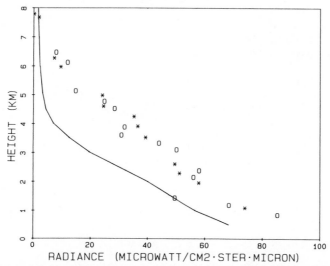

FIG. 15 Atmospheric thermal radiance as a function of height in the region of water-vapor continuum emission at $\lambda = 11.20\ \mu m$: *, O, measured radiance during ascent and descent, respectively; ——, LOWTRAN calculation using radiosonde data from Payerne. Flight 23, April 16, 1982.

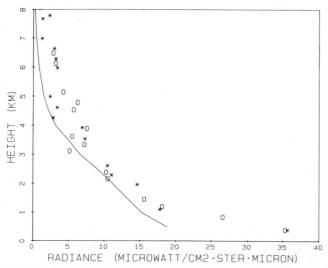

FIG. 16 Atmospheric radiance as a function of height in the region of water-vapor continuum emission at $\lambda = 4.68\ \mu m$: *, O, measured radiance during ascent and descent, respectively; ——, LOWTRAN calculation using radiosonde data from Payerne. Flight 23, April 16, 1982.

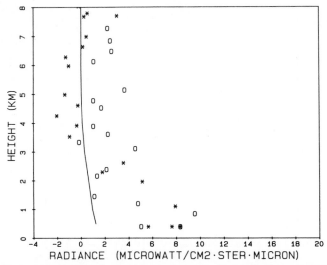

FIG. 17   Atmospheric thermal radiance as a function of height in the region of water-vapor continuum emission at $\lambda = 3.81$ μm: *, O, measured radiance during ascent and descent, respectively; ——, LOWTRAN calculation using radiosonde data from Payerne.   Flight 23, April 16, 1982.

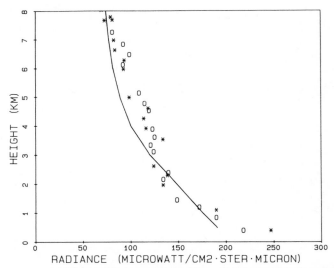

FIG. 18   Atmospheric thermal radiance as a function of height in the region of the ν_3 ozone band at $\lambda = 9.71$ μm: *, O, measured radiance during ascent and descent, respectively; ——, LOWTRAN calculation using radiosonde data from Payerne.   Flight 23, April 16, 1982.

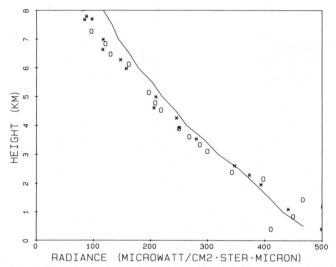

FIG. 19 Atmospheric thermal radiance as a function of height in the region of the ν_2 CO_2 band at $\lambda = 13.70$ μm: *, O, measured radiance during ascent and descent, respectively; ——, LOWTRAN calculation using radiosonde data from Payerne. Flight 23, April 16, 1982.

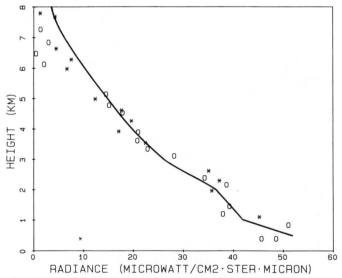

FIG. 20 Atmospheric thermal radiance as a function of height in the region of the ν_3 CO_2 band at $\lambda = 4.30$ μm: *, O, measured radiance during ascent and descent, respectively; ——, LOWTRAN calculation using radiosonde data from Payerne. Flight 23, April 16, 1982.

The height profile of atmospheric thermal radiance in the wing of the ν_2 vibration band of water-vapor at $\lambda = 8.20$ μm is shown in Fig. 13. Because of reduced detector sensitivity and transmission of the CVF, the scattering of measured radiance data is increased. However, measured and calculated height profiles are in good agreement within experimental errors, again confirming that the procedures of measurement and calibration as well as the data processing were performed properly.

The wavelength channel at $\lambda = 11.20$ μm is most suitable for the examination of the water-vapor continuum emission for the following two reasons. First, water-vapor continuum absorption is by far the most important absorption process at this wavelength because of very low monomolecular absorption of infrared-active atmospheric gases. This is demonstrated by the LOWTRAN transmission calculation plotted in Fig. 1, which shows the contribution of different absorbing gases. Second, at this wavelength the responsivity of the radiometer reaches its maximum value of R ($\lambda = 11.21$ μm) $= 20$ kV $\cdot$ cm^2 $\cdot$ sr $\cdot$ μm/W.

The height profile of atmospheric thermal emission for the wavelength channel at $\lambda = 11.20$ μm is shown in Fig. 15. The discrepancy between the calculated and observed radiance is considerable. At heights between 4000 and 5000 m, the LOWTRAN calculation yields a radiance that is 25 W/ cm^2 $\cdot$ sr $\cdot$ m below the experimentally observed value. If we consider the height profile of relative humidity (Fig. 6), it is intriguing that at this height the relative humidity also reaches maximum values close to 100%. This suggests that a more refined model should take into account the degree of saturation of water vapor, especially for conditions near saturation. Carlon (1982) drew the same conclusion from his laboratory experiments.

In the wavelength channel $\lambda = 11.76$ μm, the deviations of measured continuum emission from the LOWTRAN prediction are smaller than the deviations at $\lambda = 11.20$ μm, as can be seen in Fig. 14. Again, maximum differences are observed at heights above 4000 m.

Water-vapor continuum emission was also examined in the other important atmospheric window at wavelengths from $\lambda = 3$ to 5 μm. Calculated and measured radiance profiles for the wavelength channels $\lambda = 4.68$ μm and $\lambda = 3.81$ μm are plotted in Figs. 16 and 17. Because we are on the short-wavelength side of the Planck curve, intensities of thermal radiation decrease exponentially with decreasing wavelength, which is evident from the radiance scales in Figs. 16 and 17. In comparison with the continuum model for the atmospheric window $\lambda = 8 - 14$ μm, the water-vapor continuum emission at $\lambda = 4.68$ μm is better modeled by the LOWTRAN calculation, as is apparent in Fig. 16. Measured radiances tend to be about 2 μW/ cm^2 $\cdot$ sr $\cdot$ μm higher than calculated. This difference is the same amount as the experimental uncertainty. At the wavelength channel $\lambda = 3.81$ μm shown in Fig. 17, radiance levels of continuum emission are at the limit of the sensitivity of the radiometer system. Thus, no reliable conclusions can be drawn from this height profile.

The height profile of atmospheric thermal radiance for the wavelength channel $\lambda = 9.71 \, \mu m$ in the center of the ν_3 vibration band of ozone is represented in Fig. 18. The measured radiance agrees well with the calculated radiance at low and high altitudes. At a height of 4000 m the measured radiance exceeds the LOWTRAN prediction by 20 $\mu W/cm^2 \cdot sr \cdot \mu m$. This deviation is somewhat smaller yet of the same order of magnitude as observed for the water-vapor continuum emission in the neighboring wavelength channel at $\lambda = 11.2 \, \mu m$ (Fig. 15). This is expected if the deviations are due to water-vapor continuum emission, which is decreasing with decreasing wavelength (Fig. 1). Nevertheless, apart from the uncertain background of continuum emission the height profile of the ozone emission is correctly predicted by the LOWTRAN calculation.

The same arguments hold for the water-vapor emission band in the wavelength channel at $\lambda = 12.62 \, \mu m$, shown in Fig. 12. Deviations between measured and calculated radiance can be explained by the water-vapor continuum background, measured in the neighboring wavelength channel at $\lambda = 11.76 \, \mu m$ (Fig. 14).

Finally, to check the calibration, we present in Figs. 19 and 20 height profiles of atmospheric thermal emission in two emission bands of carbon dioxide. Figure 19 shows radiance profiles in the wing of the ν_2 deformation–vibration of CO_2 at $\lambda = 13.70 \, \mu m$. Observed radiance is fairly well predicted by the LOWTRAN model up to a height of 6000 m, above which experimental values are somewhat below calculated radiance.

The measurement at $\lambda = 13.70 \, \mu m$ is critical for two reasons. First, the wavelength calibration must be accurate, because this wavelength lies in the steep wing of the CO_2 vibration band. Second, the detector is very sensitive to temperature changes of the LN_2 bath of the detector Dewar, because this wavelength is above the cutoff wavelength of the mercury cadmium telluride (MCT) detector, corresponding to photons with energies slightly smaller than the band gap of the MCT semiconductor. The agreement of the measured and the calculated radiance profiles confirms that the detector stabilization and the precision of calibration guarantee reliability of the measurement of the excess continuum water-vapor emission.

The height profile (Fig. 20) of the very intense emission of the asymmetric stretching vibration of carbon dioxide in the middle of the atmospheric window $\lambda = 3 - 5 \, \mu m$ was measured to test calibration and stabilization of the InSb detector at short wavelengths. The absorption in the center of the ν_3 vibration band at $\lambda = 4.3 \, \mu m$ is so strong that the atmosphere appears as a blackbody emitting according to Planck's law at ambient air temperature.

Agreement between measurement and calculation is satisfactory. This demonstrates the reliability of our continuum measurement at $\lambda = 4.68$ μm.

Because we were aware of the problems involved in the humidity measurements taken with the aid of radiosondes at Payerne, as discussed in Section V.A, we estimated that the uncertainty of the predicted water-vapor continuum emission in the wavelength range between 11.2 μm (Fig. 15) and 11.76 μm (Fig. 14) introduced an error of 5% of the relative humidity. The result is 7% uncertainty, which cannot account for the difference between calculated and observed continuum emission.

VI.   Interpretation of the Water-Vapor Continuum

A.   THE TWO BASIC MODELS

Our experimental results demonstrate that the measured water-vapor continuum emission in the $\lambda = 8-14$ μm atmospheric window is even stronger than the excess emission predicted on the basis of the semiempirical formula of Roberts et al. (1976). Although there is no definite answer to the question of the physical mechanisms of this continuum absorption, different interpretations suggested by many investigators are offered in the literature. They may be grouped into two categories: water clusters, both dimers and larger aggregates; and novel line-shape theories for distant wings far from the line center. In the following we will briefly discuss these two interpretations.

B.   THE WATER-CLUSTER INTERPRETATION

Laboratory experiments on the absorption coefficiens of pure water vapor in the spectral region between 600 and 1000 cm^{-1} were performed by Varanasi et al. (1968). He suggested that the excess absorption, which cannot be related to the known spectral properties of the water molecule, might be due to hydrogen-bonded water dimers. Absorption was measured at a water-vapor pressure of $p_{H_2O} = 2$ atm with temperatures varying between $T = 400$ K and $T = 450$ K. Assuming that the dimer concentration is proportional to $\exp(-E_2/kT)$, Varanasi derived a hydrogen-bond energy of $E_2 = 5$ kcal/mol. This spectroscopic estimate of the hydrogen-bond energy is in reasonable agreement with *ab initio* calculations of the water dimer (Curtiss and Pople, 1975). Subsequent laboratory absorption measurements by Bignell (1970), Adiks et al. (1975), Burch (1971), and Gryvnak et al. (1976) confirmed the negative temperature dependence. These data are incorporated in the formula of Roberts et al. (1976), which we used for our calculations. Multipath-absorption measurements performed with a CO$_2$ laser (McCoy et al., 1969; Peterson, 1978) and laser diodes (Eng and Mantz, 1980; Montgomery, 1978) can also be interpreted in terms of dimer absorption. Near-millimeter-wave absorption measurements using both lasers (Birch *et al.*, 1969) and Fourier transform spectroscopy (Bohlander *et al.*, 1980) as well as millimeter-wave measurements in an untuned cavity (Llewellyn-Jones *et al.*, 1978), were also carried out in a search for evidence supporting the dimer hypothesis. In addition, field experiments from ground-based sky-emission measurements of Grassl (1973) and in-flight emission measurements by Coffey (1977) in the $\lambda = 14$ μm region were also interpreted by the authors to be caused by water-dimer emission.

Apart from the experiments cited here, considerable theoretical effort was made to describe the water dimers, which are of interest not only as a possible cause of infrared-continuum absorption but also as a reference system for the study of hydrogen bonding and for the theory of liquid water (Coker *et al.*, 1982; Owicki *et al.*, 1977; Stillinger and David, 1978). Also, dimers play an important role in the process of prenucleation, because they are the first step in a series of stable species (Dace *et al.*, 1972). For these reasons extensive calculations were carried out to find the structure and binding energy E_2 of the dimer.

A quantum-mechanical LCAO–MO calculation yields the 12-dimensional energy surface of the water dimer. Minima of the total electronic energy correspond to equilibrium positions of the nuclei. These calculations performed on the self-consistent field (SCF) (Curtiss and Pople, 1975; Hankins *et al.*, 1970; Popkie *et al.*, 1973) or on the more elaborate configuration-interaction (CI) level (Matsuoka *et al.*, 1976) are exceedingly tedious, especially for water clusters larger than the dimer. For this reason, empirical potentials (EPEN) have been constructed that take point charges as representations of the nuclei, lone-electron pairs, and bonding-electron pairs of the rigid water molecules that constitute the dimer (Owicki *et al.*, 1975). Another potential based on the polarization model (Stillinger and David, 1978) allows for deformable water molecules.

The first spectroscopic measurements definitely confirming the existence of stable dimers were carried out by Dyke *et al.* (1977) and Odutola and Dyke (1980). They studied the water dimer in a molecular beam using electric resonance spectroscopy.

Results of calculations and experiments on the equilibrium structure and the binding energy are listed in Table I. General agreement exists on the trans-near-linear configuration of the dimer, which is represented in Fig. 21. The proton-donor molecule lies in the x–y plane and the proton-acceptor molecule in a plane perpendicular to the x–y plane. The angles formed by the x axis and the symmetry axis of the proton-donor and proton-acceptor molecules are θ_1 and θ_2, respectively. The indicated angle δ differs from zero for nonlinear hydrogen bonds. R_{O-O} is the distance between the oxygen atoms that lie on the x axis. It can be seen from Table I that consider-

TABLE I

CALCULATED BINDING ENERGY AND GEOMETRY OF THE WATER DIMER

Reference	Method[a]	ϵ (kcal/mol)	R_{O-O} (Å)	θ_1 (deg)	θ_2 (deg)	δ (deg)
Curtiss and Pople (1975)	SCF	5.60	2.97	57.3	61.0	4.5
Popkie et al. (1973)	SCF	4.60	3.00	53.0	45.0	0.0
Hankins et al. (1970)	SCF	4.72	3.00	53.0	40.0	0.0
Matsuoka et al. (1976)	CI	5.55	2.91	45.0	26.0	−7.26
Owicki et al. (1975)	EPEN	5.44	2.88	49.46	61.9	2.8
Stillinger and David (1978)	Polarization	6.95	2.90	55.53	20.5	4.6
Coker et al. (1982)	RKW	—	2.745	53.75	53.45	1.88
Dyke et al. (1977)	Microwave spectroscopy	—	2.976	51 ± 10	57 ± 10	0 ± 10

[a] SCF, self-consistent field; CI, configuration interaction; EPEN, empirical potentials; RKW, Reimers–Klein–Watts potential.

able uncertainty exists on the angles θ_1, θ_2 and the dimer binding energy E_2, which ranges from 4.6 to 6.85 kcal/mol depending on the calculation procedure. Because the binding energy of the dimer is only a fraction of 5×10^{-5} of the total dimer energy, calculations of energy eigenvalues must surpass an accuracy of 10^{-5} for reasonable estimates of the hydrogen-bond energy. Cooperative effects of coulomb interaction, exchange repulsion, charge transfer, polarization energy, configuration interaction, and dispersion become important and none of them may be neglected.

The twelve internal degrees of freedom of $(H_2O)_2$ can be separated into six high-frequency intramolecular and six low-frequency intermolecular vibrations, as shown by a vibrational analysis of the energy surface of the dimer. Except for the OH vibration of the OH bond involved in the hydrogen bonding, intramolecular vibrations are only slightly disturbed. They have been studied by matrix spectroscopy (Hagen and Tielens, 1981; Trusi and

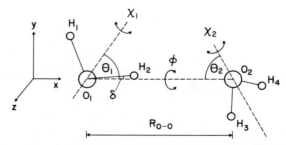

FIG. 21 The configuration and geometrical parameters of the equilibrium water dimer.

Nixon, 1970). The remaining six intermolecular vibrations and their frequencies are listed in Table II. The modes are out-of-plane and in-plane hydrogen-bond bends and shears, and a hydrogen-bond stretch and torsion. The normal modes show considerable mixing of various internal coordinates. According to Curtiss and Pople (1975), the normal modes can be ascribed mainly to the oscillation of one coordinate, given in the second column of Table II and indicated in Fig. 21. The authors performed a vibrational analysis of their SCF calculation. Their frequencies are given in Table II. The normal modes for the EPEN (Owicki *et al.*, 1975) as well as for the Shipman–Scheraga potential (Shipman and Scheraga, 1974) and the Analytical–Fit to Hartree–Fock (AFHF) potential (Kistenmacher *et al.*, 1974) have been calculated by Owicki *et al.* (1975). Recently Coker *et al.* (1982) obtained the dimer spectrum from classical trajectory studies using a Monte Carlo algorithm. The spectral features below 1200 cm^{-1} can be attributed to the intermolecular-dimer modes identified from the normal-mode analysis of the Reimers–Klein–Watts (RKW) potential, which are also listed in Table II.

As indicated in Table II, considerable discrepancies exist among the frequencies of intermolecular normal modes of different potentials. Nevertheless, all potentials unanimously predict large bending-force constants for the hydrogen-bond bend (θ_1 and χ_1 motion) with high vibration frequencies, whereas hydrogen-bond torsion and bond stretch have significantly lower frequencies. Therefore, the most probable explanation of the observed $\lambda = 8 - 14 \ \mu$m continuum is the wing of the librational band of $(H_2O)_2$ corresponding to the out-of-plane hydrogen-bond bend (Wolynes and Roberts,

TABLE II

INTERMOLECULAR NORMAL-MODE VIBRATION FREQUENCIES AND
INTEGRATED INFRARED ABSORPTIVITIES

Mode	Potential Coordinate	Symmetry type	Frequency^a (cm^{-1}) SCF	EPEN	SS	AFHF	RKW
Out-of-plane H-bond bend	χ_1	A″	536	593	681	537	780
In-plane H-bond bend	θ_1	A′	452	496	451	319	483
In-plane H-bond shear	θ_2	A′	185	189	106	432	169
H-bond stretch	R_{O-O}	A′	204	168	183	161	272
Out-of-plane H-bond shear	χ_2, Φ	A″	118	161	63	127	219
H-bond torsion	χ_2	A″	81	98	173	105	115

^{a} SCF, Curtiss and Pople (1975); EPEN, Owicki *et al.* (1975); SS, Shipman and Scheraga (1974); AFHF, Kistenmacher *et al.* (1974); RKW, Coker *et al.* (1982).

1978). It can be assigned to a libration about the symmetry axis of molecule 1 (χ_1 motion), as indicated in Fig. 21.

The extent to which continuum absorption can be attributed to water aggregates is crucially dependent on the equilibrium concentration of clusters. Water vapor can be regarded as a multicomponent system of ideal gases, namely, water monomers and water clusters of size n. The different gases undergo chemical reactions with each other. By use of statistical mechanics, the equilibrium concentration C_n of clusters of size n can be evaluated if the total binding energy of the electronic ground state E_n and the vibrational and rotational energy levels of the clusters are known (Suck et al., 1979). An expression for C_n has been derived by Suck et al. (1982). It can be reduced to the following equation:

$$C_n \lesseqgtr A_n (p_{H_2O}/kT)^n (kT)^{3(1-n)} \exp[-(E_n + E_z)/kT], \qquad (16)$$

where

$$E_z = \tfrac{1}{2} \sum_{j=1}^{6(n-1)} h\nu_j^{\text{inter}}.$$

At temperatures $kT \ll h\nu_j^{\text{inter}}$, only the ground states of the intermolecular vibrations are occupied, and the equal sign holds in this expression. E_z is the zero-point energy of intermolecular vibrations, and k is the Boltzmann constant. The factor A_n contains the principal moments of inertia of water monomers and water clusters of size n. The concentration is proportional to the nth power of water-vapor pressure p_{H_2O}. The exponential temperature dependence is determined by the zero-point energy E_z and the binding energy E_n that has a negative sign. For weak hydrogen bonds, the zero-point energy of the dimers calculated by Owicki et al. (1975) is $E_z = 2.44$ kcal/mol. Neglect of anharmonicities and coupling between different kinds of internal degrees of freedom introduce errors in E_z. Because of the exponential dependence of cluster concentrations on binding and zero-point energy, which cannot be calculated accurately, the predicted equilibrium concentrations are uncertain. At $p_{H_2O} = 21$ Torr and $T = 296$ K the ratio of dimer-to-monomer concentration is estimated to be $C_2/C_1 = 4.4 \times 10^{-5}$ for $E_2 = -4.6$ kcal/mol and $C_2/C_1 = 2.4 \times 10^{-3}$ for $E_2 = -6.95$ kcal/mol and Owicki's value of E_z.

For the wavelength $\lambda = 10$ μm, Suck et al. (1982) recently calculated an attenuation coefficient $\sigma_{H_2O \, \text{cont}} = 0.17$/km, which is attributed to the wing of the out-of-plane bond-bend libration of the dimer at $\nu_j^{\text{inter}} = 624$ cm^{-1} with halfwidth $\alpha_j^{\text{inter}} = 200$ cm^{-1}. For their estimate they used $(\partial M / \partial Q_k^{\text{inter}})^2 = 0.83$ D^2/amu$\cdot$Å^{-2}, where M is the dipole moment and Q_k the corresponding normal coordinate. The dimer concentration assumed in this estimate is 7.6×10^{14}/cm^3 at $p_{H_2O} = 14$ Torr and $T = 296$ K. The librational-dimer-attenuation coefficient is in reasonable agreement with the value $\sigma_{H_2O\ cont} = 0.153/km$ obtained from the semiempirical model of Roberts *et al.* (1976) that we used in our LOWTRAN programs. Suck *et al.* (1982) argue that equilibrium water clusters larger than the dimer are of no importance for the $8 - 14\ \mu m$ continuum absorption, because the concentration ratios C_n/C_1 for water trimers and larger clusters become less than 10^{-6}.

Contrary to this conclusion, absorption measurements of Emery *et al.* (1979) on a horizontal path in the free atmosphere and atmospheric emission measurements by Coffey (1977) indicate a negative temperature dependence that is too strong to be explained by thermal dissociation of equilibrium dimers. These measurements were carried out for temperatures below 300 K in the $8 - 14\ \mu m$ atmospheric window. The same temperature dependence was observed for the absorption at $7.1\ cm^{-1}$ in the low-temperature region (Llewellyn-Jones *et al.*, 1978). Furthermore, Gebbie analyzed millimeter-wave absorption observed in the atmosphere and in the laboratory (Gebbie, 1980, 1982). He suggests that large metastable water polymers containing about 50 molecules cause the strong temperature dependence. Anomalous spectral features near saturation are attributed to collective acoustic resonances of the Fröhlich type. Clathratelike structures maximize the number of possible hydrogen bonds, as shown in Fig. 22. They are rather stable and can also absorb radiation in the infrared windows (Kassner *et al.*, 1980). On the other hand, Carlon proposes that anomalous absorption in moist air is caused by both neutral water clusters and hydrated ions containing between 10 and 30 water molecules (Carlon, 1982). The concentration of water clusters is expected to be enhanced in supersaturated water vapor. Absorption measurements in supersaturation, which is produced by an expansion chamber, show a time dependence that is difficult to explain (W. Johnson, Appleton Lab., pers. comm.). At present, photoacoustic measurements with a CO_2 laser in a diffusion chamber are performed in our own laboratory to examine H_2O continuum absorption under well-defined conditions of supersaturation (Hinderling *et al.*, 1983).

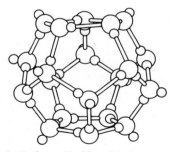

FIG. 22   Energetically favorable 20-molecule water–clathrate cluster.

In view of the variety of experimental and theoretical studies on the cluster interpretation, it must be noted that so far no spectroscopic feature in water-vapor absorption spectra has been found that can be unambiguously identified with a transition of a water aggregate.

C. THE LINE-SHAPE INTERPRETATION

In 1938 Elsasser proposed that the $8-14$-μm water-vapor continuum results from the absorption in the far wings of strong nearby rotational and ν_2 vibrational lines of water monomers (Elsasser, 1938). Because the cluster interpretation discussed in Section VI.B has not yet been confirmed by a spectroscopic assignment of specific dimer-cluster lines, the line-shape interpretation has recently been revitalized by different authors, including Clough *et al.* (1980), Thomas and Nordstrom (1982), and Birnbaum (1979).

The standard line shapes given in Section III.A neglect the finite duration τ_c of collisions. In this impact approximation, the validity of the line shapes is restricted to frequencies near the line center with $|\nu - \nu_0| < (c \cdot \tau_c)^{-1}$. For typical collision times of $\tau_c \simeq 10^{-12}$ sec, this implies an off-resonance frequency limit of 30 cm^{-1}.

To develop an adequate line-shape theory that is also far from the resonance condition, strong encounters between absorbing and broadening molecules must be taken into account. The quantum-mechanical description of the absorption process in the presence of disturbing molecules can be formulated in terms of time-dependent perturbation theory, as shown by Anderson (1949) and Tsao and Curnutte (1962). The dynamics of the collision between the absorbing and the broadening molecule is described by the collision interaction $H_c(t)$, which was discussed for two water molecules in Section VI.B. From the time development of the scattering process, which is related to the dipole moment $\mu(t)$ of the system, the power spectrum is obtained from linear response theory, as follows:

$$I(\nu) = \int_0^\infty < \mu(0)\mu(t) > \exp(-i\omega t)\, dt. \qquad (17)$$

The spectrum $I(\nu)$ is the Fourier transform of the dipole autocorrelation function $\langle \mu(0)\mu(t) \rangle$, as stated by the Wiener–Khintchine theorem. The brackets in Eq. (17) indicate an average over a Boltzmann distribution of initial conditions. A detailed analysis of autocorrelation functions corresponding to vibration–rotation spectra of diatomic, linear, symmetric-top, and asymmetric molecules was performed by Keller and Kneubühl (1971, 1972). The far wings of a spectral line in the frequency domain correspond to the short-time behavior of the autocorrelation function in the time domain. This behavior is determined by the mechanism of strong collisional perturbations of the absorbing molecule.   In a first step of the development of an adequate line shape for the far wings of H_2O, Clough *et al.* (1980) incorporated the finite duration of the collisions τ_c into their autocorrelation function and thus removed the discontinuity of all derivatives of the auto-correlation function at $t = 0$.   This statement also applies to the work by Birnbaum (1979) and by Boulet and Robert (1982).   Clough *et al.* (1980) find a Gaussian for the far wings of their line shapes, whereas Birnbaum (1979) as well as Boulet and Robert (1982) obtain exponential-like far wings.   In their theory, Clough *et al.* (1980) did not consider the temperature dependence.   Yet, they expect far-wing temperature effects similar to the making and breaking of bonds in dimer formation that are predicted by a proper treatment of finite collision times τ_c.

In a more explicit calculation, Thomas and Nordstrom (1982) took into account both the self-broadening and the foreign-broadening contributions to the water-vapor continuum absorption by explicit interaction Hamiltonians H_c.   Self-broadening results from collisions between two water mole-cules.   It is dominated by the strong dipole–dipole interaction:

$$H_{c,\,s}(t) = \langle \mu^2/R^3(t) \rangle_{\mathrm{Rot}}, \tag{18}$$

where R indicates the intermolecular distance and μ the dipole moment. The duration of the collision interaction is of the order of 10^{-12} sec.   There-fore, the colliding water molecule can rotate many times during the colli-sion.   If the rotational average is performed, the collision interaction for self-broadening becomes (Elsenberg and Kauzmann, 1969)

$$H_{c,\,s}(t) = (2\mu^4/3kT)[1/R^6(t)]. \tag{19}$$

The negative temperature dependence of this interaction indicates a breakup of attractive orientations between water molecules.   It is similar to the breakup of hydrogen bonds of water dimers at elevated temperatures. The dipole–quadrupole interaction is the dominant contribution to foreign broadening.   It is caused by collisions between absorbing water molecules and disturbing N_2 or O_2 molecules.   Because the rotation of N_2 molecules is slow compared with τ_c, no rotational average needs to be taken:

$$H_{c,\,f}(t) = \mu Q/R^4(t). \tag{20}$$

Theoretical considerations by Fomin and Tvorogov (1973) show that in the phase-shift approximation the statistical line shape takes the far-wing form represented by Eq. (21) if the dependence of the collision Hamiltonian on intermolecular distance R is proportional to R^{-m}.   Thus we have

$$k(v) = \Lambda/(v - v_0)^{1+(3/m)}, \tag{21}$$

where Λ is a parameter that depends on the potential.   The far-wing absorption decreases with $(v - v_0)^{-1.5}$ for self-broadening $(m = 6)$ and with $(v - v_0)^{-1.75}$ for foreign broadening $(m = 4)$. Thus, self-broadening will dominate far from the line centers in the atmospheric-window region. Near the line centers, the line shape is Lorentzian. This results from the exponential decay of the autocorrelation function for $\tau \gg \tau_c$ where the impact approximation holds. An adequate theory in the intermediate-frequency range is missing; therefore line shapes are smoothly interpolated from the near-line-center region to the far-wing region. The line-shape formalism describes the pressure dependence of the continuum absorption correctly. On the other hand, it cannot quantitatively model the strong negative temperature dependence that has been observed experimentally (Burch and Gryvnak, 1979; Montgomery, 1978). However, qualitative agreement is achieved by the introduction of the temperature-dependent dipole interaction given in Eq. (19).

Finally, it should be noticed that the quadratic pressure dependence of the continuum absorption can be explained by the line shape as well as by the water-dimer hypothesis. Yet, to the best knowledge of the author, none of the present line-shape theories can account for the strength of the temperature dependence of continuum absorption (Boulet and Robert, 1982). In this context a computational simulation of the molecular beam experiment of Odutola and Dyke (1980) was carried out by Coker *et al.* (1982), which starts with Eq. (17). With a tentative vibrational temperature of 50 K, a classical trajectory study using Monte Carlo methods for averaging Eq. (17) demonstrates that the resulting spectrum can be interpreted in terms of low-frequency dimer modes. This proves that the line-shape formulation yields results similar to those of the dimer hypothesis.

VII. Conclusion

The spectral thermal emission of the terrestrial atmosphere was observed from an aircraft with upward-looking radiometers while simultaneous radiosonde flights were probing temperature, humidity, and ozone profiles. Measured spectra and radiation profiles at selected wavelength channels were compared with LOWTRAN calculations based on the radiosonde data. The good agreement of measured and calculated spectra in spectral regions of the well-known H_2O, CO_2, and O_3 absorption bands confirms both the quality of our absolute in-flight calibration of the radiometer and the capability of the LOWTRAN model to predict atmospheric thermal emission and absorption. Furthermore, radiance data measured during ascent of the aircraft to a height of 8200 m are consistent with data taken during descent. This proves that neither systematic errors nor drift of radiometer responsivity seriously affect the measurements. Thus, we conclude that the observed deviation of continuum emission in the $\lambda = 8-14\ \mu m$ atmospheric window, which exceeds the prediction based on the formula of Roberts [Eq. (12)] up to 25 $\mu W/cm^2 \cdot sr \cdot \mu m$, represents a real atmospheric effect.

The water-vapor continuum emission that we observed during the flights originated from atmospheric layers with temperatures ranging between 10 and $-57°C$. Moreover, the humidity approached saturation during the relevant flight at a height of 4000 m. However, the continuum model of Roberts used for our calculations is based on laboratory and field experiments that were carried out at temperatures well above 0°C at moderate relative humidities. Therefore, we attribute the poor agreement between calculated and measured continuum emission to an inadequate extrapolation of the semiempirical LOWTRAN formula (12) to low temperatures and humidities near saturation. Experimental evidence supporting this argument is derived from the previous measurements by Coffey, who installed the SCR radiometer developed for the *Nimbus* 5 satellite in an aircraft for a measurement of the emission in the $\lambda = 10-12\ \mu m$ window. In the temperature range between -15 and $26°C$, he derived from his experiment a negative temperature dependence of the continuum absorption corresponding to the characteristic temperature $T_B \simeq 4000$ K, whereas the temperature dependence of the continuum absorption due to equilibrium dimers as incorporated in our calculations is characterized by $T_B = 1800$ K [see Eq. (11), Section III]. Furthermore, field measurements by Emery *et al.* (1979) along a horizontal absorption path, as well as laboratory measurements at a wavelength of $\lambda = 1.41$ mm by Llewellyn–Jones *et al.* (1978) showed that the temperature exponent of continuum absorption increases with decreasing temperature. Additional complications arise when the absorption coefficient $K_{H_2O\ cont}$ depends on both the saturation and the ion content of air, as Carlon concluded from his absorption measurements of air humidified by boiling water (Carlon, 1982). Experimental observations, including our own results, suggest that the LOWTRAN formula (12) for water-vapor continuum absorption should be revised at low temperatures and large relative humidities. This revision may be performed by replacing the LOWTRAN formula (12) by an equation that is in better agreement (Hinderling, pers. comm.) with the analysis by Roberts *et al.* (1976):

$$\sigma_{H_2O\ cont} = C(v)\ \exp[T_B(1/T - 1/T_0)]\{p_{H_2O} + \gamma(p - p_{H_2O})\}W_{H_2O}. \quad (22)$$

In the literature, molecular theories on the continuum absorption have been developed in the cluster and in the line-shape formalism. Because the photon energy at $\lambda = 10\ \mu m$ is of the same order of magnitude as the energy required to break up weak hydrogen bonds of water aggregates, it is difficult to decide at the present state of theoretical models whether bound–bound,

bound – free, or free – free transitions in the water-bimolecular system are the dominant mechanisms of continuum absorption. In any case, the strong negative temperature dependence implies some kind of a bound state as initial state.

To retrieve accurate absorption coefficients from the radiance gradient of the height profile of measured radiance, the sensitivity of the radiometer system should be improved. This can only be achieved by cooling the optics and the reference blackbody to cryogenic temperatures. To gain more insight into the mechanisms of continuum absorption, we now supplement the reported field experiments with photoacoustic spectroscopy of supersaturated water vapor in a diffusion chamber (Hinderling *et al.,* 1982).

ACKNOWLEDGMENTS

The authors would like to thank G. Birnbaum, Washington, D. C.; R. A. Bohlander, Atlanta, Georgia; H. R. Carlon, Aberdeen, Maryland; H. U. Dütsch, Zürich; R. J. Emery, Noordwijk; H. A. Gebbie, Slough; J. T. Houghton, Oxford; W. Johnson, Slough; and R. J. Nordstrom, Columbus, Ohio for many helpful discussions and suggestions. Furthermore, we are indebted to J. Joss, Locarno, and G. Scherrer, Bern, for lending the radiometers and to A. Greuter, Zürich, for providing the computer code LOWTRAN 3. We would also like to express our thanks to H. Häfliger and his crew at the air base in Emmen for assistance in mounting the equipment and for piloting the aircraft. J. Ricker, Station Aérologique, Payerne and H. U. Dütsch, Zürich, kindly provided the radiosonde data. Finally, we express our thanks to B. Bachofner for help with the design, construction, and flying of the instrument and to D. Keusch for typing the manuscript.

This work was funded by the ETH Zürich, the GRD of the EMD, and the Swiss National Science Foundation, National Program Energy, Project 4.089-076.04.

REFERENCES

Adiks, T. G., Aref'ev V. N., and Dianov-Klokov, V. I. (1975). *Sov. J. Quantum. Electron.* **5,** 481.

Anderson, P. W. (1949). *Phys. Rev.* **76,** 647.

Bean, B. L., and White, K. O. (1981). *6th Int. Conf. Infrared Millimeter Waves Dig.,* Miami Beach, Florida.

Ben Shalom, A., *et al.* (1980). *Appl. Opt.* **19,** 838.

Berdahl, P., and Fromberg, R. (1982). *Solar Energy* **29,** 299.

Bignell, K. J. (1970). *Q. J. Roy. Meteorol. Soc.* **96,** 390.

Birch, J. R., Burroughs, W. J., and Emery, R. J. (1969). *Infrared Phys.* **9,** 75.

Birnbaum, G. (1979). *J. Quant. Spectros. Radia. Transfer* **21,** 597.

Bohlander, R. A. (1979). "Spectroscopy of Water Vapour." Ph. D. Thesis, Imperial College of Science and Technology, University of London, London, United Kingdom.

Bohlander, R. A., *et al.* (1980). *In* "Atmospheric Water Vapour" (A. Deepak, T. D. Wilkerson, and L. H. Ruhnke, eds.), p. 241. Academic Press, New York.

Boulet, C., and Robert, D. (1982). *J. Chem. Phys.* **77,** 4288.

Burch, D. E. (1971). "Investigation of the Absorption of Infrared Radiation by Atmospheric Gases," Aeronautic Report U-4784. Philco Ford Corp., Newport Beach, California.

Burch D. E., and Gryvnak D. A. (1979). "Method of Calculating H_2O Transmission between 333 and 633 cm^{-1}," AFGL-TR-79-0054. U.S. Air Force, Hanscom, Massachusetts.

Burch, D. E., and Gryvnak, D. A. (1980). In "Atmospheric Water Vapour" (A. Deepak, T. D. Wikerson, and L. H. Ruhnke, eds.), p. 47. Academic Press, New York.

Burch, D. E., Gryvnak, D. A., and Pembrook, J. D. (1971). Aeronautic Report U-4897, ASTLA (AD 882876). Philco Ford Corp., Newport Beach, California.

Carli, B. (1977). J. Opt. Soc. Am. 67, 908.

Carlon, H. R. (1982). Infrared Phys. 22, 43.

Chandrasekhar, S. (1950). "Radiative Transfer," Oxford Univ. Press, Oxford (Dover Publ., New York, 1960).

Clough, S. A., Kneizys, F. X., Davies, R., Gamache, R., and Tipping, R. (1980). In "Atmospheric Water Vapour" (A. Deepak, T. D. Wilkerson, and L. H. Ruhnke, eds.), p. 25. Academic Press, New York.

Coffey, M. T. (1977). Q. J. Roy. Meteorol. Soc. 103, 685.

Coker, D. F., Reimers, J. R., and Watts, R. O. (1982). Aust. J. Phys. 35, 623.

Curtis, P. D., Houghton, J. T., Peskett, G. D., and Rodgers, C. D. (1974). Proc. Roy. Soc. London, Ser. A. 135.

Curtiss, L. A., and Pople, J. A. (1975). J. Mol. Spectrosc. 55, 1.

Daee, M., Lund, L. H., Plummer, P. L. M., Kassner, J. L., and Hale, B. N. (1972). J. Colloid and Interface Sci. 39, 65.

Dianov-Klokov, V. I., Ivanov, V. M., Aref'ev, V. N., and Sizov, N. J. (1981). J. Quant. Spectrosc. Radiat. Transfer 25, 83.

Dütsch H. U. (1984). "Regular Ozone Soundings at the Aerological Station of the Swiss Meteorological Office at Payerne, Switzerland," Publ. Lab. Atmos. Phys., ETH, LAPETH, Zürich, Switzerland.

Dyke, T. R., Mack, K. M., and Muenter, J. S. (1977). J. Chem. Phys. 66, 498.

Eisenberg, D., and Kauzmann, W. (1969). "The Structure and Properties of Water," Oxford Univ. Press, Oxford.

Ellis, P., et al. (1973). Proc. Roy. Soc. London, Ser. A. 334, 149.

Elsasser, W. M., (1938). Phys. Rev. 53, 768.

Emery, R. J., Zavody, A., and Gebbie, H. A. (1979). Nature 277, 462.

Eng, R. S., and Mantz, A. W. (1980). In "Atmospheric Water Vapour" (A. Deepak, T. D. Wilkerson, and L. H. Ruhnke, eds.), p. 101. Academic Press, New York.

Fomin, V. V., and Tvorogov, S. D. (1973). Appl. Opt. 12, 584.

Gebbie, H. A. (1980). In "Atmospheric Water Vapour" (A. Deepak, T. D. Wilkerson, and L. H. Ruhnke, eds.), p. 133. Academic Press, New York.

Gebbie, H. A. (1982). Nature 296, 1.

Goody, R. (1952). Q. J. Roy. Meteorol. Soc. 78, 165.

Goody, R. (1964). "Atmospheric Radiation, Vol. I, Theoretical Basis," Oxford Univ. Press, Oxford.

Grassl, H. (1973). Beiträge zur Physik der Atmosphäre 46, 75.

Gross, E. P. (1955). Phys. Rev. 97, 395.

Gruenzel, R. R. (1978). Appl. Opt. 17, 2591.

Gryvnak, D. A., Burch, D. E., Alt, R. L., and Zgnoc, D. K. (1976). "Infrared Absorption by CH_4, H_2O and CO_2," AFGL-TR-76-0246. U.S. Air Force, Hanscom, Massachusetts.

Hagen, W., and Tielens, A. G. G. M. (1981). J. Chem. Phys. 75, 4198.

Hankins, D., Moskowitz, J. W., and Stillinger, F. H. (1970). J. Chem. Phys. 53, 4544.

Harrison, A. W. (1976). Can. J. Phys. 54, 1442.

Hinderling, J., Sigrist, M. W., and Kneubühl, F. K. (1982). "Proc. Lasers '82." STS Press, McLean, Virginia.

Hinderling, J., Sigrist, M. W., and Kneubühl, F. K. (1983). *Proc. 3rd Int. Topical Meeting on Photoacoustic and Photothermal Spec., J. Phys.* 44-*Colloq* **C6**, 559. Paris, France.
Houghton, J. T. (1977). "The Physics of Atmospheres." Cambridge Univ. Press, Cambridge, Massachusetts.
Hummel, J. R. (1982). *J. Atmos. Sci.* **39**, 879.
Huppi, E. R., Rogers, J. W., and Stair, A. T. (1974). *Appl. Opt.* **13**, 1466.
Kassner, J. L. *et al.* (1980). *In* "Atmospheric Water Vapour" (A. Deepak, T. D. Wilkerson, and L. H. Ruhnke, eds.), p. 613. Academic Press, New York.
Keller, B., and Kneubühl, F. K. (1971). *Chem. Phys. Lett.* **9**, 178.
Keller, B., and Kneubühl, F. K. (1972). *Helv. Phys. Acta* **45**, 1127.
Kistenmacher, H., Lie, G. C., Popkie, H., and Clementi, E. (1974). *J. Chem. Phys.* **61**, 546.
Kneizys, F. X., *et al.* (1980). "Atmospheric Transmittance/Radiance: Computer Code LOW-TRAN 5," AFGL-TR-80-0067. U.S. Air Force, Hanscom, Massachusetts.
La Rocca, A. J. (1975). *Proc. IEEE* **63**, 75.
Lee, A. C. (1973). *Q. J. Roy. Meteorol. Soc.* **99**, 490.
Llewellyn-Jones, D. T., Knight, R. J., and Gebbie, H. A. (1978). *Nature* **274**, 876.
Lorentz, H. A. (1906). *Proc. Amst. Acad. Sci.* **8**, 591.
Matsuoka, O., Clementi, E., and Yoshimine, M. (1976). *J. Chem. Phys.* **64**, 1351.
McClatchey, R. A., Fenn, R. W., Selbey, J. E. A., Volz, F. E., and Garing, J. S. (1972). "Optical Properties of the Atmosphere," AFCRL-72-0497. U.S. Air Force, Hanscom, Massachusetts.
McClatchey, R. A., *et al.* (1973). "AFCRL Atmospheric Absorption Line Parameter Compila-tion," AFCRL-TR-73-0096. U.S. Air Force, Hanscom, Massachusetts.
McCoy, J. H., Rensch, D. B., and Long, R. K. (1969). *Appl. Opt.* **8**, 1471.
Montgomery, P. (1978). *Appl. Opt.* **17**, 2299.
Murcray, D. G., Goldman, A., Kosters, J. J., Murcuray, F. H., and Williams, W. J. (1978). "Spectral Radiometric Measurements of Sub-Arctic Stratospheric Constituents," DAAD-05-76-C-0740.
Odutola, J. A., and Dyke, T. R. (1980). *J. Chem. Phys.* **72**, 5062.
Oetjen, R. A., Bell, E. E., Young, J., and Eisner, L. (1960). *J. Opt. Soc. Am.* **50**, 1313.
Owicki, J. C., Shipman, L. L., and Scheraga, H. A. (1975). *J. Phys. Chem.* **79**, 1794.
Peterson, J. C. (1978). "A Study of Water Vapour Absorption at CO_2 Laser Frequencies Using a Differential Spectrophone and White Cell," Ph.D. Thesis, Ohio State University, Physics Dept., Columbus, Ohio.
Phillips, P. D., Richner, H., Joss, J., and Ohmura, A. (1981). *PAGEOPH* **119**, 259.
Platt, C. M. R. (1971). *J. Appl. Meteorol.* **10**, 1307.
Popkie, H., Kistenmacher, H., and Clementi, E. (1973). *J. Chem. Phys.* **59**, 1325.
Ricker, J. (1980–1982). Institute Suisse de Météorologie, Station Aerologique, Payerne, pers. comm.
Roberts, R. E., Selby, J. E. A., and Bibermann, L. M. (1976). *Appl. Opt.* **15**, 2085.
Selby, J. E. A., and McClatchey, R. A. (1972). "Atmospheric Transmittance from 0.25 to 28.5 μm: Computer Code LOWTRAN 2," AFCRL-72-0745. U.S. Air Force, Hanscom, Massa-chusetts.
Selby, J. E. A., and McClatchey, R. A. (1975). "Atmospheric Transmittance from 0.25 to 28.5 μm: Computer Code LOWTRAN 3," AFCRL-TR-75-0255. U.S. Air Force, Hanscom, Massachusetts.
Selby, J. E. A., Shettle, E. P., and McClatchey, R. A. (1976). "Atmospheric Transmittance from 0.25 to 28.5 μm: Supplement LOWTRAN 3B," AFGL-TR-76-0258. U.S. Air Force, Hanscom, Massachusetts.
Selby, J. E. A., Kneizys, F. X., Chetwynd, J. H., and McClatchey, R. A. (1978). "Atmospheric

Transmittance/Radiance Computer Code LOWTRAN 4," AFGL-TR-78-0053. U.S. Air Force, Hanscom, Massachusetts.

Shipman, L. L., and Scheraga, H. A. (1974). *J. Phys. Chem.* **78,** 909.

Shumate, M. S., Menzies, R. T., Margolis, J. S., and Rosengren, L. G. (1976). *Appl. Opt.* **15,** 2480.

Stair, A. T., *et al.* (1983). *Appl. Opt.* **22,** 1056.

Stillinger, F. H., and David, C. W. (1978). *J. Chem. Phys.* **69,** 1473.

Suck, S. H., Kassner, J. L., and Yamaguchi, Y. (1979). *Appl. Opt.* **18,** 2609.

Suck, S. H., Wetmore, A. E., Chen, T. S., and Kassner, J. L. (1982). *Appl. Opt.* **21,** 1610.

Thomas, M. E., and Nordstrom, R. J. (1982). *J. Quant. Spectrosc. Radiat. Transfer* **28,** 103.

Traub, W. A., and Stier, M. T. (1976). *Appl. Opt.* **15,** 364.

Trusi, A. J., and Nixon, E. R. (1970). *J. Chem. Phys.* **52,** 1521.

Tsao, C. J., and Curnutte, B. (1962). *J. Quant. Spectrosc. Radiat. Transfer.* **2,** 41.

Van Vleck, J. H., and Weisskopf, V. F. (1945). *Rev. Mod. Phys.* **17,** 277.

Valero, F. P. J., Warren, J. Y. G., and Giver, L. P. M. (1982). *Appl. Opt.* **21,** 831.

Varanasi, P., Chou, S., and Penner, S. S. (1968). *J. Quant. Spectrosc. Radiat. Transfer* **8,** 1537.

Viktorova, A. A., and Zhevakin, S. A. (1966). *Sov. Phys. Dokl.* **11,** 1059.

Viktorova, A. A., and Zhevakin, S. A. (1975). *Izv. VUZ, Radiophys.* **18,** 211.

Watkins, W. R., White, K. O., Bower, L. R., and Sojka, B. Z. (1979). *Appl. Opt.* **18,** 1149.

White, K. O., Watkins, W. R., Tuer, T. W., Smith, F. G., and Meredith, R. E. (1975). *J. Opt. Soc. Am.* **65,** 1201.

Wolynes, P. G., and Roberts, R. E. (1978). *Appl. Opt.* **17,** 1484.

Woody, D. P., and Richards, P. L. (1981). *Astrophys. J.* **248,** 18.

Wyatt, C. L. (1975). *Appl. Opt.* **14,** 3086.

Zürcher, C., *et al.* (1982). *Infrared Phys.* **22,** 277.

CHAPTER 5

Frequency Tuning and Efficiency Enhancement of High-Power Far-Infrared Lasers

B. G. Danly, S. G. Evangelides, R. J. Temkin, and B. Lax

*Plasma Fusion Center
and National Magnet Laboratory
Massachusetts Institute of Technology
Cambridge, Massachusetts*

I. Introduction

The generation of coherent radiation in the far-infrared region of the electromagnetic spectrum is an important problem for modern physics. While powerful sources such as microwave tubes and optical lasers have existed on both sides of this spectral region for some time, the development of powerful sources in the far infrared (FIR) remains a largely unsolved problem. Present FIR sources include gyrotrons, backward-wave oscillators, Ledatrons, free-electron lasers, and far-infrared lasers (DeTemple, 1979; Hirshfield, 1979; Kantorowicz and Palluel, 1979; Mizuno and Ono, 1979; Sprangle *et al.,* 1979). The harmonic generation of FIR radiation is a well-established technique (Gordy, 1960). The nonlinear mixing of the radiation from two infrared lasers is also known to produce radiation in the FIR (Aggarwal and Lax, 1977).

This investigation is generally concerned with the underlying physics of optically pumped, high-power far-infrared lasers. In particular, it is concerned with the theory and experimental development of frequency-tunable far-infrared lasers and possible methods for efficiency enhancement of these lasers. This investigation represents the first comprehensive study of CO_2-laser-pumped tunable FIR lasers.

A. MOTIVATION

There are many applications of coherent sources in the far-infrared or submillimeter ($\lambda = 0.1 - 1$ mm) spectral range (Lax, 1982). Many molecules with permanent-dipole moments and modest mass possess rotational energy levels in the far infrared. Coherent sources in the FIR have many spectroscopic applications. However, all continuous-wave (cw) FIR laser sources and all previously developed pulsed FIR laser sources have been limited to a large set of discrete frequencies (Rosenbluh *et al.,* 1976). There have been advances in cw FIR lasers that are optically pumped by cw-waveguide CO_2 lasers (Strumia and Inguscio, 1981). These FIR lasers are tunable by $\sim \pm 500$ MHz about the line-center emission frequencies. Pulsed FIR lasers that are tunable by ± 2 GHz about line center have also been reported (Fetterman *et al.,* 1979), but, in general, for any desired FIR frequency, one must hope for a coincidence with a number of available discrete laser lines. It is clear that a broadly tunable source of coherent far-infrared radiation would be a valuable tool for research and application in this spectral region.

The development of a tunable FIR laser source would lead to significant improvements in both molecular and solid-state spectroscopy. The range of atmospheric contaminants that could be remotely sensed would increase substantially with a high-power, tunable source in the FIR. In the military arena, FIR lasers are of interest for communications and high-resolution radar as well as for scaled-down measurements of model radar targets.

One of the more important applications of high-power FIR lasers is in the field of plasma diagnostics (Jassby *et al.,* 1974). In future deuterium–tritium plasmas in tokamak fusion reactors, ion-Thomson scattering could be the only unambiguous method of determining the ion temperature.

Powerful laser sources in the far infrared are needed for this diagnostic. For typical plasma densities and temperatures and reasonable scattering angles, the scattering is primarily caused by the collective motion of the electrons with the ions for radiation only in the FIR region of the spectrum. Although a powerful megawatt-level far-infrared laser has been developed for the first demonstration of this diagnostic on the Alcator C tokamak (Woskoboini-kow et al., 1981), the use of a tunable FIR laser would solve some of the problems associated with this diagnostic. In the Alcator C experiment, a gas-absorption cell at the receiver has been shown to operate effectively as a narrow-bandwidth stray-light filter. One difficulty lies in finding a coinci-dence between an absorption line in this gas and a powerful FIR laser line. With a tunable laser the choice of a suitable gas-absorption line would be simplified. The use of a broadly tunable source would also allow some degree of flexibility in avoiding the broad electron cyclotron-emission har-monics that extend down into the FIR and introduce noise at the receivers.

B. PREVIOUS FAR-INFRARED LASER WORK

Ever since the first demonstration of an optically pumped FIR laser by Chang and Bridges (1970), a large effort has been put into the search for other transitions that could be optically pumped to obtain FIR laser emission. Several good reviews of progress in this field have appeared (Chang, 1977; DeTemple, 1979). The many observations of new FIR laser transitions and accompanying theoretical developments will not be reviewed here, except as they apply to this specific investigation. While these many developments have produced both cw and pulsed FIR lasers at discrete and narrowly tunable (<2 GHz) frequencies (Strumia and Inguscio, 1981; Woskoboin-kow, 1981), no previous work has produced such broadly tunable high-power FIR laser emission. The theory and experimental results for a con-tinuously tunable, laser-pumped, molecular FIR laser form the main body of the investigation. The mechanism of stimulated multiphoton emission in FIR lasers as a method of efficiency enhancement will also be discussed. The organization of this discussion is as follows.

C. ORGANIZATION

A review of Raman transitions in pulsed, optically pumped FIR lasers is presented in Section II. The theory of single and double Raman resonance is discussed in both the perturbation and strong-field limits. General oper-ating principles of tunable FIR lasers are described.

Section III deals with the tunable Raman laser experiment. The develop-ment and operation of a high-pressure ($10-12$ atm) transversely excited CO_2 laser for use as the pump laser is discussed first, followed by a discussion of the far-infrared waveguide laser. The experimental method and results on *P* and *R* branch tuning in $^{12}CH_3F$ are then presented.

The prospect of laser efficiency enhancement by stimulated multiphoton emission is considered in Section IV. This includes a general-level analysis, a density-matrix treatment of the problem of two-photon emission, and a discussion of saturation and startup conditions. Several multiphoton emission experiments are suggested.

Section V provides a summary of the final results and a discussion of possible follow-up work.

II. Tunable Raman Laser Theory

The basic operating principles of pulsed, optically pumped far-infrared (FIR) lasers can be understood by considering the interaction between two laser fields and three quantum levels (Fig. 1). The population inversion is created by a pump laser rather than by an incoherent process such as electron-impact excitation. In the far infrared, the pump laser is often a transversely-excited atmospheric-pressure (TEA) CO_2 laser. The pump laser induces transitions from one rotational level in the vibrational ground state to another rotational level in an excited vibrational band. The far-infrared laser transition then occurs between two adjacent rotational levels within the excited vibrational band. Optical pumping is preferred because it is more highly selective than other forms of excitation such as electron impact, chemical, or vibrational–vibrational energy transfer. Far-infrared energies are typically less than 1 kT at room temperature. Whereas optical pumping results in a selectivity of $\sim 10^{-4}$ kT, the other commonly used mechanisms

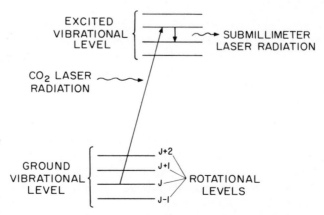

FIG. 1   Excited and ground-state levels for a submillimeter laser.

are much less selective ($\gtrsim 2$ kT), making it difficult to obtain population inversions on FIR transitions.

Although this theoretical description of tunable FIR lasers is generally applicable to any optically pumped FIR laser, the experimental work reported here involves CO_2 laser pumping of methyl fluoride (CH_3F) gas. The first optically pumped FIR laser was the CO_2-laser-pumped CH_3F laser (Chang and Bridges, 1970). The spectroscopy of the symmetric-top molecule CH_3F is well known, and the dynamics of the CH_3F laser have been studied extensively. For these reasons, when specific examples are useful, the salient features of the pulsed, optically pumped FIR laser will be discussed in terms of this laser. It should be emphasized, however, that the concept and theory of tunable Raman far-infrared lasers is not restricted to either the CH_3F laser or symmetric-top molecules.

In the CH_3F laser studied by Chang and Bridges, the 9.55-μm $P(20)$ line of the CO_2 TEA laser was used to pump the $Q(12)$, $g \rightarrow v_3$ vibrational transition in CH_3F. The ground vibrational state is denoted g. The v_3 vibrational state corresponds to the $C-F$ stretch mode of the molecule, and the notation $Q(12)$ indicates that $J = 12$ in the initial state and that the dipole transition is a $\Delta J = 0$ or Q transition.

Initially there are few molecules in the excited vibrational state; the CO_2 laser pumps molecules into the $J = 12$, v_3 state. This creates a population inversion, and the resulting FIR transition $J = 12 \rightarrow J = 11$ (R transition) produces radiation at 496.1 μm. An induced-dipole moment is responsible for the pump transition, but the FIR transition results from the presence of a permanent-dipole moment in the excited vibrational state.

The spectroscopy of symmetric-top molecules is further complicated by K-level splittings. Associated with any rotational state specified by the total angular-momentum quantum number J are two additional quantum numbers K and M. These quantum numbers specify the projection of the total angular momentum on the molecular-symmetry axis and an arbitrary space-fixed axis, respectively (Townes and Schalow, 1955). Both K and M may take the $2J + 1$ values from 0 to $\pm J$. The energy of a particular J, K, M level is given by $E/h = BJ(J + 1) + (C - B)K^2$, where B and C are molecular constants, so that the K-level splittings are twofold degenerate for $K > 0$. For a prolate symmetric top such as CH_3F, $C - B > 0$, and the energy increases with increasing K. For CH_3F, $C - B \simeq 13$ MHz. Although the energy does not depend on M, the transition-dipole moments do depend on the orientation of the molecule relative to the laser field polarization; this dependence is treated later. For the present the K and M-level substructure of a given rotational energy level is neglected, and each level is taken to be uniquely determined by the single quantum number J.

The selection rules for dipole transitions in symmetric-top molecules are

$\Delta J = 0, \pm 1$ and $\Delta K = 0$. The initial and final states also must be of opposite parity. However, in noninverting symmetric-top molecules the energy levels involved are twofold degenerate for $K \neq 0$, and states of a given energy are of mixed parity.

There are a wide variety of relaxation mechanisms important to the CH_3F laser. These are depicted in Fig. 2 and are more fully discussed elsewhere (Chang, 1977). The pump laser at frequency ω_p excites the molecule to the ν_3 vibrational manifold; the FIR laser transition at frequency ω_s occurs between adjacent rotational states. The fastest relaxation rate τ_R^{-1} results from J-changing collisions that thermalize the populations among all rotational levels within the ν_3 band. This rate is assumed to apply only to $\Delta K = 0$ collisions; the $\Delta K \neq 0$, J-changing collisions are generally slower (Oka, 1973). In CH_3F, $\tau_R p \simeq 8$ nsec · Torr for low J values, but increases for increasing J (Trappeniers and Elenbaas-Bunshoten, 1979). Thermalization of the ν_3 vibrational band with the $2\nu_3, 3\nu_3, \ldots$ bands, the ν_6 bands, which are nearly resonant with ν_3, and the ground state occurs in a characteristic time $\tau_{vv} p \simeq 2 - 9 \times 10^{-6}$ sec · Torr (Hertzberg, 1945). The relaxation mechanism that eventually returns the system to the ground state in thermal equilibrium with the surroundings is that of either vibrational-translational relaxation (V – T/R) transfer or collisions with the resonator walls. For high gas pressures and large volumes the V – T/R processes dominate, and the relaxation time is approximately $\tau_v p \simeq 1.7 \times 10^{-3}$ sec · Torr (Weitz *et al.*, 1972). De-excitation via wall collisions, which dominate for smaller-bore resonators and lower pressues, is determined by the diffusion rate to the wall. For a resonator of radius r, $\tau_v / pr \simeq 6 \times 10^{-3}$ sec/(Torr · cm) (Weitz *et al.*, 1972).

Pulsed CH_3F lasers are most often operated in the homogeneously broadened regions at pressures of 3 – 30 Torr. At $p = 10$ Torr, the different relaxation times are approximately $\tau_R \simeq 1$ nsec, $\tau_{vv} \simeq 500$ nsec, $\tau_v \simeq 200$ μsec.

A.   RAMAN TRANSITIONS IN OPTICALLY PUMPED LASERS

Theoretical treatments of the laser-pumped laser generally fall into two categories: rate-equation analyses, in which the pumping process is treated phenomenologically as an incoherent pumping rate; and semiclassical density-matrix analysis. The rate-equation approach models the dynamics of the laser-pumped laser as a two-step process. The pump-laser field induces transitions to the intermediate level in the first step; in the second step, the far-infrared (FIR) emission is produced via an uncorrelated, single-photon emission process. Although the rate-equation approach provides a thorough description of many aspects of the optically pumped laser, including the relaxation processes of Fig. 2, it is inadequate to describe two-photon coherent processes (Chang, 1977; Panock and Temkin, 1977; Temkin and

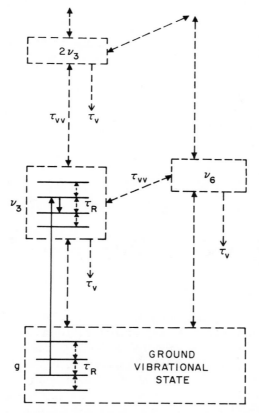

FIG. 2 Relaxation mechanisms important to the CH_3F laser.

Cohn, 1976). Since this investigation is concerned with two-photon Raman processes, the density-matrix approach is adopted.

In a Raman process, the molecule simultaneously absorbs one photon from the pump-laser field and emits one photon into the FIR laser field. There is no intermediate loss of phase as with the two-step process, and there need be no population inversion between the intermediate and the final states. Figure 3 defines the notation adopted here: $E_p(E_s)$ and $\omega_p(\omega_s)$ denote the pump (FIR signal or Stokes) field amplitude and frequency, respectively. The dipole moments connecting the states are μ_{13} and μ_{32}, and the pump and Stokes offsets from resonance are defined by $\delta_p \equiv \omega_p - \omega_{31}$ and $\delta_s \equiv \omega_s - \omega_{32}$. The β_{ij} are a measure of the field strength and are half the value of the Rabi frequency.

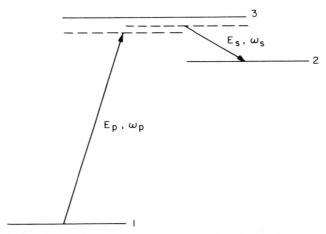

FIG. 3   Definition of notation used in a two-photon Raman process: $E_p(E_s)$ and $\omega_p(\omega_s)$ denote the pump (Stokes) field amplitude and frequency, respectively. The β_{ij} are a measure of the field strength and dipole moment: $\beta_{13} \equiv \pi|\mu_{13}|E_p/h$; $\beta_{32} \equiv \pi|\mu_{32}|E_s/h$; $\delta_p \equiv \omega_p - \omega_{31}$; $\delta_s \equiv \omega_s - \omega_{32}$.

The most important relaxation process for pulsed laser-pumped CH_3F lasers results from rotational thermalizing collisions at the rate τ_R^{-1}. For typical pump-laser pulse lengths ($t_p \simeq 100$ nsec), only the rotational levels within the ν_3 vibrational band of CH_3F have time to equilibrate. This rate is related to the homogeneous, pressure-broadened linewidth $\Delta\nu_h$ by $\tau_R = (\pi\Delta\nu_h)^{-1}$. Henceforth, τ_R, the collision time due to J-changing collisions, is represented by τ.

The quantum theory of two-photon Raman transitions was first derived by Javan (1957) in the limit of a small-signal field, $\beta_{32}^2\tau^2 \ll 1$. This result forms the foundation of the tunable Raman laser theory presented here and provides a useful review of the density-matrix method. The gain (or absorption) α at frequency ω is defined by the equation

$$dI(\omega)/dz = \alpha(\omega)I(\omega),$$

where $I(\omega)$ is the intensity of the laser field at frequency ω. For $\alpha > 0$, there is amplification as the wave travels in the z direction, whereas for $\alpha < 0$ there is absorption. The gain is related to the imaginary part of the electric susceptibility $\chi \equiv \chi' + i\chi''$ by

$$\alpha(\omega) = (4\pi\omega/c)\chi''(\omega)$$

in electrostatic units. The susceptibility $\chi''(\omega)$ is evaluated using the density-matrix formalism.

In the semiclassical approach, the molecular system is treated quantum mechanically, and the fields are treated classically.   Although a classical treatment of the fields cannot account for spontaneous emission or the statistical properties of the emitted radiation (Loudon, 1973), this method can account for most of the observed properties of far-infrared lasers.   The treatment by Panock and Temkin (1977) of this three-level problem is reviewed here.   The density-matrix equations, after ensemble averaging over collisions, can be expressed (Marcuse, 1980) as

$$dp_{mm}/dt = -\tau_{mm}^{-1}(\rho_{mm} - \rho_{mm}^{\circ}) + (i/\hbar)[\rho, H]_{mm},$$
$$dp_{mn}/dt = -\tau_{mn}^{-1}\rho_{mn} + (i/\hbar)[\rho, H]_{mn},$$
(1)

where for the three-level system (Fig. 3), m and n vary between 1 and 3.   The elements of the density matrix are related by the constraints $\rho_{mn} = \rho_{nm}^{*}$ and $\Sigma_i \rho_{ii} = 1$.   The ρ_{mm}° are the zero-field equilibrium values of the diagonal elements, i.e., the Boltzmann factors.   The rates τ_{mm}^{-1} and τ_{mn}^{-1} represent the equilibration of the molecular levels with a thermal bath at temperature T. For $\hbar\omega_{32}/kT < 1$, the relaxation processes are adequately described by interaction of the levels with this thermal bath.   For $\hbar\omega_{32}/kT > 1$, the relaxation processes are more accurately described by an additional mechanism describing the direct collisional coupling of the optically active states (Osche, 1978; Polanyi and Woodall, 1972).   The thermal-bath relaxation mechanism is adequate for room temperatue and the typical level spacings in the CH_3F molecule.   Osche (1978) has extended the results of Panock and Temkin (1977) for the case $\hbar\omega_{32}/kT > 1$.

The collision rates for off-diagonal relaxation τ_{mn}^{-1} are usually denoted T_2 rates, whereas those for the population relaxation τ_{mm}^{-1} are T_1 rates.   Provided all the collisions in a gas are J-changing collisions, the dephasing T_2 rates are equal to the T_1 rates.   The collision times are assumed equal, $T_1 = T_2 \equiv \tau$; this is a reasonable approximation for homogeneously broadened transitions, provided purely M-changing and velocity-changing collisions are not probable relative to J-changing collisions.   The M- and v-changing collisions are not important for high J, low K states in CH_3F (Shoemaker et al., 1974; Johns et al., 1975).   For other molecules, such as NH_3, $T_1 \neq T_2$.   The addition of K-changing collisions, which occur on a time scale approximately one order of magnitude longer than J-changing collisions, would not affect this approximation.   Although a K-changing collision would alter the phase of the off-diagonal elements, it would also alter the population.

The Hamiltonian is taken to be the sum of the molecular Hamiltonian H_0 and the dipole interaction term $V(t)$, as given here:

$$H = H_0 + V(t)$$
$$= H_0 - \mu \cdot \mathbf{E}(t),$$

where $H_{nm} = \langle n|H|m \rangle = E_m \delta_{mn} + V_{nm}$. The interaction matrix elements are taken to be $V_{nm} = -\mu_{nm}E(t)$, where we have assumed a linearly polarized total-electric field and evaluated the dipole moment along the space-fixed axis defined by this polarization.

The resulting set of equations for the elements of the density matrix is obtained by substitution of the Hamiltonian into Eq. (1):

$$(d/dt)r_{13} = -(r_{13} - r_{13}^\circ)\tau^{-1} + (i/\hbar)[2(\mu_{13}\rho_{31} - \mu_{31}\rho_{13})$$
$$- (\mu_{32}\rho_{23} - \mu_{23}\rho_{32})] E(t),$$

$$(d/dt)r_{32} = -(r_{32} - r_{32}^\circ)\tau^{-1} + (i/\hbar)[(\mu_{31}\rho_{13} - \mu_{13}\rho_{31})$$
$$+ 2(\mu_{32}\rho_{23} - \mu_{23}\rho_{32})] E(t); \tag{2}$$

$$(d/dt + \tau^{-1} - i\omega_{21})\rho_{12} = (i/\hbar)(\mu_{13}\rho_{32} - \mu_{32}\rho_{13}) E(t),$$

$$(d/dt + \tau^{-1} - i\omega_{31})\rho_{13} = -(i/\hbar)(\mu_{13}r_{13} + \mu_{23}\rho_{12}) E(t), \tag{3}$$

$$(d/dt + \tau^{-1} + i\omega_{32})\rho_{32} = -(i/\hbar)(\mu_{32}r_{32} - \mu_{31}\rho_{12}) E(t);$$

where $r_{ij} \equiv \rho_{ii} - \rho_{jj}$ are population differences and $\omega_{mn} \equiv (E_m - E_n)/\hbar$. The dipole moments μ_{ii} and μ_{12} are omitted from these equations; this simplification will be discussed later.

The electric field is taken to be the sum of two fields: the pump-laser field at frequency ω_p and the far-infrared emission field at frequency ω_s, as given here:

$$E(t) = \tfrac{1}{2}A_p \exp(i\omega_p t) + \tfrac{1}{2}A_s \exp(i\omega_s t) + \text{c.c.}$$

The complex amplitudes are defined by

$$A_p = E_p \exp(i\varphi_p) \quad \text{and} \quad A_s = E_s \exp(i\varphi_s),$$

where φ_p and φ_s are phase factors. Both the pump and FIR Stokes fields are assumed to be monochromatic. The interaction of a three-level system with a two-mode laser field and a multimode laser field has been discussed in the literature (Dupertuis et al., 1983a,b; Finkelstein, 1982).

To obtain a solution to the density-matrix equations, we use the standard rotating wave approximation and write

$$\rho_{13} = \lambda_{13} \exp(i\omega_p t),$$

$$\rho_{32} = \lambda_{32} \exp(-i\omega_s t), \tag{4}$$

$$\rho_{12} = \lambda_{12} \exp[i(\omega_p - \omega_s)t],$$

where the λ_{ij} are independent of time. These trial solutions are then substituted into Eqs. (2) and (3). Terms with nonresonant denominators are set to zero and the steady-state approximation is assumed. In the steady state, the populations (on-diagonal elements) have no time dependence. The time derivatives of r_{13} and r_{32} are set to zero in Eq. (2). This is valid provided the pump-laser pulse length t_p is much longer than the rotational equilibration time τ. With the trial solutions [Eq. (4)] and these simplifying approximations, solutions for the populations and the ρ_{ij} can be found (Panock and Temkin, 1977).

The imaginary part of the susceptibility is related to the induced electric polarization by $P = (\chi' + i\chi'')E$. The ensemble-averaged polarization is given by the trace of the product of the dipole-moment matrix and the density matrix: $P = Tr(\mu\rho)$. For the levels and dipole moments relevant to this problem (Fig. 3), this gives

$$P = \sum_i \mu_{ii}\rho_{ii} + 2 \, \text{Re}(\mu_{21}\rho_{12}) + 2 \, \text{Re}(\mu_{32}\rho_{23}) + 2 \, \text{Re}(\mu_{31}\rho_{13}).$$

For molecules with permanent-dipole moments, such as CH_3F, the $\mu_{ii} \neq 0$. However, because the contribution to the polarization for the $\mu_{ii}\rho_{ii}$ terms has no time dependence near the relevant frequencies ω_p and ω_s, these terms do not contribute directly toward the absorption or emission of radiation at these frequencies. Although the inclusion of the permanent-dipole moments and their resulting interaction energy could give rise to modified collision processes and enhanced polarization effects, the thermal energy and dipole-field interaction energy are much larger for reasonable gas densities. We therefore assume $\mu_{ii} = 0$.

Furthermore, μ_{12} is not necessarily zero for levels with mixed parity (because of K-level degeneracy). However, we assume that the spontaneous-emission rate is small at this frequency and that there is no field present near frequency ω_{12}. This term is then unimportant. The ensemble-averaged polarization is given by

$$P = 2 \, \text{Re}(\mu_{31}\rho_{13}) + 2 \, \text{Re}(\mu_{32}\rho_{23})$$
$$= P(\omega_p) + P(\omega_s),$$

where the contributions to the polarization at the two different frequencies have been identified from the ω_p time dependence of ρ_{13} and the ω_s time dependence of ρ_{23}.

The imaginary part of the susceptibility can also be written as the sum of two terms $\chi''(\omega) = \chi''(\omega_p) + \chi''(\omega_s)$, where $P(\omega_s) = \chi''(\omega_s)E(\omega_s)$ and $P(\omega_p) = \chi''(\omega_p)E(\omega_p)$. The real and imaginary parts of χ are the proportionality constants for the components of the polarization that are in phase and out of phase, respectively, with the driving field. For ω_s we have

$$2 \, \text{Re}[\mu_{32}\lambda_{23} \exp(i\omega_s t)] = \text{Re}(\chi' + i\chi'')A_s \exp(i\omega_s t).$$

The susceptibilities are thus related to the polarizations and matrix elements by the following equations:

$$\chi''(\omega_s) = (2/A_s) \operatorname{Im}(\mu_{32}\lambda_{23}),$$

$$\chi''(\omega_p) = (2/A_p) \operatorname{Im}(\mu_{31}\lambda_{13}),$$

$$\chi'(\omega_s) = (2/A_s) \operatorname{Re}(\mu_{32}\lambda_{23}),$$

$$\chi'(\omega_p) = (2/A_p) \operatorname{Re}(\mu_{31}\lambda_{13}).$$

In the limits of small-signal (FIR) fields, $\beta_{32}^2\tau^2 \ll 1$, the solutions are

$$\chi''(\omega_p) = -\frac{|\mu_{13}|^2}{\hbar}\tau\frac{r_{13}^\circ}{1 + \delta_p^2\tau^2 + 4\beta_{13}^2\tau^2}, \tag{5}$$

$$\chi''(\omega_s) = \frac{|\mu_{23}|^2}{4\hbar}\tau r_{13}^\circ\frac{\beta_{13}^2}{\gamma^2}\left\{\frac{1}{1 + (\gamma - \Omega)^2\tau^2} + \frac{1}{1 + (\gamma + \Omega)^2\tau^2}\right.$$

$$\left. + \frac{2(\gamma^2 - \Omega^2)\tau^2(1 + 2\gamma^2\tau^2) - 2}{[1 + 4\gamma^2\tau^2][1 + (\gamma - \Omega)^2\tau^2][1 + (\gamma + \Omega)^2\tau^2]}\right\}, \tag{6}$$

where $\gamma = \frac{1}{2}[\delta_p^2 + 4\beta_{13}^2]^{1/2}$, $\Omega = \frac{1}{2}\delta_p - \delta_s$.

The solution for the real part of the electric susceptibility χ' also follows in a similar manner from the solutions of Eqs. (2) and (3), in the same limit $(\beta_{32}^2\tau^2 \ll 1)$, as given here:

$$\chi'(\omega_p) = \frac{|\mu_{13}|^2}{\hbar}\tau r_{13}^\circ\frac{\delta_p\tau}{1 + \delta_p^2\tau^2 + 4\beta_{13}^2\tau^2},$$

$$\chi'(\omega_s) = -\frac{|\mu_{23}|^2}{4\hbar}\tau r_{13}^\circ\frac{\beta_{13}^2}{\gamma^2}\left\{\frac{(\Omega - \gamma)\tau}{1 + (\Omega - \gamma)^2\tau^2} + \frac{(\Omega + \gamma)\tau}{1 + (\Omega + \gamma)^2\tau^2}\right.$$

$$\left. + \frac{2\Omega\tau[4\gamma^2\tau^2 + (\gamma^2 - \Omega^2)\tau^2 - 1]}{(1 + 4\gamma^2\tau^2)[1 + (\Omega - \gamma)^2\tau^2][1 + (\Omega + \gamma)^2\tau^2]}\right\}.$$

The real part of the susceptibility is important in the description of effects such as self-focusing and cavity frequency pulling in laser-pumped lasers (Danly and Temkin, 1980; Siegrist *et al.*, 1978; Vass *et al.*, 1982). Solutions for $\chi''(\omega_s)$ and $\chi''(\omega_p)$ have been obtained in the strong-field limit $(\beta_{32}\tau \gtrsim 1)$ by this method and are given in the literature (Brewer and Hahn, 1975; Panock and Temkin, 1977; Takami, 1976a). The solutions for the real part of the susceptibility in the strong-field limit are given by (Danly and Temkin, 1980)

$$\chi'(\omega_s) = \frac{2\pi|\mu_{23}|^2}{h} \frac{\tau}{1 + \delta_s^2\tau^2} \left\{ r_{32}\left[\delta_s\tau - \beta_{13}^2\tau^2\left(\frac{\alpha_4}{1 + \delta_s^2\tau^2} - \frac{\alpha_5}{1 + \delta_p^2\tau^2} \right) \right] \right.$$

$$\left. - r_{12}\beta_{13}^2\tau^2 \frac{\alpha_5}{1 + \delta_p^2\tau^2} \right\},$$

$$\chi'(\omega_p) = -\frac{2\pi|\mu_{13}|^2}{h} \frac{\tau}{1 + \delta_p^2\tau^2} \left\{ r_{13}\left[\delta_p\tau - \beta_{32}^2\tau^2\left(\frac{\alpha_6}{1 + \delta_p^2\tau^2} - \frac{\alpha_5}{1 + \delta_s^2\tau^2} \right) \right] \right.$$

$$\left. + r_{12}\beta_{32}^2\tau^2 \frac{\alpha_5}{1 + \delta_s^2\tau^2} \right\},$$

where the population differences (r_{13}, r_{32}, r_{12}) are given by Eq. (8) of Panock and Temkin (1977); α_4, α_5, and α_6 are given by

$$\alpha_4 = 2\delta_s\tau R_2 + (1 - \delta_s^2\tau^2) R_1,$$

$$\alpha_5 = (\delta_p - \delta_s)\tau R_2 - (1 + \delta_p\delta_s\tau^2) R_1,$$

$$\alpha_6 = 2\delta_p\tau R_2 - (1 - \delta_p^2\tau^2) R_1,$$

where R_1 and R_2 are defined in Eq. (6) of Panock and Temkin (1977). These solutions for χ' and χ'' account for saturation and ac Stark effects in lasers with strong pump and emission fields.

The results of Eq. (6) indicate that for a weak, off-resonant pump laser, $|\delta_p\tau| \gg 1 \gg \beta_{13}\tau$, the gain spectrum has two maxima: one at line center, $\delta_s = 0$, and one at the Raman frequency given by $\delta_s = \delta_p$. For larger pump intensities, the ac Stark shift becomes important, and the maxima occur at $(\delta_s\tau)^\pm = \frac{1}{2}\delta_p\tau \pm \frac{1}{2}(\delta_p^2\tau^2 + 4\beta_{13}^2\tau^2)^{1/2}$. The Stark shifts introduce a correction to the location of the maxima near line center and the Raman frequency. In the small Stokes field limit $(\beta_{32}^2\tau^2 \ll 1)$ these maxima are of equal height, but for strong Stokes fields $(\beta_{32}\tau \gtrsim 1)$ the gain at the Raman resonance frequency is larger than that at line center (Panock and Temkin, 1977). The saturation intensity at the Raman frequency is higher than that at line center.

Experimental results for a variety of optically pumped laser systems indicate that for an off-resonant pump laser, emission at $\delta_s = \delta_p$ resulting from the Raman process dominates the line-center process at all but the lowest pressures (Fetterman et al., 1979; Rolland et al., 1983; Wiggins et al., 1978). This has been explained by noting that although emission at both the Raman and line-center frequencies is initially present, the gain at the line-center frequency saturates earlier than that at the Raman frequency. Consequently, the Raman emission at $\delta_s = \delta_p$ dominates the output. Dupertuis et al. (1983b, 1984) have considered the case of a multimode FIR field and have concluded that the observed preference of the Raman line is the result of an inversion of the line center gain for a moderate to high intensity Raman mode.

The importance of the ac Stark effect in determining the emission frequency of optically pumped FIR lasers is still under investigation. It has been observed in experimental studies of the gain spectrum of the 385-μm D_2O laser (Drozdowicz, 1978; Drozdowicz *et al.*, 1979b). It has not been observed in several other Raman tuning experiments, even with large pump intensities (DeMartino *et al.*, 1978; Fetterman *et al.*, 1979; Frey *et al.*, 1977; Woskoboinikow, pers. comm.). This may result from the temporal profile of the pump-laser pulse. The FIR emission is initiated very early during the pump pulse at a frequency determined by the Raman resonance condition. At the time the FIR emission is initialized, the pump intensity is low and the ac Stark shift is small. Consequently, the emission field grows at the frequency determined by the condition $\delta_s = \delta_p$. For the present, the FIR laser emission is assumed to occur at the frequency given by the Raman resonance condition.

B.   DOUBLE RAMAN PROCESSES

Optically pumped lasers operating via the two-photon Raman process have a number of advantages over lasers operating via single-photon processes. The Raman process has a maximum quantum efficiency twice that of the ordinary two-step absorption–emission process (Panock and Temkin, 1977). The higher efficiency of the Raman process occurs because the intermediate level is not occupied; for the two-step line-center emission process, a population inversion between the intermediate and final states must be created. As previously noted, the Raman process saturates at a higher intensity than the line-center single-photon process. However, by far the most important advantage of Raman transitions is that tunable pump-laser radiation in one frequency band can be downconverted into tunable radiation in a different frequency band.

This method of generating tunable radiation is not new to nonlinear optics. The technique applies equally well in principle to different regions of the spectrum, from the ultraviolet to the far infrared. In practice, the range of applicability is limited to some extent by the availability of tunable pump lasers and appropriate quantum systems.

In the visible and near-infrared region of the spectrum, hot metal vapors, which exhibit very large susceptibilities near resonance, have been used in conjunction with tunable dye lasers to produce tunable radiation (Sorokin *et al.*, 1974; Wynne and Sorokin, 1977). Stimulated Raman scattering in hydrogen has produced tunable radiation in the 0.7–7.0 μm wavelength range (Berry *et al.*, 1982; Hartig and Schmidt, 1979).

One of the first extensive uses of the Raman process for the generation of broadly tunable radiation in the near and far inrared was carried out by Frey, Pradere, DeMartino, and Ducuing for diatomic molecules (DeMartino *et al.*, 1978; DeMartino *et al.*, 1980; Frey *et al.*, 1977). They used a high-power (400 – 1000 MW) broadly tunable (0.72 – 1.09 μm) ruby-pumped dye-laser system as the pump laser. The output of this system was Raman shifted in gaseous H_2 to produce 1 – 200 MW powers in the 1 – 10 μm wavelength range. This 1 – 10 μm emission was then Raman shifted a second time in liquid N_2 to produce emission in the 10 – 18 μm region, which was then Raman shifted a third time in gaseous HF into the 40 – 250 μm region of the spectrum. Gaseous HCl was also employed as a scattering medium for dye-laser radiation, which was Raman shifted twice in H_2.

It is by now well known that many CO_2 laser-pumped FIR lasers operate via Raman processes (Fetterman *et al.*, 1979; Wiggins *et al.*, 1978). Frequency tuning of $\sim \pm 1.5$ GHz has been obtained on the 385-μm D_2O laser transition, pumped by the 9.26-μm R_{22} line of the CO_2 laser (Fetterman, 1979; Woskoboinokow, pers. comm.). The pump laser in this case was an atmosphoric-pressure TEA laser, tunable $\sim \pm 1.5$ GHz about the $9R22$ line center.

Biron *et al.* (1981) have pumped a CH_3F laser with a line-tunable TEA CO_2 laser and obtained far-infrared Raman emission on 25 transitions in $^{12}CH_3F$ and 23 transitions in $^{13}CH_3F$. Although their output was not tunable, their work proved to be an important step toward the goal of a continuously tunable FIR laser. Their results demonstrated that through the use of intense, highly focused pump beams, Raman emission is obtainable on transitions that are as much as 30 GHz from resonance with the excited state. The work of Biron *et al.* on Raman generation in polyatomic molecular gases is extended in this investigation. A high-pressure (10 – 12 atm) continously tunable CO_2 TE laser is used to pump a CH_3F waveguide Raman laser. The theory of frequency tuning in these lasers will be reviewed and then compared with experimental results from the tunable FIR Raman laser.

Prior to the work of Biron *et al.* and DeMartino *et al.*, the largest pump offset δ_p for which laser emission in the far infrared had been observed was approximately ± 2 GHz for most gases. Wiggins *et al.* (1978) observed Raman emission in NH_3 that was off-resonance by 5 GHz. For these values of small pump offset, no other molecular levels are near resonance, and the three-level treatment can adequately describe the problem. However, for large pump offsets, a second process becomes important. This process involves the absorption and emission of pump and Stokes photons of the same frequency as for the ordinary three-level process, but in this case the Stokes photon is emitted in the ground vibrational state. These two processes are shown in Fig. 4. Because both processes begin at level 1 and end at

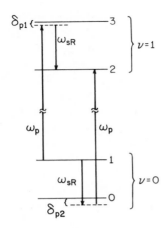

FIG. 4 A diagram of the double Raman process for a tunable far-infrared laser.

level 2, they both contribute to the gain at the frequency ω_{sR}, which is the emission frequency at the Raman resonance condition. There are now two pump offsets, $\delta_{p1} \equiv \omega_p - \omega_{31}$ and $\delta_{p2} \equiv \omega_p - \omega_{20}$, and two signal or Stokes offsets, $\delta_{s1} = \omega_s - \omega_{32}$ and $\delta_{s2} = \omega_s - \omega_{10}$.

For large excited-state pump offsets δ_{p1}, the ground-state pump offset δ_{p2} can approach zero, and the ground-state Raman emission process can contribute significantly to the gain. In fact, the interference of the two processes shown in Fig. 4 causes a large asymmetry in the frequency-tuning curves. Consequently, this double Raman process is crucial to a complete theory of the tunable FIR Raman laser.

The perturbation theory for the double Raman resonance has been derived and discussed with regard to Raman emission in HF and HCl (Ducuing *et al.*, 1976; Frey *et al.*, 1977) and in CH_3F (Biron, 1980; Biron *et al.*, 1981). The growth of the Stokes wave at frequency ω_s is given by

$$dI_s/dz = \hbar\omega_s r^\circ_{12} R_{1\rightarrow 2},$$

where $R_{1\rightarrow 2}$ is the transition rate for the two Raman processes contributing to gain at ω_s. The rate can be calculated from standard second-order perturbation theory (Loudon, 1973), wherein the interaction Hamiltonian has no time dependence but involves photon creation and annihilation operators at ω_p and ω_s. The perturbation theory result is given by

$$R_{1\rightarrow 2} = \frac{2\pi E_p^2 E_s^2}{\hbar^4} \left| \frac{\mu_{23}\mu_{31}}{\delta_{p1}} - \frac{\mu_{20}\mu_{01}}{\delta_{p2}} \right|^2 \delta(\omega_p - \omega_s - \omega_{21}).$$

When the energy delta function is replaced by a collision-broadened resonance function, the Raman gain at ω_s can be written

$$\alpha(\omega_s) = \frac{8\pi^2\omega_s\tau r_{12}^\circ}{c^2\hbar^3}\left|\frac{\mu_{23}\mu_{31}}{\delta_{p1}} - \frac{\mu_{20}\mu_{01}}{\delta_{p2}}\right|^2 \frac{I_p}{1 + (\delta_{p1} - \delta_{s1})^2\tau^2}, \qquad (7)$$

where $\delta_{p1} - \delta_{s1} = \delta_{p2} - \delta_{s2}$. Although this solution is valid only in the perturbation limit for nonresonant pumping, it provides valuable insight into
many of the properties of tunable Raman lasers. The gain is proportional to
the pump intensity and inversely proportional to the square of the pump
offsets. The contribution to the gain from the first term inside the absolute
value signs results from the excited-state Raman emission process; the contribution from the second term results from the ground-state Raman emis-
sion process. The cross term is proportional to $(\delta_{p1}\delta_{p2})^{-1}$ and arises from the
interference of the two Raman processes. For large values of either pump
offset, Eq. (7) reduces to the standard three-level result.

The gain caused by the double Raman resonant process shown in Fig. 4
can be calculated for arbitrary pump and FIR field strengths using the
density-matrix approach (Biron, 1980; Biron et al., 1981). The same as-
sumptions discussed previously for the three-level system are made, and the
only nonzero dipole moments are taken to be $\mu_{31}, \mu_{32}, \mu_{02}, \mu_{10}$. The density-matrix equations for the off-diagonal elements of the four-level problem
can be written

$$[d/dt + \tau^{-1} - i\omega_{21}]\rho_{12} = -(i/\hbar)[\mu_{32}\rho_{13} - \mu_{13}\rho_{32} + \mu_{02}\rho_{10} - \mu_{10}\rho_{02}]\,E(t),$$

$$[d/dt + \tau^{-1} - i\omega_{31}]\rho_{13} = -(i/\hbar)[\mu_{13}r_{13} + \mu_{23}\rho_{12} - \mu_{10}\rho_{03}]\,E(t),$$

$$[d/dt + \tau^{-1} + i\omega_{32}]\rho_{32} = -(i/\hbar)[\mu_{32}r_{32} - \mu_{31}\rho_{12} + \mu_{02}\rho_{30}]\,E(t),$$

$$[d/dt + \tau^{-1} + i\omega_{10}]\rho_{10} = -(i/\hbar)[\mu_{10}r_{10} + \mu_{20}\rho_{12} - \mu_{13}\rho_{30}]\,E(t), \qquad (8)$$

$$[d/dt + \tau^{-1} - i\omega_{20}]\rho_{02} = -(i/\hbar)[\mu_{02}r_{02} + \mu_{32}\rho_{03} - \mu_{01}\rho_{12}]\,E(t),$$

$$[d/dt + \tau^{-1} - i\omega_{30}]\rho_{03} = -(i/\hbar)[\mu_{13}\rho_{01} + \mu_{23}\rho_{02} - \mu_{01}\rho_{13} - \mu_{02}\rho_{23}]\,E(t),$$

where, as before, $r_{ij} \equiv \rho_{ii} - \rho_{jj}$ and $\omega_{ij} \equiv (E_i - E_j)/\hbar$. The equations for the
diagonal elements are written in terms of these population differences r_{ij}:

$$\frac{d}{dt}r_{13} = -(r_{13} - r_{13}^\circ)\,\tau^{-1} + \frac{i}{\hbar}[2(\mu_{13}\rho_{31} - \mu_{31}\rho_{13}) - (\mu_{32}\rho_{23} - \mu_{23}\rho_{32})$$

$$+ (\mu_{10}\rho_{01} - \mu_{01}\rho_{10})]\,E(t),$$

$$\frac{d}{dt}r_{32} = -(r_{32} - r_{32}^\circ)\,\tau^{-1} + \frac{i}{\hbar}[(\mu_{31}\rho_{13} - \mu_{13}\rho_{31}) + (\mu_{02}\rho_{20} - \mu_{20}\rho_{02})$$

$$+ 2(\mu_{32}\rho_{23} - \mu_{23}\rho_{32})]\,E(t),$$

$$\frac{d}{dt} r_{10} = -(r_{10} - r_{10}^\circ)\, \tau^{-1} + \frac{i}{\hbar} [(\mu_{13}\rho_{31} - \mu_{31}\rho_{13}) + 2(\mu_{10}\rho_{01} - \mu_{01}\rho_{10}) \tag{9}$$

$$+ (\mu_{20}\rho_{02} - \mu_{02}\rho_{20})]\, E(t),$$

$$\frac{d}{dt} r_{02} = -(r_{02} - r_{02}^\circ)\, \tau^{-1} + \frac{i}{\hbar} [(\mu_{32}\rho_{23} - \mu_{23}\rho_{32}) + 2(\mu_{02}\rho_{20} - \mu_{20}\rho_{02})$$

$$+ (\mu_{01}\rho_{10} - \mu_{10}\rho_{01})]\, E(t).$$

These equations reduce to the density-matrix equations for the three-level problem, provided all the terms with a subscript 0 are set to zero.

The electric field is taken to be the sum of two fields at ω_p and ω_s:

$$E(t) = \tfrac{1}{2} A_p \exp(i\omega_p t) + \tfrac{1}{2} A_s \exp(i\omega_s t) + \text{c.c.},$$

where the complex field amplitude is $A_i = E_i \exp(i\varphi_i)$. The rotating-wave approximation is employed and the off-diagonal density-matrix elements are written as

$$\begin{aligned}
\rho_{13} &= \lambda_{13} \exp(i\omega_p t), & \rho_{02} &= \lambda_{02} \exp(i\omega_p t), \\
\rho_{32} &= \lambda_{32} \exp(-i\omega_s t), & \rho_{10} &= \lambda_{10} \exp(-i\omega_s t), & (10) \\
\rho_{12} &= \lambda_{12} \exp[i(\omega_p - \omega_s)t], & \rho_{03} &= \lambda_{03} \exp[i(\omega_p + \omega_s)t],
\end{aligned}$$

where the λ_{ij} are time independent and $\lambda_{ij} = \lambda_{ji}^*$. The field amplitude parameters β_{ij} are now defined by

$$\beta_{13} = \frac{\mu_{13} A_p}{2\hbar}, \qquad \beta_{02} = \frac{\mu_{02} A_p}{2\hbar}, \qquad \beta_{23} = \frac{\mu_{23} A_s}{2\hbar}, \qquad \beta_{01} = \frac{\mu_{01} A_s}{2\hbar},$$

where $\beta_{ij}^* = \beta_{ji}$ and $\mu_{ij}^* = \mu_{ji}$.

When the ansatz given by Eq. (10) is substituted into Eq. (8), only terms with the same time dependence are retained; terms with nonresonant denominators are neglected. This results in a set of six linear equations in six unknowns for the off-diagonal elements, given here:

$$\begin{aligned}
L_{32}\lambda_{32} &= \beta_{32}r_{32} - \beta_{31}\lambda_{12} + \beta_{02}\lambda_{30}, \\
L_{10}\lambda_{10} &= \beta_{10}r_{10} + \beta_{20}\lambda_{12} - \beta_{13}\lambda_{30}, \\
L_{12}\lambda_{12} &= \beta_{02}\lambda_{10} - \beta_{13}\lambda_{32} + \beta_{32}\lambda_{13} - \beta_{10}\lambda_{02}, \\
L_{03}\lambda_{03} &= \beta_{13}\lambda_{01} - \beta_{02}\lambda_{23} - \beta_{01}\lambda_{13} + \beta_{23}\lambda_{02}, \qquad (11) \\
L_{13}\lambda_{13} &= \beta_{13}r_{13} + \beta_{23}\lambda_{12} - \beta_{10}\lambda_{03}, \\
L_{02}\lambda_{02} &= \beta_{02}r_{02} - \beta_{01}\lambda_{12} + \beta_{32}\lambda_{03}.
\end{aligned}$$

The resonance functions are defined by

$$
\begin{aligned}
L_{13} &= -\delta_{p1} + i\tau^{-1}, & L_{10} &= \delta_{s2} + i\tau^{-1}, \\
L_{12} &= \delta_{s1} - \delta_{p1} + i\tau^{-1}, & L_{02} &= -\delta_{p2} + i\tau^{-1}, \quad (12)\\
L_{32} &= \delta_{s1} + i\tau^{-1}, & L_{03} &= -\delta_{s1} - \delta_{p2} + i\tau^{-1},
\end{aligned}
$$

where the detuning parameters are defined by

$$
\begin{aligned}
\delta_{p1} &= \omega_p - \omega_{31}, & \delta_{p2} &= \omega_p - \omega_{20}, \\
\delta_{s1} &= \omega_s - \omega_{32}, & \delta_{s2} &= \omega_s - \omega_{10}.
\end{aligned}
$$

The imaginary part of the electric susceptibility near ω_s is given by

$$
\chi''(\omega_s) = -(2/A_s^*)\,\mathrm{Im}(\mu_{23}\lambda_{32} + \mu_{01}\lambda_{10}).
$$

Substitution of the equations for λ_{32} and λ_{10} [Eq. (11)] into the expression for $\chi''(\omega_s)$ yields

$$
\begin{aligned}
\chi''(\omega_s) = &\frac{|\mu_{23}|^2}{\hbar}\frac{r_{32}\tau}{(1+\delta_{s1}^2\tau^2)} + \frac{|\mu_{01}|^2}{\hbar}\frac{r_{10}\tau}{(1+\delta_{s2}^2\tau^2)}\\
&+ \frac{2}{A_s^*}\,\mathrm{Im}\left[\lambda_{12}\left(\frac{\beta_{20}\mu_{01}}{L_{10}} - \frac{\beta_{31}\mu_{23}}{L_{32}}\right)\right]\\
&+ \frac{2}{A_s^*}\,\mathrm{Im}\left[\lambda_{30}\left(\frac{\beta_{13}\mu_{01}}{L_{10}} - \frac{\beta_{02}\mu_{23}}{L_{32}}\right)\right]. \quad (13)
\end{aligned}
$$

The first two terms of Eq. (13) represent the single-photon contribution to the gain at ω_s. The term proportional to λ_{12} represents the coherent two-photon contribution to the gain. The term proportional to λ_{30} represents a coherent two-photon process that transfers molecules from level 0 to level 3 by the simultaneous absorption of one ω_s photon and one ω_p photon. When this two-photon absorption is resonant, the gain will be reduced by this term. At this point we assume two-photon absorption is not resonant and set $\lambda_{30} = 0$. The qualitative effect of two-photon absorption will be discussed in Section II.D. For L_{03} nonresonant, the ac Stark shifts caused by the λ_{30} term are small compared with the other shifts.

The solution for λ_{12} can now be obtained by simple substitution and Eq. (13) becomes

$$
\begin{aligned}
\tfrac{1}{2}A_s^*\chi''(\omega_s) = &\frac{\mu_{23}}{L_{32}}\left[r_{32} - \frac{\beta_{13}^2}{L_{12}'}\left(\frac{r_{13}}{L_{13}} - \frac{r_{32}}{L_{32}}\right)\right]\beta_{32}\\
&+ \frac{\mu_{01}}{L_{10}}\left[r_{10} - \frac{\beta_{02}^2}{L_{12}'}\left(\frac{r_{10}}{L_{10}} - \frac{r_{02}}{L_{02}}\right)\right]\beta_{10}
\end{aligned}
$$

$$+ \frac{\mu_{01}\beta_{13}\beta_{20}}{L_{10}L'_{12}} \left(\frac{r_{13}}{L_{13}} - \frac{r_{32}}{L_{32}} \right) \beta_{32}$$

$$- \frac{\mu_{23}\beta_{02}\beta_{31}}{L_{32}L'_{12}} \left(\frac{r_{10}}{L_{10}} - \frac{r_{02}}{L_{02}} \right) \beta_{10}, \tag{14}$$

where the resonance L'_{12} is given by

$$L'_{12} = L_{12}\left[1 - \frac{|\beta_{23}|^2}{L_{12}L_{13}} - \frac{|\beta_{02}|^2}{L_{12}L_{10}} - \frac{|\beta_{13}|^2}{L_{12}L_{32}} - \frac{|\beta_{10}|^2}{L_{12}L_{02}} \right].$$

The first two terms of Eq. (14) are of the same form as Eq. (5a) in the analysis by Panock and Temkin (1977). Writing $\chi''(\omega_s)$ in Eq. (14) as $\chi''_1 + \chi''_2 + \chi''_3 + \chi''_4$, Biron (1981) has obtained a solution for $\chi''(\omega_s)$, as follows:

$$\chi''(\omega_s) = \chi''_1 + \chi''_2 + \chi''_3 + \chi''_4; \tag{15}$$

where

$$\chi''_1(\omega_s) = \frac{|\mu_{23}|^2\tau}{\hbar(1 + \delta_{s1}^2\tau^2)}\left[r_{32}\left(1 + \frac{\beta_{13}^2\tau^2\alpha_3}{1 + \delta_{s1}^2\tau^2} \right) + \frac{r_{13}\beta_{13}^2\tau^2\alpha_2}{1 + \delta_{p1}^2\tau^2} \right],$$

$$\chi''_2(\omega_s) = -\frac{|\mu_{01}|^2\tau}{\hbar(1 + \delta_{s2}^2\tau^2)}\left[r_{01}\left(1 + \frac{\beta_{02}^2\tau^2\alpha'_3}{1 + \delta_{s2}^2\tau^2} \right) + \frac{r_{20}\beta_{02}^2\tau^2\alpha'_2}{1 + \delta_{p2}^2\tau^2} \right],$$

$$\chi''_3(\omega_s) = -\frac{|\mu_{01}|^2\beta_{13}\beta_{20}\tau}{\hbar(1 + \delta_{s2}^2\tau^2)}\left[\frac{r_{32}\gamma_3}{1 + \delta_{s1}^2\tau^2} + \frac{r_{13}\gamma_2}{1 + \delta_{p1}^2\tau^2} \right],$$

$$\chi''_4(\omega_s) = -\frac{|\mu_{23}|^2\beta_{02}\beta_{31}\tau}{\hbar(1 + \delta_{s1}^2\tau^2)}\left[\frac{r_{10}\gamma_3}{1 + \delta_{s2}^2\tau^2} + \frac{r_{02}\gamma'_2}{1 + \delta_{p2}^2\tau^2} \right].$$

The following definitions have been made:

$$\alpha_1 = 2\delta_{p1}\tau R_1 + (1 - \delta_{p1}^2\tau^2)\,R_2$$

$$\alpha_2 = (\delta_{p1} - \delta_{s1})\,\tau R_1 + (1 + \delta_{p1}\delta_{s1}\tau^2)\,R_2,$$

$$\alpha_3 = 2\delta_{s1}\tau R_1 - (1 - \delta_{s1}^2\tau^2)\,R_2,$$

$$\alpha'_1 = 2\delta_{p2}\tau R_1 + (1 - \delta_{p2}^2\tau^2)\,R_2,$$

$$\alpha'_2 = (\delta_{p2} - \delta_{s2})\,\tau R_1 + (1 + \delta_{p2}\delta_{s2}\tau^2)\,R_2,$$

$$\alpha'_3 = 2\delta_{s2}\tau R_1 - (1 - \delta_{s2}^2\tau^2)\,R_2,$$

$$\gamma_1 = (\delta_{p1} + \delta_{p2})\,\tau R_1 + (1 - \delta_{p1}\delta_{p2}\tau^2)\,R_2,$$

$$\gamma_2 = (\delta_{p1} - \delta_{s2})\,\tau R_1 + (1 + \delta_{p1}\delta_{s2}\tau^2)\,R_2,$$

$$\gamma'_2 = (\delta_{p2} - \delta_{s1})\,\tau R_1 + (1 + \delta_{p2}\delta_{s1}\tau^2)\,R_2,$$

$$\gamma_3 = (\delta_{s1} + \delta_{s2})\,\tau R_1 - (1 - \delta_{s1}\delta_{s2}\tau^2)\,R_2.$$

The constants R_1 and R_2 are related to the L'_{12} resonance by $1/L'_{12} = (R_1 - iR_2)\tau$; R_1 and R_2 are given by $R_1 = A/(A^2 + B^2)$, $R_2 = B/(A^2 + B^2)$; and A and B are given by

$$A = (\delta_{s1} - \delta_{p1})\,\tau + \frac{\beta_{23}^2\tau^2\delta_{p1}\tau}{1 + \delta_{p1}^2\tau^2} - \frac{\beta_{13}^2\tau^2\delta_{s1}\tau}{1 + \delta_{s1}^2\tau^2}$$

$$+ \frac{\beta_{01}^2\tau^2\delta_{p2}\tau}{1 + \delta_{p2}^2\tau^2} - \frac{\beta_{02}^2\tau^2\delta_{s2}\tau}{1 + \delta_{s2}^2\tau^2},$$

$$B = 1 + \frac{\beta_{01}^2\tau^2}{1 + \delta_{p2}^2\tau^2} + \frac{\beta_{13}^2\tau^2}{1 + \delta_{s1}^2\tau^2} + \frac{\beta_{02}^2\tau^2}{1 + \delta_{s2}^2\tau^2} + \frac{\beta_{23}^2\tau^2}{1 + \delta_{p1}^2\tau^2}.$$

Equation (15) for $\chi''(\omega_s)$ describes the gain for arbitrary field strengths in terms of the equilibrium population differences r_{ij}. The populations are next determined by substitution of the solutions for λ_{ij} into Eqs. (9). At this point it is assumed that the emission field is weak ($\beta_{01}^2\tau^2 \ll 1$, $\beta_{23}^2\tau^2 \ll 1$) and does not affect the populations. The r_{ij} are then given in terms of the zero-field values r_{ij}° by

$$r_{13} = r_{13}^\circ \frac{1 + \delta_{p1}^2\tau^2}{1 + \delta_{p1}^2\tau^2 + 4\beta_{13}^2\tau^2},$$

$$r_{32} = r_{32}^\circ + r_{13}^\circ \frac{2\beta_{13}^2\tau^2}{1 + \delta_{p1}^2\tau^2 + 4\beta_{13}^2\tau^2} - r_{02}^\circ \frac{2\beta_{02}^2\tau^2}{1 + \delta_{p2}^2\tau^2 + 4\beta_{02}^2\tau^2},$$

$$r_{02} = r_{02}^\circ \frac{1 + \delta_{p2}^2\tau^2}{1 + \delta_{p2}^2\tau^2 + 4\beta_{02}^2\tau^2}, \tag{16}$$

$$r_{10} = r_{10}^\circ - r_{13}^\circ \frac{2\beta_{13}^2\tau^2}{1 + \delta_{p1}^2\tau^2 + 4\beta_{13}^2\tau^2} + r_{02}^\circ \frac{2\beta_{02}^2\tau^2}{1 + \delta_{p2}^2\tau^2 + 4\beta_{02}^2\tau^2}.$$

The solution for $\chi''(\omega_s)$ with the above expression for the populations is valid for arbitrary pump intensity and small emission fields, $\beta_{01}^2\tau^2 \ll 1$, $\beta_{23}^2\tau^2 \ll 1$. However, because the frequency of the emission field is determined by the small-signal gain spectrum, this solution [Eqs. (15) and (16)] adequately describes most aspects of a single-pass Raman laser.

For pump offsets large compared with the Rabi frequency, $|\delta_{p1}| \gg \beta_{13}$, $|\delta_{p2}| \gg \beta_{02}$, the gain near the Raman resonance $\delta_{s1} = \delta_{p1}$ is given by $\alpha(\omega_s) = 4\pi\omega_s\chi''(\omega_s)/c$ with

$$\chi''(\omega_s) = -\frac{|\mu_{01}|^2\tau r_{01}^\circ}{\hbar(1 + \delta_{p2}^2\tau^2)} - \frac{|\mu_{23}|^2\tau r_{23}^\circ}{\hbar(1 + \delta_{p1}^2\tau^2)} + \frac{\tau r_{12}^\circ E_p^2}{4\hbar^3[1 + (\delta_{p1} - \delta_{s1})^2\tau^2]}$$

$$\times \left[\frac{|\mu_{13}|^2|\mu_{32}|^2}{\delta_{p1}^2} + \frac{|\mu_{02}|^2|\mu_{10}|^2}{\delta_{p2}^2} - \frac{\mu_{10}\mu_{20}\mu_{32}\mu_{13}}{\delta_{p1}\delta_{p2}} - \frac{\mu_{23}\mu_{01}\mu_{02}\mu_{31}}{\delta_{p1}\delta_{p2}} \right], \tag{17}$$

where we have used the relation $\delta_{p1} - \delta_{s1} = \delta_{p2} - \delta_{s2}$. The first term corresponds to single-photon absorption in the ground state, and the second term corresponds to single-photon absorption in the excited state. We set $r_{32}^\circ = 0$ and assume $\delta_{p2}^2 \tau^2 \gg 1$. The remaining term in Eq. (17) represents the Raman gain at $\omega_{sR} \equiv \delta_p + \omega_{32}$ resulting from the double Raman emission process. This expression is equivalent to that obtained by the perturbation theory [Eq. (7)] and is valid for arbitrary pump intensity.

The approximation $|\delta_{p1}| \gg \beta_{13}$ and $|\delta_{p2}| \gg \beta_{02}$ must be justified by comparison with real optically pumped lasers. For the CO_2-laser-pumped CH_3F laser, the important dimensionless parameters, $\delta_{p1}\tau, \beta_{13}\tau, \delta_{p2}\tau, \beta_{02}\tau$, can be calculated as follows. These parameters do not vary greatly with J; they are evaluated here for a $J = 20$, R-pump transition.

The rotational-collision time for CH_3F is given by

$$\tau = (\pi \Delta \nu_h)^{-1} = 10.6 \text{ nsec}/p \text{ [Torr]},$$

where $\Delta \nu_h \simeq 30$ MHz/Torr is the homogeneous linewidth appropriate for $J = 20$ (Trappeniers and Elenbaas-Bunshoten, 1979). For pump offsets of $\delta_p'(= \delta_p/2\pi)$, the pump detuning parameter is

$$\delta_p \tau = 66.6\delta_p'/p$$

for δ_p' in gigahertz and p in torr. At a typical pressure $p = 10$ Torr, $\delta_p^2\tau^2 > 1$ for $\delta_p' > 0.15$ GHz.

The pump field-amplitude parameters ($\beta_{13} \simeq \beta_{02}$) are obtained by spatially averaging the dipole moment:

$$\beta_{13av}^2 = \frac{|\mu_{13}|_{av}^2 E_p^2}{4\hbar^2} = \frac{2\pi}{\hbar^2 c} \left[\frac{\mu_0^2[(J+1)^2 - K^2]}{3(2J+1)(J+1)} \right] I_p.$$

In this equation, μ_0 is the induced dipole moment. For $J \simeq 20$ and low K, $|\mu_{13}|_{av}^2 \simeq \mu_0^2/6$, this equation reduces to

$$\beta_{13av}^2\tau^2 = 1.5 \times 10^3 I_p \text{ [MW/cm}^2\text{]}/p^2 \text{[Torr]}^2.$$

Typical pump intensities used in the experimental work discussed here are $I_p = 1-4$ MW/cm². This yields $\beta_{13}^2\tau^2 \simeq 15-60$ for a pressure of 10 Torr. Consequently, the condition $|\delta_{p1}| \gg \beta_{13}, |\delta_{p2}| \gg \beta_{02}$ is fulfilled only when $|\delta_{p1}'|, |\delta_{p2}'| > 1.2$ GHz. Equation (17) is strictly valid only for pump fields off-resonance by more than this amount.

The far-infrared Rabi frequency can be calculated in a similar manner. For an $R(20)$ pump transition, the Stokes field-amplitude parameters ($\beta_{32} \simeq \beta_{01}$) are obtained by spatially averaging the dipole moment μ_{32}, as given here:

$$\beta_{32av}^2 = \frac{|\mu_{32}|_{av}^2 E_s^2}{4\hbar^2} = \frac{2\pi}{\hbar^2 c} \left(\frac{\mu_0^2}{6} \right) I_s.$$

In this expression, μ_0 is the permanent-dipole moment of CH_3F. The Stokes field parameter $\beta_{32}^2\tau^2$ is then given by

$$\beta_{32av}^2\tau^2 = 0.13I_s \, [\text{W/cm}^2]/p^2 \, [\text{Torr}]^2.$$

This yields an FIR saturation intensity of approximately 192 W/cm^2 at a pressure of 10 Torr.

To apply Eq. (17) to a real symmetric-top molecule, the K- and M-level substructure must be included. In a real gas the molecules are randomly oriented in space and the strength of their interaction with the laser fields will be angle dependent. This dependence can be incorporated into the theory by making the interaction matrix elements $\mu_{fi} \equiv \langle f|\vec{\mu} \cdot \hat{e}|i \rangle$ dependent on the M quantum numbers. This method has been discussed by several authors (Biron et al., 1981; Drozdowicz et al., 1979a).

Because of the selection rule $\Delta K = 0$ and the assumption that K-changing collisions are unimportant, the contributions to the total Raman gain from each K level can be treated separately. Provided the different M-level subsystems do not interact with each other, we can write the total response as a sum over the contribution from each M subsystem. For noninverting symmetric-top molecules, the $K \neq 0$ levels are doubly degenerate and of mixed parity. Although this makes M-changing collisions dipole allowed, we assume that these collisions occur on a time scale slower than the J-changing collision (Johns et al., 1975). Then the total gain for level (J, K) can be written

$$\alpha(J, K) = \sum_M \alpha(J, K, M), \tag{18}$$

where the dependence on K and M in $\alpha(J, K, M)$ is through the dipole transition matrix elements

$$\langle J', K', M'|\mu \cdot \hat{e}|J, K, M \rangle = \mu^\circ \phi_{JJ}\phi_{JKJ'K'}\phi_{JMJ'M'}. \tag{19}$$

The ϕ are the direction-cosine matrix elements tabulated in the literature (Table 4-4 in Townes and Schalow, 1955) and $\hat{e}$ is the electric-field-polarization vector that defines the spaced-fixed axis associated with the M quantum numbers.

For arbitrary field intensities, the summation in Eq. (18) must be carried out numerically. However, for Eq. (17) it is possible to perform this sum analytically. Neglecting the ground and excited-state absorptions in Eq. (17), the gain for level J, K at the frequency $\delta_{s1} = \delta_{p1}$ is given by

$$\alpha(J, K) = \sum_M \frac{8\pi^2\omega_s\tau N_0}{c^2\hbar^3}I_p f_{JKM}\left[\frac{|\mu_{13}|^2|\mu_{32}|^2}{\delta_{\delta 1}^2} + \frac{|\mu_{02}|^2|\mu_{10}|^2}{\delta_{p2}^2}\right.$$
$$\left. - \frac{2\,\text{Re}(\mu_{10}\mu_{20}\mu_{32}\mu_{13})}{\delta_{p1}\delta_{p2}}\right], \tag{20}$$

TABLE I

GAIN COEFFICIENTS FOR SYMMETRIC-TOP MOLECULES

Transition branch	$A(J, K)$	$B(J, K)$	$C(J, K)$
$R(\parallel)$	$\dfrac{[(J+1)^2 - K^2]^2(4J^2 + 8J + 5)}{15(J+1)^3(2J+1)^2(2J+3)}$	$\dfrac{(J^2 - K^2)^2(4J^2 + 1)}{15J^3(2J+1)^2(2J-1)}$	$\dfrac{4(J^2 - K^2)[(J+1)^2 - K^2]}{15J(J+1)(2J+1)^2}$
$Q(\perp)$	$\dfrac{(J^2 - K^2)K^2(4J+1)}{30J^3(J+1)(2J+1)}$	$\dfrac{(J^2 - K^2)K^2(4J-1)}{30J^3(2J+1)(J-1)}$	$\dfrac{(J^2 - K^2)K^2(4J+1)}{15J^3(J-1)(2J+1)}$
$P(\parallel)$	$\dfrac{2(J^2 - K^2)[(J-1)^2 - K^2]}{15J(2J+1)(J-1)(2J-1)}$	$\dfrac{2(J^2 - K^2)[(J-1)^2 - K^2]}{15J(2J+1)(J-1)(2J-1)}$	$\dfrac{4(J^2 - K^2)[(J-1)^2 - K^2]}{15J(2J+1)(J-1)(2J-1)}$

where $I_p = cE_p^2/8\pi$ is the pump intensity and $r_{12}^\circ = N_0 f_{JKM}$. N_0 is the number density and f_{JKM} is the Boltzmann factor for level J, K, M ($f_{JK} = (2J + 1)f_{JKM}$). Each transition matrix element is written in the form given by Eq. (19), and the summation is performed. The J, K dependence of each term is grouped into the coefficients A, B and C, defined by

$$\alpha(J, K) = \frac{8\pi^2\omega_s\tau N_0}{c^2\hbar^3} I_p f_{JK}$$

$$\times \left[A(J, K)\frac{\mu_{IR}^2\mu_0^2}{\delta_{p1}^2} + B(J, K)\frac{\mu_{IR}^2\mu_0^2}{\delta_{p2}^2} - C(J, K)\frac{\mu_{IR}^2\mu_0^2}{\delta_{p1}\delta_{p2}} \right], \quad (21)$$

where $\mu_{IR} \equiv \mu_{13}^\circ = \mu_{02}^\circ$ (=0.205 D in CH_3F) is the induced-dipole moment and $\mu_0 \equiv \mu_{23}^\circ = \mu_{01}^\circ$ (= 1.905 D in CH_3F) is the permanent-dipole moment.

The coefficients A, B, and C depend on the relative polarization of the pump and Stokes fields. For P and R pump transitions, parallel polarization dominates the output, while for Q pump transitions, perpendicular relative polarization dominates (Henningsen, 1977; Drozdowicz *et al.*, 1979a). A, B, and C are summarized in Table I for P, Q and R-branch pump transitions. These coefficients, together with Eq. (21), describe the gain of a tunable Raman laser.

C. GENERAL TUNING THEORY FOR SYMMETRIC-TOP MOLECULES

The general theory of Raman tuning in optically pumped molecular lasers is obtained by application of the result [Eq. (21)] to a particular set of energy levels. For noninverting, symmetric-top molecules, such as CH_3F, this is straightforward. The assumption implicit in Eq. (21) is that for any given pump-laser frequency, the emission is at the Raman frequency ($\delta_s = \delta_p$); we assume there are no ac Stark shifts.

Although the ordinary three-level theory of Raman processes predicts that the gain at ω_s is identical above ($\delta_s = \delta_p > 0$) and below ($\delta_s = \delta_p < 0$) line center for equal offsets, on closer examination we have found it necessary to include the contribution to the gain from a second near-resonant Raman process. With this contribution, the gain is no longer symmetric about the line center. Because asymmetric gain produces a preference for Raman tuning on one side of line center, the addition of this second process can have profound effects on the frequency-tuning behavior of these tunable Raman lasers. The resulting asymmetry will be discussed for the three separate cases of R, P, and Q-branch pump transitions in symmetric-top molecules.

A detailed description of the energy-level structure of symmetric-top molecules can be found in the literature (Townes and Schalow, 1955). Any rotational state is completely specified by the three quantum numbers J, K, and M. The M levels are degenerate and the K levels add a substructure to each of the J levels. For any rotational state J, there are $J + 1$ separate energy levels, each corresponding to a separate value of $|K|$. The rotational energies are given by

$$E_g/h = B°J(J + 1) + (A° - B°)K^2 - D_J°J^2(J + 1)^2$$
$$- D_{JK}°J(J + 1)K^2 - D_K°K^4 \tag{22}$$

for the ground vibrational state and

$$E_{v3}/h = v_0 + BJ(J + 1) + (A - B)K^2 - D_JJ^2(J + 1)^2$$
$$- D_{JK}J(J + 1)K^2 - D_KK^4 \tag{23}$$

for rotational levels within the v_3 excited vibrational state. The values of these constants for CH_3F are shown in Table II.

TABLE II

$^{12}CH_3F$ MOLECULAR CONSTANTS

Quantity	Value	Reference
v_0	31,436,560.7 MHz	Arimondo *et al.* (1979)
$A - A°$	−294.64 MHz	Arimondo *et al.* (1979)
B	25,197.545 MHz	Arimondo *et al.* (1979)
$B°$	25,536.1494 MHz	Arimondo *et al.* (1979)
D_{JK}	0.5186 MHz	Arimondo *et al.* (1979)
$D_{JK}°$	0.4399 MHz	Arimondo *et al.* (1979)
D_J	0.05741 MHz	Arimondo *et al.* (1979)
$D_J°$	0.06020 MHz	Arimondo *et al.* (1979)
$D_K - D_K°$	−0.098 MHz	Arimondo *et al.* (1979)
$A°$	154,000 MHz	Arimondo *et al.* (1979)
μ_{IR}	0.205 D	Hodges *et al.* (1976)
μ_{FIR}	1.905 D	Freund *et al.* (1974)

For a molecule with a symmetry axis, the permanent-dipole moment must lie along that axis. Dipole transitions between rotational states then obey the selection rules $\Delta J = \pm 1$, $\Delta K = 0$. For laser emission the transitions are always $J \to J - 1$, $\Delta K = 0$, and the line-center emission frequencies are given by

$$\nu_e = 2J(B - D_{JK}K^2) - 4D_J J^3 \simeq 2BJ,$$

$$\nu_g = 2J(B^\circ - D_{JK}^\circ K^2) - 4D_J^\circ J^3 \simeq 2B^\circ J$$

for the excited and ground vibrational states. The approximations are useful for the general analysis; the only effect of the D_J and D_{JK} terms is to shift the absolute emission frequency by a small amount. The spacing of the ground-state rotational levels is larger than the excited-state spacing in CH_3F ($B^\circ \gtrsim B$).

The pump laser induces transitions from a particular rotational level J, K in the ground state to a rotational level J', K in the excited state. These line-center transitions are either $P(J' = J - 1)$, $Q(J' = J)$, or $R(J' = J + 1)$ transitions, and they are denoted $P(J), Q(J), R(J)$. From Eqs. (22) and (23), the absorption frequencies are given by

$$\begin{aligned}
R(J) = \nu_0 &+ K^2[(A - A^\circ) - (B - B^\circ)] - K^4[D_K - D_K^\circ] \\
&+ (J + 1)[(B + B^\circ) - K^2(D_{JK} - D_{JK}^\circ)] \\
&+ (J + 1)^2[(B - B^\circ) - K^2(D_{JK} - D_{JK}^\circ) - (D_J - D_J^\circ)] \\
&- 2(J + 1)^3[D_J + D_J^\circ] - (J + 1)^4[D_J - D_J^\circ];
\end{aligned} \tag{24}$$

$$\begin{aligned}
P(J) = \nu_0 &+ K^2[(A - A^\circ) - (B - B^\circ)] - K^4[D_K - D_K^\circ] \\
&- J[(B + B^\circ) - K^2(D_{JK} + D_{JK}^\circ)] \\
&+ J^2[(B - B^\circ) - K^2(D_{JK} - D_{JK}^\circ) - (D_J - D_J^\circ)] \\
&+ 2J^3[D_J + D_J^\circ] - J^4[D_J - D_J^\circ];
\end{aligned} \tag{25}$$

$$\begin{aligned}
Q(J) = \nu_0 &+ K^2[(A - A^\circ) - (B - B^\circ)] - K^4[D_K - D_K^\circ] + J(J + 1) \\
&\times [(B - B^\circ) - K^2(D_{JK} - D_{JK}^\circ)] - J^2(J + 1)^2[D_J - D_J^\circ].
\end{aligned} \tag{26}$$

As the pump-laser frequency is tuned from one frequency to another, it sweeps through absorption resonances originating on different J values for each band. These resonances are separated by different amounts for each band:

$$R(J) - R(J - 1) = 2B - 2J(B^\circ - B),$$

$$P(J) - P(J - 1) = -2B - 2J(B^\circ - B),$$

$$Q(J) - Q(J - 1) = -2J(B^\circ - B).$$

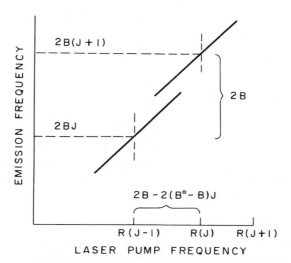

FIG. 5   The general features of a frequency-tuning curve are shown diagrammatically for R-branch step tuning.

For pumping on the Q-branch absorptions, the spacing between resonances is very small. The spacing between P absorptions is larger than the spacing between R absorptions. However, for P-branch absorptions the spacing decreases with increasing J; for R-branch absorptions it increases with increasing J.

Frequency-tuning curves are graphs of the emission frequency ($\nu_s = \omega_s/2\pi$) as a function of the pump-laser frequency ν_p. For a Raman process operating on a particular P, Q, or R-branch pump transition, $\delta_s = \delta_p$, and the tuning curve is a straight line of slope unity. The general features of a frequency-tuning curve are shown in Fig. 5. Because of the frequency difference between the spacing of line-center absorptions and line-center emissions, there are discontinuities or steps introduced into the tuning curve whenever the lasing transition switches to a new set of levels. The magnitude of this discontinuity is different for each of the three types of pump transitions. The tuning theory for R, P, and Q-branch pump transitions will now be examined in detail.

1. R-Branch Tuning

For R transitions, the pump laser induces a vibrational transition from $J \rightarrow J + 1$. Far-infrared emission is obtained on the $J + 1 \rightarrow J$ rotational transition in the excited vibrational state. In the small-signal field limit, the gain at the Raman resonance condition is given by Eq. (21):

$$\alpha(J, K) = \frac{8\pi^2 \omega_s \tau N_0 f_{JK}}{c^2 \hbar^3}$$

$$\times \left[A(J, K) \frac{|\mu_{23}^\circ|^2 |\mu_{31}^\circ|^2}{\delta_{p1}^2} + B(J, K) \frac{|\mu_{01}^\circ|^2 |\mu_{02}^\circ|^2}{\delta_{p2}^2} \right.$$

$$\left. - C(J, K) \frac{\mu_{23}^\circ \mu_{31}^\circ \mu_{20}^\circ \mu_{01}^\circ}{\delta_{p1} \delta_{p2}} \right] I_p, \qquad (27)$$

where A, B, and C are the polarization coefficients defined in Table I for an R absorption with parallel relative polarization of the pump and Stokes fields. For R absorptions, the offsets are defined by $\delta_{p1} \equiv \omega_p - 2\pi R(J)$ and $\delta_{p2} \equiv \omega_p - 2\pi R(J-1)$.

The level configurations for three different pump frequencies are shown in Fig. 6. In Case I, the pump frequency $\omega_p > 2\pi R(J)$ and $\delta_{p1} > 0$. The ground-state Raman offset is very large and positive ($\delta_{p2} > 0$); the only important term in the gain is the first term of Eq. (27), which varies as δ_{p1}^{-2}. As the pump frequency is decreased the pump field becomes resonant with

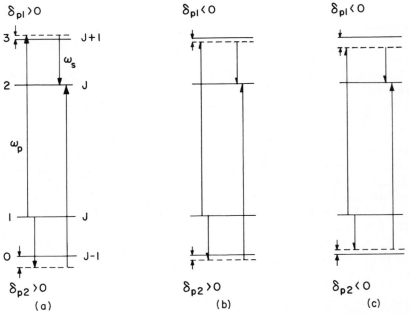

FIG. 6 Diagrams of the R-branch double Raman process: (a) case I, (b) case II, and (c) case III, where $\omega_{pI} > \omega_{pII} > \omega_{pIII}$.

the $R(J)$ transition. Although Eq. (27) is not valid for a resonant pump field, the major contribution to the gain is from the simple three-level process. As ω_p is decreased further (Case II), the sign of the offset δ_{p1} changes, but the sign of δ_{p2} remains positive. At this point, the interference term changes sign; this term now contributes positively toward the gain. Negative values of δ_{p1} yield higher gain than positive values of the same magnitude. As the pump frequency is decreased still further, it passes through the ground-state Raman resonance ($\delta_{p2} = 0$), and the sign of δ_{p2} changes. The interference term once again assumes a negative value.

The positive contribution of the interference term for $\delta_{p1} < 0$ and $\delta_{p2} > 0$ gives rise to the gain asymmetry. This gain asymmetry in turn manifests itself in the frequency-tuning curves by affecting the pump frequency for which the laser switches operation from one transition [$R(J)$] to an adjacent transition [$R(J-1)$ or $R(J+1)$]. For R-branch pump transitions, the emission remains on the $R(J)$ Raman transition for large negative δ_{p1}. For small positive δ_{p1}, the emission switches to Raman emission on the $R(J+1)$ transition.

The gain for several adjacent R transitions is shown in Fig. 7; it is calculated from Eq. (27) by summing over K levels for the R pump transitions originating on $J = 21$, 22, and 23 in CH_3F. The emission is assumed to occur at the Raman frequency that is resonant with the $K = 3$ level; this level has the largest population and contributes the largest amount to the gain. Equation (27) is not valid near the resonances $\delta_{p1} = 0$ or $\delta_{p2} = 0$; the gain

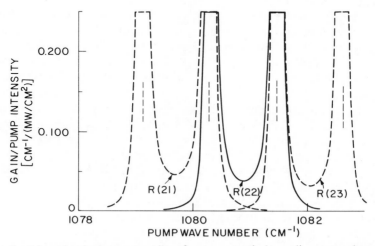

FIG. 7 R-branch gain versus pump-laser frequency: vertical $---$, line-center absorption in $^{12}CH_3F$.

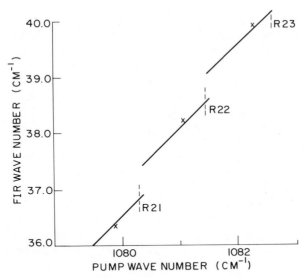

FIG. 8   R-branch frequency-tuning curve: x, experimental data (from Biron *et al.,* 1981) and vertical – – –, line-center absorption in $^{12}CH_3F$.

must saturate for $\delta_{p1}, \delta_{p2} \to 0$.   Gain versus ν_p for the R transition originating on $J = 22$ is shown by the solid line.   The gain peak at the higher frequency corresponds to the resonance at $\delta_{p1} = 0$.   The peak at the lower frequency corresponds to $\delta_{p2} = 0$.   Between these two maxima, the interference term contributes positively toward the gain.

To determine the location of the steps in the tuning curves, the gain for Raman transitions on the neighboring rotational levels must be compared with the gain resulting from Raman transitions on $R(22)$.   Gain curves for the $R(21)$ and $R(23)$ transitions are also shown in Fig. 7.   For a given pump frequency, the transition with the highest gain will produce the Raman emission.

The frequency-tuning curve in the region of the $R(21)$ to $R(23)$ resonances is shown in Fig. 8.   The offsets δ_{p1} and δ_{p2} are defined relative to the $R(22)$ transition.   For positive pump offsets, $\delta_{p1} > 0 \,(\nu_p \gtrsim 1082 \text{ cm}^{-1})$, the gain for the Raman transition on the $R(23)$ transition is much larger than that for the $R(22)$ process.   As δ_{p1} approaches zero, the $R(22)$ line-center process becomes resonant and the Raman process shifts from operation on the $R(23)$ Raman transition to operation on the $R(22)$ transition.   This shift produces a discontinuity in the far-infrared emission frequency.   As the pump frequency is decreased further, $\delta_{p1} < 0$ and $\delta_{p2} > 0$.   The gain remains high for the $R(22)$ Raman process because the interference term is contributing positively toward the gain.   Finally, as δ_{p2} approaches zero the Raman process again shifts from operation on $R(22)$ to operation on $R(21)$. The exact location of these discontinuities cannot be predicted by Eq. (27) because the discontinuities occur slightly above but near resonance. However, this treatment is adequate to predict the marked asymmetry in the Raman tuning relative to line center. This asymmetry favors negative values of δ_{p1} for R transitions.

The magnitude of the step in the R-branch tuning curve is $2(B° − B) J$. The fraction of coverage of emission frequencies between $2BJ_{min}$ and $2BJ_{max}$ is given by

$$1 - \frac{\sum_{J_{min}}^{J_{max}} 2(B° − B) J}{2B(J_{max} − J_{min})} = 1 - \tfrac{1}{2}\alpha(J_{max} + J_{min} + 1),$$

where $\alpha \equiv (B° − B)/B$. In CH_3F, $B° − B \simeq \tfrac{1}{3}$ GHz, $\alpha \simeq \tfrac{1}{75}$, and R-branch tuning can cover ~75% of the range from 200 to 600 μm.

2. P-Branch Tuning

For P transitions, the pump laser induces a vibrational transition from $J \to J - 1$. The emission is then obtained on the $J - 1 \to J - 2$ transition. The gain at the Raman resonance condition is again given by Eq. (27) with the appropriate values of the A, B, and C coefficients. As shown in Fig. 9, for P transitions the ground-state resonance occurs at a higher pump frequency than the excited-state resonance. For $\delta_{p1} < 0$, the interference term has an overall negative sign (Case I, Fig. 9). As the pump frequency is increased, the resonance at $\delta_{p1} = 0$ occurs; this is followed by the region for which $\delta_{p1} > 0$, $\delta_{p2} < 0$, and the interference term contributes positively to the gain (Case II, Fig. 9). Once the pump frequency has increased past the ground-state resonance ($\delta_{p2} = 0$) the interference term is again negative, and the gain falls off rapidly for $\delta_{p2} > 0$.

The Raman gain as a function of pump frequency is shown for the $P(30)$ transition of $^{13}CH_3F$ (solid line) in Fig. 10. In contrast to the R-branch gain curves, the resonance $\delta_{p1} = 0$ occurs on the low-frequency (~ 968 cm^{-1}) side of the double-peaked gain curve for P-branch transitions. The $\delta_{p2} = 0$ resonance occurs at a higher pump frequency (~ 971 cm^{-1}) than the $\delta_{p1} = 0$ resonance. The interference term contributes positively to the gain in the region between these two resonances.

The gain curves for the competing Raman processes $P(29)$ and $P(31)$ are also shown in Fig. 10; the resulting frequency-tuning curve is shown in Fig. 11. The experimental points are those obtained with a line-tunable TEA CO_2 laser (Biron, 1980; Biron et al., 1981). As the pump-laser frequency is increased from the $P(30)$ $\delta_{p1} = 0$ resonance, the emission frequency in-

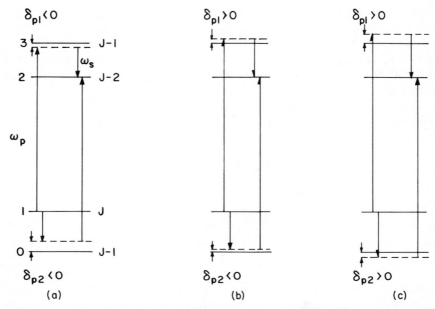

FIG. 9   Diagrams of the *P*-branch double Raman process: (a) case I, (b) case II, and (c) case III, where $\omega_{\text{pI}} < \omega_{\text{pII}} < \omega_{\text{pIII}}$.

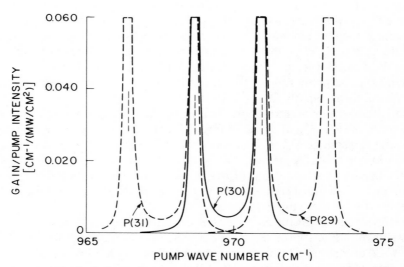

FIG. 10   *P*-branch gain versus pump-laser frequency: vertical – – –, line-center absorption in $^{13}CH_3F$.

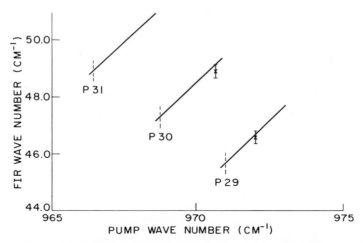

FIG. 11 P-branch frequency-tuning curve: x, experimental data (from Biron *et al.*, 1981) and vertical – – –, line-center absorption in $^{13}CH_3F$.

creases linearly. At a point just below the $P(30)$ $\delta_{p2} = 0$ resonance, the gain for the Raman process originating on the $P(29)$ transitions exceeds that from the Raman process on $P(30)$, and the laser transition switches to the Raman process originating on $P(29)$. The emission frequency is changed discontinuously to a lower value. The step size for the P-branch transitions is $4B + 2J(B° - B)$. Because the spacing between absorptions is greater than the spacing between line-center emission frequencies, all far-infrared emission frequencies between two line centers are obtainable.

The resonance at $\delta_{p2} = 0$ and the region in which the interference term contributes positively toward the gain are both at higher frequencies than the resonance at $\delta_{p1} = 0$. Consequently, for any given P transition, positive values of δ_{p1} have much larger gain than negative values, and the tuning curves exhibit a preference for positive δ_{p1}.

3. Q-Branch Tuning

The pump laser induces a $J \rightarrow J$ vibrational transition for Q transitions. The emission is then on a $J \rightarrow J - 1$ transition within the excited state. Figure 12 depicts the level diagram for a Q-branch pump transition at three different pump frequencies. Initially, for $\delta_{p1} < 0$, $\delta_{p2} < 0$, the interference term has an overall negative sign. For resonance within the excited state $\delta_{p1} = 0$, the ground-state emission process is also nearly resonant, with $\delta_{p2} \lesssim 0$. For equal ground-state and excited-state rotational constants, the $\delta_{p1} = 0$ resonance would occur at exactly the same pump frequency as the $\delta_{p2} = 0$ resonance. The Raman gain would then be totally symmetric about

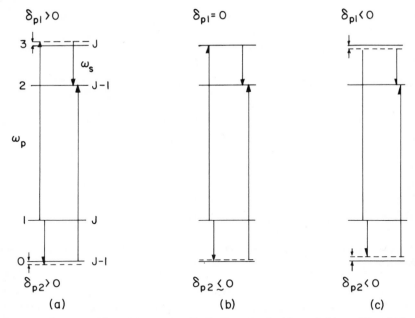

FIG. 12　Diagrams of the Q-branch double Raman process: (a) case I, (b) case II, and (c) case III, where $\omega_{pI} > \omega_{pII} > \omega_{pIII}$.

$\delta_{p1} = \delta_{p2} = 0$. In CH_3F, the frequency difference between the δ_{p1} resonance and the δ_{p2} resonance is $2J(B° - B)$, or approximately 13 GHz for $J = 20$. This gives rise to a slightly asymmetric tuning curve. In the narrow region $\delta_{p1} \gtrsim 0$, $\delta_{p2} \lesssim 0$ between these two resonances, the interference term contributes positively to the gain. Consequently, the Q-branch tuning is qualitatively similar to the P-branch tuning. The step size is $2B + 2J(B° - B)$ for the Q branch.

D.　ABSORPTION PROCESSES

Although the simple theory of frequency tuning on P, Q, and R branches in symmetric-top molecules, as discussed in Section II.C, is adequate to explain the overall behavior and asymmetries of the frequency-tuning curves, it does not include many important effects. These include saturation effects, ac Stark shifts, and single and multiple photon-absorption processes. The threshold behavior of these tunable far-infrared lasers must also be discussed.

Saturation effects caused by strong-pump and Stokes fields may change the gain and the threshold pump intensity necessary for Raman emission. The ac Stark shifts can give rise to nonlinear Raman tuning curves; single and multiple photon-absorption processes can give rise to localized depressions or holes in the gain for particular values of the emission frequency. Some of these absorption processes are included in the four-level density-matrix treatment; some are not, and must be accounted for in a separate consideration of additional absorption resonances. Even for those absorption processes included in the full density-matrix treatment, it is useful to study where they are resonant. An account of these processes together with the small-signal gain curves provides a deeper understanding of the expected frequency-tuning behavior in far-infrared Raman lasers.

For the following analysis, the ac Stark shifts are assumed small and unimportant. Then the emission frequency is given by the Raman resonance condition $\delta_{s1} = \delta_{p1}$ (or, equivalently, $\delta_{s2} = \delta_{p2}$). We discuss the location of the relevant absorption processes in terms of the pump-frequency offsets δ_{p1}. The inclusion of Stark shifts would simply add pump and Stokes intensity-dependent corrections to the location of these absorption resonances; ac Stark shifts are discussed in Section II.E.

Single-photon absorption processes can occur in both the ground and excited states and are shown in Fig. 13. Ground-state absorption reso-

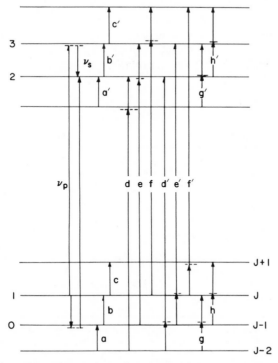

FIG. 13 Single- and two-photon absorption processes in both the ground and excited states.

nances originating on some level J' occur for $v_3 = \omega_s/2\pi = 2B°J'$, or for pump offsets given by

$$\delta_{p1}/2\pi = 2B°J' - v_{32}$$

$$= 2B°J' - \begin{cases} 2B(J+1) & (R) \\ 2BJ & (Q) \\ 2B(J-1) & (P) \end{cases}$$

for R, Q, and P-branch pump transitions originating on level J. The relevant absorption resonances are summarized in Table III. The locations of these resonances are shown in Fig. 14, where they are plotted as a function of the normalized pump frequency $\delta_{p1}^* \equiv \delta_{p1}/(2\pi)(2B)$. In CH_3F, $\alpha \equiv (B_0 - B)/B = 0.013$; the dependence on α and J of the most important resonances is shown in Fig. 14. The qualitative effect of these resonances on the Raman gain curves of Section II.C is also depicted.

The single-photon ground-state absorption process b (Fig. 13) corresponds to the $\delta_{p2} = 0$ resonance and is included in the density-matrix treatment. The gain at $\delta_{p2} = 0$ is smaller than that at $\delta_{p1} = 0$ because of the large populations in levels 0 and 1. The ground-state absorption processes α and c are not included in the four-level treatment. While process c is not important because it occurs at a pump frequency outside the tuning range for the R branch, the resonance corresponding to process α occurs within the Raman tuning range for P-branch pumping. This process can contribute additional gain or absorption depending on the relative populations of the two levels connecting the transition.

TABLE III

SINGLE-PHOTON ABSORPTION RESONANCES

Pump transition[a]	Pump frequency $(\delta_{p1}/2\pi)$	Process[b]
		Ground-state absorption
R	$2(B° - B)(J+1)$	c
R	$2(B° - B)J - 2B$	b
Q	$2(B° - B)J$	b
P	$2(B° - B)J + 2B$	b
P	$2(B° - B)(J-1)$	a
		Excited-state absorption
P	$2B$	c'

[a] All pump transitions originate on level J in the ground state.
[b] Letters refer to absorption processes shown schematically in Figs. 13 and 14.

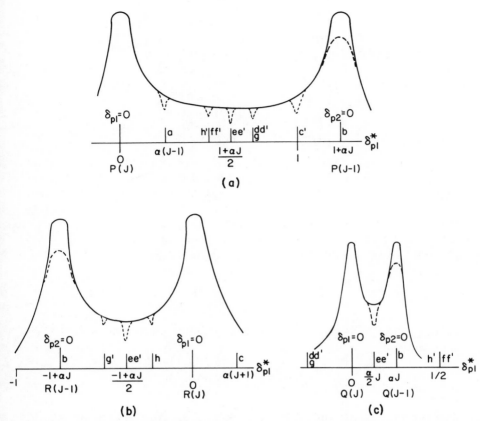

FIG. 14 Locations of absorption resonances for the *P*-branch (a), the *R*-branch (b), and the *Q*-branch (c), plotted as functions of normalized pump frequency. The qualitative effect on the gain is shown by the dashed lines.

Single-photon resonances also occur in the excited state. The $\delta_{p1} = 0$ resonance (b′) is, of course, the line-center process included in the four-level treatment. For $\delta_{p1} = \pm 2B$ (process c′ and a′ in Fig. 13), the emission field can be resonant with transitions in the excited state. Only process c′ is important. It occurs for *P*-branch pumping and can cause absorption of the emitted radiation.

Two-photon absorption processes constitute a second class of important resonances. For certain pump frequencies, the pump field and the Raman emission field will combine to drive a two-photon absorption process at frequency $\omega_p + \omega_s$. A two-photon absorption resonance occurs when, for some J',

$$
v_p + v_s = \begin{cases} R(J') + 2\dot{B}(J' + 2), \\ 2B°(J' + 1) + R(J' + 1), \end{cases} \qquad R \text{ branch}
$$

$$
v_p + v_s = \begin{cases} Q(J') + 2B(J' + 1), \\ 2B°(J' + 1) + Q(J' + 1), \end{cases} \qquad Q \text{ branch}
$$

$$
v_p + v_s = \begin{cases} P(J') + 2BJ', \\ 2B°(J' + 1) + P(J' + 1). \end{cases} \qquad P \text{ branch}
$$

The first condition for each type of transition corresponds to excited-state absorption of the Stokes photon; the second condition for each type of transition corresponds to ground-state absorption of the Stokes photon. Although the location of the resonances for these two processes are the same for equal J', the offset of the two-photon absorption from the intermediate level δ_t is different for ground and excited-state absorption of the Stokes photon. These two-photon absorption offsets are given by

$$
\delta_t/2\pi = v_p - \begin{cases} R(J') \\ Q(J') \\ P(J) \end{cases}
$$

for R, Q, and P transitions where the Stokes photon is absorbed in the excited state, and

$$
\delta_t/2\pi = v_p - \begin{cases} R(J' + 1) \\ Q(J' + 1) \\ P(J' + 1) \end{cases}
$$

for R, Q, and P transitions where the Stokes photon is absorbed in the ground state. The important resonances are noted in Table IV and Fig. 14; the corresponding processes are depicted in Fig. 13.

For R, P, and Q transitions, an important two-photon absorption occurs for $J' = J - 1$ and causes a decrease in the Raman gain exactly halfway between two pump-absorption resonances. This process (e, e′) is the same as that accounted for by the ρ_{03} element of the density matrix; it has been discussed previously (Biron *et al.*, 1981; Lin and Gong, 1982), and was observed for R-branch tuning in CH_3F (Mathieu and Izatt, 1981).

For P and Q-branch pump transitions, there are additional absorptions that are important. Two-photon absorption originating on $J' = J$ and $J' = J - 2$ occur at the locations $\delta_{p1}^* = \frac{1}{2}, -\frac{1}{2} + \frac{1}{2}\alpha(2J - 1)$ for the Q branch (Fig. 14). For low J values, these resonances are outside the normal Raman tuning range, but for $\alpha J \gtrsim \frac{1}{2}$ ($J \gtrsim 38$ in CH_3F) these resonances become important. For P-branch pumping there are in fact many two-photon absorption resonances occurring between $\delta_{p1} = 0$ and $\delta_{p2} = 0$ for any J

TABLE IV

MULTIPHOTON ABSORPTION RESONANCES

Pump transition[a]	J'	$\omega_s{}^b$	Pump frequency $(\delta_{pl}/2\pi)$	Absorption offset[c] $(\delta_t/2\pi)$	Process[d]
					$\nu_p + \nu_s$ absorption
R	$J-1$	E	$-B + J(B^\circ - B)$	$B - J(B^\circ - B)$	e
R	$J-1$	G	$-B + J(B^\circ - B)$	$-B + J(B^\circ - B)$	e'
Q	$J-2$	E	$-B + (2J-1)(B^\circ - B)$	$-B - (2J-1)(B^\circ - B)$	d
Q	$J-2$	G	$-B + (2J-1)(B^\circ - B)$	$-B - (B^\circ - B)$	d'
Q	$J-1$	E	$J(B^\circ - B)$	$-J(B^\circ - B)$	e
Q	$J-1$	G	$J(B^\circ - B)$	$J(B^\circ - B)$	e'
Q	J	E	B	B	f
Q	J	G	B	$B + 2(J+1)(B^\circ - B)$	f'
P	$J-2$	E	$B + (2J-1)(B^\circ - B)$	$-3B - (2J-1)(B^\circ - B)$	d
P	$J-2$	G	$B + (2J-1)(B^\circ - B)$	$-B - (B^\circ - B)$	d'
P	$J-1$	E	$B + J(B^\circ - B)$	$-B - J(B^\circ - B)$	e
P	$J-1$	G	$B + J(B^\circ - B)$	$B + J(B^\circ - B)$	e'
P	J	E	B	B	f
P	J	G	B	$3B + 2(J+1)(B^\circ - B)$	f'
					$2\nu_s$ absorption
R	$J-1$	G	$-B + (2J+1)(B^\circ - B)$	B°	h
R	$J-2$	G	$-3B + (2J-1)(B^\circ - B)$	B°	g
Q	$J-2$	G	$-B + (2J-1)(B^\circ - B)$	B°	g
P	$J-2$	G	$B + (2J-1)(B^\circ - B)$	B°	g
R	$J-1$	E	$-B$	B	g'
Q	$J-1$	E	B	B	h'
P	$J-2$	E	B	B	h'

[a] All pump transitions originate on level J in ground state.
[b] E(G) indicates ω_s photon(s) absorbed in excited (ground) state.
[c] Offset of absorption from intermediate state.
[d] Letters refer to absorption processes shown schematically in Figs. 13 and 14.

(Fig. 14). Only processes originating on $J' = J, J-1$, and $J-2$ have reasonably small two-photon absorption offsets δ_t. Although two-photon resonances for $J' = J+1, J+2, \ldots$ and $J' = J-3, J-4, \ldots$ occur within the Raman tuning region of the P-branch gain spectrum, the offsets (δ_t) are large enough to preclude absorption at those frequencies except for very high-power tunable lasers.

The simultaneous absorption of two Stokes photons in the ground or excited state can also be important for large Stokes intensities. For $2\nu_s$ absorption originating on level J', the conditions are $2\nu_s = 2B^\circ(J' + 1) + 2B^\circ(J' + 2)$ and $2\nu_s = 2B(J' + 1) + 2B(J' + 2)$. The relevant resonances are listed in Table IV and shown in Figs. 13 and 14.

As with the single-photon absorption, the effect of the two-photon absorption holes is to suppress the Raman gain in the region of the resonance. Provided the Raman gain from an adjacent transition (originating on $J - 1$ or $J + 1$) is large enough, the lasing transition can switch to this adjacent transition. This has been observed in HF and HCl Raman lasers (Frey *et al.*, 1977; DeMartino *et al.*, 1978).

Other processes such as $3v_s$, $v_p + 2v_s$ absorption and $v_p - 2v_s$ three-wave processes can affect the gain, but are of higher order than those already discussed. The absorption of two pump photons, yielding excitation to the $2v_3$ vibrational level, can also play a role in reducing the Raman gain at certain pump frequencies. A two-step version of this absorption process has been observed in $^{13}CH_3F$ and $^{12}CH_3F$ (Biron, 1980; Peebles *et al.*, 1980).

E.   ac STARK SHIFTS

In the previous discussion it has been assumed that the emission frequency is always related to the pump frequency by the Raman resonance condition $\delta_s = \delta_p$. However, for large pump or Stokes fields, the gain maxima occur at positions that are shifted from the small-field locations (Panock and Temkin, 1977). For a small Stokes field $\beta_{32}\tau \ll 1$ and a large pump field $\beta_{13}\tau > 1$, the gain maxima for off-resonant pumping occur at

$$\delta_s^{\pm} = \tfrac{1}{2}\delta_p \pm \tfrac{1}{2}(\delta_p^2 + 4\beta_{13}^2)^{1/2}. \tag{28}$$

For $\delta_p < 0$, the Raman emission peak is identified with the δ_s^- peak, whereas for $\delta_p > 0$ the Raman peak corresponds to the δ_s^+ peak. The maxima at $\delta_s^{\pm}$ are plotted as a function of δ_p for $\beta_{13}\tau = 2$ in Fig. 15.

Provided the emission occurs at the frequency corresponding to the gain maximum farthest from line center, one would expect the emission frequency to vary linearly with δ_p for $\delta_p^2 \gg 4\beta_{13}^2$. Near line center there would be very nonlinear tuning and perhaps even a discontinuous jump from one ac Stark peak to the other. The frequency tuning would then correspond to the solid line in Fig. 15.

The ac Stark shift of both the Raman and line-center gain peaks has been observed in optically pumped lasers by Drozdowicz *et al.* (1979b). They measured the gain as a function of δ_s for the 385-μm D_2O laser; the gain maxima occurred at $\delta_s^{\pm}$, as given by Eq. (28).

In a number of other investigations the ac Stark shift has not been present in the laser emission. Fetterman *et al.* (1979) studied Raman tuning on the 385-μm transition in D_2O and, within their tuning range of $\sim \pm 2$ GHz, they observed linear tuning with $\delta_s = \delta_p$ and no Stark shift. In fact, the far infrared emission was observed to track linearly with the pump frequency right through line center. In Raman tuning experiments on HF and HCl, Frey *et al.* (1977) and DeMartino *et al.* (1978, 1980) also observed perfectly

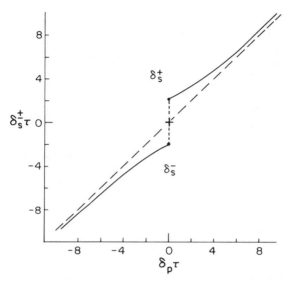

FIG. 15 Effect of ac Stark shift on tuning curves. Raman resonance maxima for $\beta_{13}\tau = 2$: ——, frequency tuning. The broken 45° line indicates tuning for zero Stark shift.

linear tuning. The experimental results on Raman tuning in CH_3F presented here and in the literature do not indicate the presence of any significant ac Stark shifts. Consequently, in this investigation, we assume linear tuning given by $\delta_s = \delta_p$ and note that the presence of Stark shifts produces nonlinearities in the tuning curves (as described) and small shifts in the locations of the resonances discussed in Section II.D (Danly *et al.*, 1981, 1983; Izatt and Mathieu, 1981; Mathieu and Izatt, 1981).

F. THRESHOLD CONDITION

The buildup of the emission field at the Raman frequency results from stimulated Raman scattering (SRS) (Bloembergen, 1965, 1967; Shen and Bloembergen, 1965); the derivation of the threshold gain and threshold pump intensity for the operation of a tunable Raman laser can be derived from the gain expression, Eq. (21). The growth of the emission field begins from spontaneous Raman scattering near the entrance to the waveguide. The small amount of Stokes radiation scattered into the mode copropagating with the pump beam is then amplified by the stimulated Raman scattering process described by Eq. (21). In the small-signal regime, $I(z) = I_0 e^{\alpha z}$, and the threshold is often experimentally defined by the condition $\alpha L \simeq 30$. For this condition, the typical spontaneous Raman scattering intensities $(10^{-12} - 10^{-13} \text{ W/cm}^2)$ reach typical observable powers of $1-10 \text{ W/cm}^2$.

This condition agrees well with observed detector thresholds for stimulated Raman scattering in optically pumped lasers (Berry *et al.*, 1982; Biron, 1980; Biron *et al.*, 1981; Frey *et al.*, 1977). In the $\alpha L \simeq 30$ condition, the net gain α is defined as the Raman gain minus the distributed losses. For free-space propagation, the losses are primarily diffraction effects. Cotter *et al.* (1975) derived both the fundamental gain threshold condition, for which the gain equals the diffractive losses, and the detector threshold condition. For propagation in dielectric waveguides, the waveguide propagation losses must be subtracted from the Raman gain. The condition $\alpha L \simeq 30$ and Eq. (21) can be used to find the threshold pump intensity I_p^t necessary for the observation of Raman emission. In general, I_p^t depends on the inverse square of the pump detuning δ_p.

III.   Tunable Raman Laser Experiment

A.   HIGH-PRESSURE CO_2 PUMP LASER

To produce continuously tunable far-infrared emission via the Raman process, a continuously tunable pump laser is required. Fortunately, there is a large overlap of the vibrational absorption bands of $^{12}CH_3F$ and $^{13}CH_3F$ with the 9 and 10-μm emission bands of the CO_2 laser. Furthermore, continuously tunable high-pressure CO_2 lasers have been widely studied and developed (Alcock *et al.*, 1973, 1975; Bagratashvili *et al.*, 1976; Taylor *et al.*, 1979; Wood, 1974).

Conventional transversely excited atmospheric (TEA) lasers are among the most powerful lasers available at any wavelength. They produce radiation on four bands in the 9- and 10-μm regions. These bands are commonly referred to as the $9P$, $9R$, $10P$, and $10R$ branches of the CO_2 laser; the P and R denote the ΔJ value of the laser transition. As many as 80 discrete emission lines are routinely obtained with wavelengths from 9.2 to 10.8 μm. The frequency spacing between R-branch transitions is typically 40 GHz, whereas the spacing between P-branch transitions is typically 55 GHz.

Pressure broadening is the basic mechanism responsible for producing the large-gain bandwidth of the continuously tunable CO_2 laser. When these lasers are operated at gas pressures of 10 atm or greater, the high pressure broadens the discrete lines of each band into a gain continuum. The total homogeneously broadened gain bandwidth, $\Delta\nu_{TOT}$, depends on the broadening contributed by each of the gas species CO_2, N_2, and He, which constitute the standard gas mix for the CO_2 laser, as shown here:

$$\Delta\nu_{TOT} = \Delta\nu_{CO_2}p_{CO_2} + \Delta\nu_{He}p_{He} + \Delta\nu_{N_2}p_{N_2}.$$

In this expression, $\Delta\nu_i$ and p_i are the pressure-broadening coefficients and partial pressures of each species i. For typical laser mixes of $8:1:1$

He:CO_2:N_2 and for $\Delta\nu_{He} = 3.72$ GHz/atm, $\Delta\nu_{CO_2} = 5.79$ GHz/atm and $\Delta\nu_{N_2} = 4.23$ GHz/atm (Abrams, 1972), the total pressure broadening at $p_{TOT} = \Sigma_i p_i = 10$ atm is $\Delta\nu \simeq 40$ GHz. Typical gain spectra at pressures of one and ten atmospheres are shown for the four CO_2 laser branches in Figs. 16 and 17 (Gibson, 1979). Above approximately 10 atm pressure, the gain can be greater than the resonator losses for frequencies intermediate between two line centers, and frequency-tunable laser output can be obtained in each of the four branches of the CO_2 laser.

Given the broad gain bandwidth of a high-pressure (> 10 atm) CO_2 discharge, an optical cavity must be designed to provide both feedback and frequency selectivity. As with other lasers, many combinations of optical components may be employed. High-pressure CO_2 lasers most frequently use a diffraction grating as the dispersive element inside the laser resonator. Typical output bandwidths are 2–4 GHz, and many longitudinal modes of the cavity are present in the output. This gives rise to a self-mode-locked pulse. The use of additional dispersive elements to produce single-mode output must involve Fabry–Perot etalons or other more elaborate cavity designs (Bernard et al., 1981; Deka et al., 1981; Smith, 1972). Although the use of a pulsed or cw low-pressure section within the cavity produces single-longitudinal-mode output for atmospheric CO_2 lasers (Gondhalekar et al., 1975; Loy and Roland, 1977), these low-pressure sections are useless for a high-pressure laser that must be tunable.

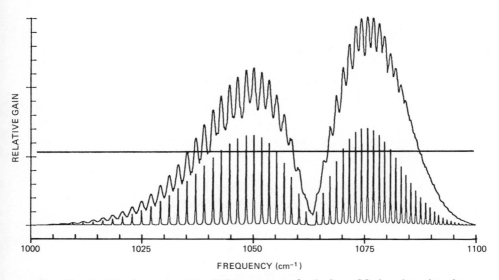

FIG. 16 Typical gain spectra at 1 and 10 atm pressure for the 9 μm CO_2-laser branches: the horizontal line indicates a typical loss level.

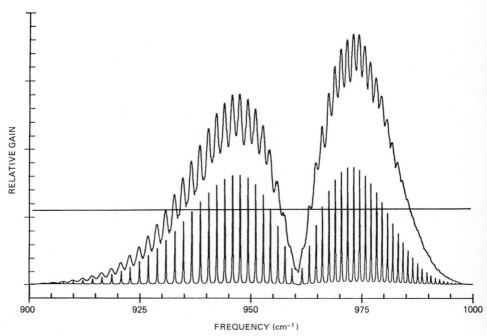

FIG. 17   Typical gain spectra at 1 and 10 atm pressure for the 10 μm CO_2-laser branches: the horizontal line indicates a typical loss level.

For this investigation a compact high-pressure CO_2 laser was designed and constructed. Such lasers are at present commercially available from one laser manufacturer (Lumonics). However, it was believed that some improvements over the commercial model could be included in this design. After experience was gained with a prototype high-pressure laser, the laser used for these far-infrared studies was designed and constructed. This 10–12-atm CO_2 laser is shown in Fig. 18; a block diagram is shown in Fig. 19. A description of the design and operational details follows.

1. Gain Section

The gain section of the high-pressure CO_2 laser is constructed from a solid aluminum block. The uniform-field electrodes are of the Rogowski type (Rogowski, 1941) with the ends filed down to prevent arcing. The electrodes are 35 cm long and 0.64 cm wide. The gap is variable from 6 to 8 mm, but the laser was always operated with a 7-mm gap. The active volume is 15.7 cm³. The high-voltage discharge was preionized by ultraviolet radiation from two pairs of spark rods or flashbars on each side of the

FIG. 18   Photograph of the high-pressure CO_2 laser used for far-infrared studies.

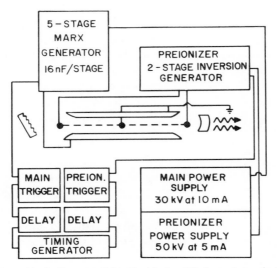

FIG. 19   A block diagram of the 10–12 atm CO_2 laser shown in Fig. 18.

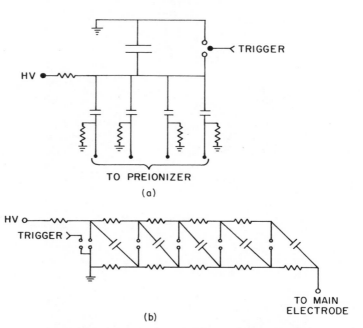

FIG. 20   Pulse-forming networks that drive the gain section of the high-pressure CO$_2$ laser:
(a) two-stage inversion generator; (b) five-stage Marx–Goodlet circuit. HV, high voltage.

main electrodes.   The position of the flashbars relative to the main electrodes is variable.   The flashbars are driven by a two-stage, spark-gap-trig-
gered inversion generator (Fig. 20a).   With charging voltages up to 25 kV,
this inversion generator drives the flashbars with 50-kV, 12-J pulses that
produce enough ultraviolet radiation to sufficiently preionize the discharge.

The main discharge is driven by a compact, N$_2$-pressurized, five-stage
Marx–Goodlet circuit (or Marx bank), as shown in Fig. 20b.   This circuit
contains five separate capacitor stages (16.2 nF per stage) that are charged in
parallel and discharged in series.   The switching is done by means of spark
gaps.   The first gap is triggered by an automotive spark plug, and the remaining four spark gaps trigger spontaneously as the voltage across them
swings to a value much larger than their hold-off voltage.   For charging
voltages of 30 kV, this Marx bank produces 150-kV, 36-J pulses with a
measured 10–90% risetime of 11 nsec, an internal inductance of 122 nH,
and an output impedance of 11.5 Ω.   This low output impedance closely
matches the impedance of the gas discharge after breakdown, allowing for
the fast energy transfer necessary for arc-free operation at these high pressures.

The preionization circuit and main discharge circuit are triggered by EG&G trigger modules (model TM – 12A) and a California Avionics time-delay generator. The time-delay generator contains a master clock that controls the repetition rate of the laser ($\frac{1}{3}$ – 2 Hz). It provides a trigger pulse to the preionizer trigger module, followed by a time-delayed trigger pulse to the main trigger module. This time delay is variable in increments of 10 nsec and is typically on the order of 500 nsec. The delay generator also provides a reference pulse coincident with the first (preionizer) trigger pulse. This reference pulse is used to trigger all the data-acquisition electronics, which will be described later. The total time-delay jitter between the preionizer-circuit discharge and the Marx-bank discharge is approximately 50 nsec. This includes contributions from the spark plugs and spark gaps in the inversion generator and Marx bank.

The Brewster windows provide optical access to the gain section and are made from a polycrystalline NaCl. Produced under the name of Polytran by Harshaw Chemical Co., these windows have the same optical properties as single-crystalline NaCl but have four times the elastic limit of ordinary NaCl. After an attempt to use ZnSe windows in the prototype laser resulted in frequent optical damage, rectangular Polytran NaCl windows (1 in. $\times$ 2 in. $\times \frac{3}{8}$ in.) were incorporated into the present laser. The reported damage resistance of the polycrystalline NaCl surfaces was 5 – 8 J/cm^2 for single 1.2-nsec pulses at 10.6 μm. Nonetheless, optical damage occurred over a period of 20 – 30 hours of operation even with these windows. Consequently, several sets of windows were used and repolished on a rotating basis.

The gas-handling system is responsible for mixing the CO_2, N_2, and He in variable concentrations and delivering this mix to the laser at pressures of 10 – 12 atm. The three gases are first mixed at an absolute pressure of 215 psi in a Matheson three-tube gas mixer that controls the relative gas concentrations. A pressure regulator then reduces the pressure to the desired operating value for the laser, and the gas enters the laser gain section through the gas-supply manifold. After traversing the electrode region, the gas exits through the exhaust manifold to an output valve that controls the total gas-flow rate. This output valve is heated to prevent the freezing of CO_2 in the valve orifice as a result of the rapid expansion of the gas. The entire gas system can sustain the laser pressure at 10 – 12 atm and at the same time flow up to 25 l/min of gas through the laser.

In many high-pressure CO_2 lasers, trace amounts of a low-ionization-potential gas are added to the laser gas mix to create more free electrons from the ultraviolet preionization. This provides a more stable discharge (Gibson et al., 1978; Morikawa, 1977). Seed gases, such as trimethylamine, tripropylamine, xylene, and others, generally make the discharge more forgiving of slow-risetime driving circuits and allow a more uniform discharge. However, they have many disadvantages, including low vapor pressures (~ 30 psi) and harmful byproducts in electrical discharges.

Although the provision for operation with seed gas was incorporated into the gas-system design, the ultraviolet preionization and the risetime of the Marx-bank driving circuit were sufficient to provide reliable operation without seed gas.

2. Optical Resonator

Once the high-voltage, high-pressure discharge region produces the broadband gain, an optical resonator is needed to provide feedback and control of the laser oscillation frequency. The optical cavity consists of a germanium output coupler and a 150-line/mm diffraction grating with a blaze wavelength of 10.6 μm. Germanium output couplers with different reflectivities (36%, 50%, 65%, and 75%) were used for different laser frequencies to maximize the power output and the linearity of the frequency tuning of the CO_2 laser. The optimum reflectivity for the output coupler was typically 50–75%.

The diffraction grating was operated in the Littrow configuration such that the $n = 1$ order was reflected back into the laser gain section. The grating has an angular dispersion of 42 Å/mr and an efficiency of 97% in the 9 to 10.6-μm wavelength range.

Two problems necessitated the introduction of telescopic beam expanders into the cavity. Optical damage to both the output coupler and the grating was a constant concern and a severe constraint on laser output power. The second problem was the presence of a large amount of frequency pulling of the laser frequency toward the line-center transition frequency.

Frequency pulling in an optical cavity results from the interaction of the cavity modes with the laser gas near resonance. One of the laser oscillation conditions requires that the cavity round-trip phase shift be an integral multiple of 2π. For a plane wave of frequency v propagating through a medium near a resonance at frequency v_0, the complex propagation constant is given by (Yariv, 1975)

$$k' \simeq k[1 + \chi'(v)/2n^2] - ik\chi''(v)/2n^2.$$

The real and imaginary parts of the electric susceptibility are χ' and χ'', $k = \omega n/c$, and n is the index of refraction of the medium. The phase condition on laser oscillation is thus $2Lk' = 2\pi m$ or

$$2Lk[1 + \chi'(v)/2n^2] = 2\pi m,$$

where L is the length of the cavity and m an integer. For a cold cavity, defined by the lack of any dispersion ($\chi' = 0$), the laser resonator frequencies

v_m are given by

$$2\pi v_m n = 2\pi m/2L,$$

$$v_m = mc/2nL.$$

The natural or cold oscillation frequency of the cavity is more accurately given in the literature (Eq. 9.1-18 in Yariv, 1975), but for a qualitative discussion of frequency pulling this is adequate. With the dispersive medium present in the cavity, the cavity oscillation frequency v can now be written in terms of the real part of the susceptibility and the cold cavity frequency v_m, as follows:

$$v[1 + \chi'(v)/2n^2] = v_m. \tag{29}$$

Only for $\chi'(v) = 0$ with the actual laser oscillation frequency equal the cavity frequency v_m.

Dispersive optical elements are introduced into the laser cavity to suppress (ideally) all but one of the cavity modes v_m. In the presence of nonzero χ' the laser will not oscillate at the frequency v_m but at the frequency v, which is shifted or pulled toward the gain maximum by an amount proportional to χ'.

The cavity oscillation frequency can also be written in terms of the full width of the passive optical resonator $\Delta v_{1/2}$ and the full width of the gain Δv, as

$$v \simeq v_m - (v_m - v_0)(\Delta v_{1/2}/\Delta v). \tag{30}$$

For a high-Q cavity ($Q = v/\Delta v_{1/2}$) for which $\Delta v_{1/2} \ll \Delta v$, $v \simeq v_m$, and there is very little frequency pulling. However, for large-gain bandwidths and low-Q cavities, there can be large amounts of frequency pulling.

For high-pressure CO_2 lasers at 10 atm, even though the individual transitions are pressure broadened into a continuum, there still exist gain maxima at line center and gain minima roughly half-way between two line centers. Consequently, there is some degree of frequency pulling as the laser is tuned by the diffraction grating away from a line-center frequency.

Initially the optical cavity consisted only of the diffraction grating and an output coupler. Optical damage and the presence of strong frequency pulling necessitated the introduction of telescopic beam expanders for both the grating and the output coupler. The use of on-axis NaCl lenses was attempted, but the reflection losses were too great. Any attempt to use antireflection-coated lenses would have met with the same optical-damage problem. The use of spherical copper mirrors for off-axis beam expansion proved to be the best solution (Fig. 21). Each beam expander consisted of one concave and one convex mirror, arranged in a close-packed Z configura-

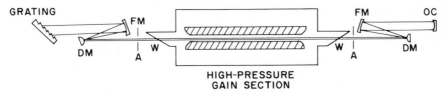

FIG. 21   The optical cavity of the high-pressure CO$_2$ laser: DM is the defocusing mirror, FM the focusing mirror, W the NaCl window, A the aperture, and OC the Ge output coupler.

tion.   Expansions of $\sim 2.5\times$ and $\sim 2\times$ for the grating and output coupler were obtained.   Although the use of off-axis spherical mirrors resulted in a certain degree of astigmatism at the output coupler, the output beam from the laser was circular and not ellipitical at the entrance to the far-infrared laser waveguide.

The beam expanders reduced the optical damage significantly; the laser routinely produces 100-mJ, 100-nsec output pulses without damage to the output coupler or grating.   The illumination of more lines of the grating increased the grating resolution and the Q of the cavity.   This decreased the frequency pulling substantially but not completely.

The elements of the optical cavity are mounted on a large aluminum table; no attempt to stabilize the length of the cavity was made.   For large-band-width multimode laser operation, the fluctuations in cavity length are not important.

3.   CO$_2$ Laser Operation

The performance and common operating parameters of the CO$_2$ laser will be described in this section.   The laser operated routinely at 10.5 atm pressure and occasionally at pressures as high as 12 atm.   Above 10.5 atm the discharge was very unstable, and arcing in the discharge region occurred often.

Stable laser operation was very dependent on the gas mixture used.   The laser operated best with a 80% He, 10% CO$_2$, 10% N$_2$ mix at a total flow rate of $\sim 10-15$ l/min.   Every attempt was made to run the laser with as much CO$_2$ in the mix as possible; both the laser output power and the amount of pressure broadening depend strongly on the CO$_2$ concentration.

Typical operating voltages were $16-24$ kV for the Marx-bank-charging voltage and $20-25$ kV for the preionizer-charging voltage.   The discharge voltages were thus $80-120$ kV for the Marx bank and $40-50$ kV for the preionizer circuit.   For Marx-charging voltages above ~ 22 kV on the $9R$ and $10R$ laser branches, there was little increase in laser output power.   This implies that there is a saturation of the CO$_2$ laser-pump transition by the electrical discharge or that for higher gains the output coupler was not of optimum reflectivity.

For high-gain operation at the centers of the $9R$ and $10R$ bands, a 50% reflectivity Ge output coupler was used. For lower-gain operation away from these band centers, a 75% reflectivity output coupler provided more output power and less frequency pulling.

Laser output energy ranged from 40 mJ to ~ 120 mJ, depending on Marx voltage and the emission frequency. For emission near the band centers of the $9R$ and $10R$ branches, the laser produced 80–120 mJ reliably. For the $9P$ and $10P$ branches, only 40–80 mJ near the line-center frequencies was obtained. No continuous tuning on either of the P branches was obtained; it was only possible to tune within ± 5 GHz about the line-center frequency. The P-branch transitions have larger spacings (55 GHz) than the R-branch transitions. At the operating pressure of 10.5 atm the pressure broadening causes less of an overlap for the P-branch transitions. Consequently, the gain between line centers is much lower for the same operating conditions, and it is difficult to obtain laser emission at these intermediate frequencies.

B. FAR-INFRARED WAVEGUIDE LASER

The far-infrared (FIR) laser used for the tunable Raman laser experiments is an amplified, spontaneous-emission waveguide laser, shown schematically in Fig. 22. Fused-quartz waveguides with inside diameters of 9, 7, 5, and 4 mm and various lengths are mounted inside a 1.5-in. i.d., pyrex-glass pipe. The windows (W1, W2), pumping ports, pressure gauges, and a photoacoustic transducer (P.A.T.) were mounted on the glass pipe containing the waveguide. The CO_2 laser beam enters the FIR laser through a NaCl window (W1). The FIR beam exits through a Teflon output window (W2). The photoacoustic signal is monitored with a low-cost, miniature, dynamic microphone, and the laser pressure is monitored by two Wallace and Tiernan absolute-pressure gauges.

The pump beam is focused into the dielectric waveguide with a focusing mirror (F.M.) external to the FIR laser. At the entrance to the waveguide, the Gaussian pump beam couples to the dielectric-waveguide modes. In a hollow dielectric waveguide, there are three types of modes: circular electric (TE_{0n}), circular magnetic (TM_{0n}), and hybrid EH_{mn} modes (Abrams, 1972; Degnan, 1976; Marcatili and Schmeltzer, 1964). For $n < 2.2$ ($n \simeq 1.4$ for fused quartz), the EH_{11} mode has the lowest loss. It is assumed, therefore, that both the pump beam and the Stokes beam propagate in this mode. The EH_{11} mode is linearly polarized to lowest order, and the electric field is given by $E(r) = J_0(ur/a)$ for $r \leq a$ where a is the guide radius and u the first zero of $J_0(x)$.

Abrams (1972) has shown that the maximum coupling (98%) between a free-space Gaussian TEM_{00} mode and the EH_{11} mode occurs for a Gaussian beam with $\omega_0/a = 0.64$ and $R = \infty$, where ω_0 and R are the $1/e$ radius of the electric field and the wavefront radius of curvature. The position of the

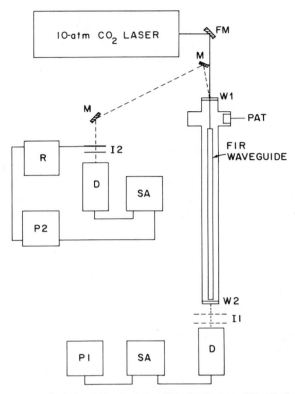

FIG. 22   The tunable far-infrared laser system: D is the detector; W_1, W_2 the windows; FM, M, mirrors; I1, I2, interferometers; P1, P2, plotters; PAT, photoacoustic transducer; SA, signal averaging electronics.

waveguide and the focal length of mirror FM in Fig. 22 are adjusted to provide the maximum coupling.

The propagation loss for the EH_{11} mode in a hollow guide with dielectric constant $n = 1.4$ is given by $a_{11} = 0.22\lambda^2/a^3$ cm^{-1} (Degnan, 1976). Although the propagation losses for the pump beam are negligible for the guide radii used, significant losses were caused by a departure from straightness of the waveguide bore. This loss mechanism places severe constraints on the maximum transmission through hollow waveguides; it has been discussed by Marcatili and Schmeltzer (1964) and Hall *et al.* (1976). For a radius of curvature of 75 m for the waveguide bore, the additional losses because of the curvature are roughly three times the losses for a perfectly straight waveguide. Although visually inspected for straightness, the quartz waveguides used in these experiments had measured CO_2 beam transmission of 100%,

74%, 70%, and 55% for the 9, 7, 5, and 4 mm waveguides of lengths 122, 122, 117, and 99 cm, respectively.

The far-infrared radiation that is generated is also present primarily in the EH_{11} mode. The other waveguide modes have higher losses; the low-loss EH_{11} mode then dominates. There is some feedback for the FIR laser resulting from the impedance mismatch at the ends of the waveguide and the reflections off the Teflon and NaCl windows. Except for the reflection from the NaCl window, this feedback is small. Biron et al. (1981) measured a decrease in pump threshold when the NaCl window is aligned normal to the optical axis. For these experiments the NaCl window was mounted at a very slight angle to the optical axis to provide a reflected portion of the pump beam for pump power and wavelength measurement. Far-infrared emission was monitored only in the forward (pump-copropagating) direction.

C. EXPERIMENTAL METHODOLOGY

The experimental apparatus used for this study of a frequency-tunable CH_3F Raman laser is shown in Fig. 22. The FIR emission is detected with a Laser Precision Pyroelectric detector (D). This detector measures the total energy of the FIR pulse with a sensitivity of approximately 10 V/mJ.

Absorption of the pump beam by the FIR laser gas is monitored by a photoacoustic transducer (PAT) mounted perpendicular to the optical axis near the entrance to the waveguide. When the pump frequency matches a line-center absorption in the laser gas, the gas is heated and produces a large acoustic signal detected by the transducer. For low pump powers there is no ac Stark splitting; thus, when the acoustic signal is maximized, both the pump and emission fields are resonant ($\delta_{p1} = 0$ and $\delta_{s1} = 0$). All frequency shifts are then measured relative to these line-center frequencies. Because the spectroscopy of CH_3F is known to high accuracy, frequency shifts measured relative to line center can be converted to absolute frequencies with the available spectroscopic data (Table II) and Eqs. (22)–(25).

An alternative, although somewhat less reliable, method for measuring the frequency-tuning behavior of the Raman laser involves the measurement of all frequency shifts relative to the well-known CO_2-laser line-center transitions. Although the CO_2-laser gain spectrum is very broad at the 10–12 atm operating pressure, there still exist gain maxima at the frequencies corresponding to line-center CO_2 transitions. The CO_2-laser output power is measured as a function of grating position; the maxima correspond to line-center CO_2 transitions. The FIR emission is then assumed to occur at the Raman frequency given by $v_s = \delta_p + v_{32}$ with $\delta_p = v_{CO_2} - v_{31}$. Frequency shifts are measured relative to this pump-laser frequency and the FIR-emission frequency. In this case, the CH_3F acoustic signal provides a check of the linearity of the CO_2 laser tuning as well as the FIR Raman tuning.

In most cases, the frequency shifts obtained experimentally were measured relative to the line-center CH_3F absorptions as determined from the photoacoustic signal. The measurement of frequency shifts for the pump and FIR beams was carried out with Fabry–Perot interferometers (I1 and I2 in Fig. 22). The FIR interferometer I1 consists of two wire-grid inductive meshes of 300, 400, or 500 wires per inch mounted on conventional mirror mounts. A geared motor drives a translation stage on which one of the mirror mounts is fixed. Translation rates are 50 and 25 μm/min for motor speeds of $\frac{1}{10}$ and $\frac{1}{20}$ rpm. The pyroelectric detector then measures the transmission through the interferometer.

Because the entire system was operated on a pulsed basis, typically at a 0.5-Hz repetition rate, a signal-averaging boxcar integrator (SA) was used to convert the pulsed data into a dc-voltage signal proportional to the power transmitted by the interferometer. The output of the signal-averaging electronics was displayed on a strip-chart recorder (P1). The entire FIR interferometer system was calibrated by the 496.1-μm $Q(12)$ emission line of CH_3F.

The pump-laser frequency shifts are determined by two methods. In the first method, the pump frequency shift is determined from the grating dial position. For grating positions between two line centers, linear tuning is assumed and the pump frequency is obtained by linear interpolation between the two line centers. This method, while straightforward, does not account for frequency pulling in the CO_2 laser. The assumption of linear grating tuning is valid only for negligible frequency pulling. However, for large frequency shifts from resonance, any frequency pulling is a small part of the total shift. When this method is used, the error in pump-frequency measurement resulting from unknown frequency-pulling effects is estimated to be $\pm 2 - 3$ GHz, based on actual measurements of frequency pulling for this laser.

An improved method for measuring the pump-laser frequency shift has been recently incorporated into the experimental system; it involves the use of a piezoelectrically driven infrared Fabry–Perot interferometer (I2 in Fig. 22). The dielectric mirrors are 96% reflecting, $\lambda/100$ germanium optics, corresponding to a reflectivity finesse of 77 and a flatness finesse of 50. The overall finesse of the instrument was measured with a cw CO_2 waveguide laser to be 42. Alignment of the germanium optics proved difficult and limited the maximum finesse to this value. Even with poorer finesse, the interferometer is adequate for measuring frequency shifts of 1 GHz or more.

The piezoelectric transducer stack of the interferometer is driven with a ramp generator (R). This generator produces the necessary $0 \rightarrow -1500$ dc voltage with a time constant sufficiently long (~ 2 min) to allow adequate signal averaging of the pump-laser interferometer scans. As with the FIR interferometer, a pyroelectric detector is used, and the output is averaged with a boxcar integrator. The integrated detector signal and a fraction of the ramp voltage are applied to the vertical and horizontal inputs of an $X-Y$ recorder (P2). Interferometer transmission as a function of plate spacing is recorded. The infrared interferometer was calibrated and aligned with a cw CO_2 waveguide laser operating on the 10.6-μm $P20$ line. A bandwidth of 4.2 GHz (FWHM) was then measured for the high-pressure CO_2 laser output.

For rough absolute measurements of the pump frequency, an Optical Engineering CO_2 spectrum analyzer is used. The precise relative measurements of the pump-frequency shifts are made with the interferometer. The basic method for measuring frequency shifts from an interferometer scan is applicable to both the infrared and far-infrared interferometers. The FIR interferometer is of the scanning type; the free spectral range is either increasing or decreasing continuously. For the infrared interferometer, the free spectral range remains fixed, and the plate separation is reset after each sweep.

Experimental results on the Raman-tuning behavior of a CH_3F waveguide laser are obtained by measuring sets of frequency shifts relative to a reference frequency, usually that of a CH_3F absorption-line center. The interferometer scans are initiated at this reference frequency, ν_A ($= \nu_{pA}$ or ν_{sA} for pump or FIR scans), and the transmission maxima are labeled A.

After several of these A peaks, the pump frequency is instantaneously changed by means of the diffraction grating to a new frequency, ν_B, giving rise to new transmission maxima, labeled B. After several of these B peaks, the grating is reset to its original position. This results in transmission peaks once again at the A position. A sequence of peaks AAABBAACCAAD-DAAA would correspond to the measurement of three frequency shifts (B, C, D) relative to the A peak frequency.

Sample interferometer scans are shown in Fig. 23 for a single frequency-shift measurement on the $R(20)$ CH_3F transition. The pump-laser interferometer scan shows the Fabry–Perot transmission maxima as a function of mirror separation for two frequencies, ν_{pA} and ν_{pB} (Fig. 23a). In this case, ν_A corresponds to the CH_3F $R(20)$ absorption-line center. The second A peak is not visible on the scan; the free spectral range was determined by performing a scan of a single-mode, cw CO_2 laser. This calibration is depicted as the horizontal distance corresponding to 2 GHz. The anomalously large pump bandwidth is caused by poor instrumental finesse.

The corresponding FIR interferometer scan is shown in Fig. 23b. The points at which the pump-laser frequency was abruptly changed are noted on this continuous scan. The scan proceeds from right to left; the positions of the transmission maxima, had the frequency not been changed from ν_{pA} to ν_{pB}, are indicated by dashed vertical lines labeled A.

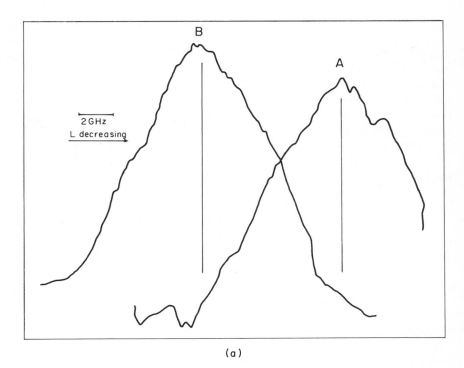

(a)

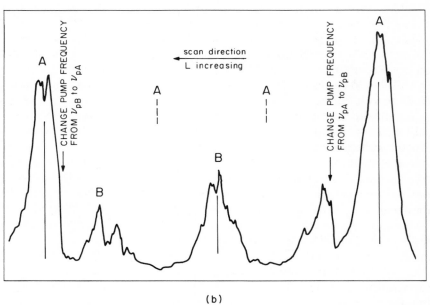

(b)

FIG. 23   Sample interferometer scans: (a) pump laser, $\nu_{\text{FSR}} = 20$ GHz; (b) far-infrared laser, $\nu_{\text{FSR}} = 17.8$ GHz.

For both of the interferometer scans, a quantitative value for the frequency shift of peak B relative to A is obtained as follows. Assume the distance between the interferometer mirrors, L, is increasing and assume that the last A peak corresponds to an order n such that $\frac{1}{2}n\lambda_A = 1$. The B peak, corresponding to frequency v_B, then occurs at a position shifted by an amount v_m from the position of the $(n + 1)^{th}$ order of the A peak. The B peak occurs at some order m such that $\frac{1}{2}m\lambda_B = L'$, where L' is the new mirror separation. With the convention that $v_m > 0$ when the B peak is shifted toward the last order of the A peak, the actual frequency shift $\Delta v = v_B - v_A$ is given by

$$\Delta v = \left[-n + \frac{mn}{(n + 1) - v_m/v_{FSR}} \right] v_{FSR}, \qquad (31)$$

where v_{FSR} is the free spectral range and $m = n, n \pm 1, n \pm 2, \ldots$.
For L decreasing (v_{FSR} increasing), the difference is given by

$$\Delta v = \left[-n + \frac{mn}{(n - 1) + v_m/v_{FSR}} \right] v_{FSR}. \qquad (32)$$

For the parameters in these experiments, $\lambda \sim 200-300 \ \mu m$, $v_{FSR} \approx 10-20$ GHz, and n and m are large. Then $|m/n| \approx 1$, and the frequency shifts are given to a good approximation by

$$\Delta v = n'v_{FSR} \pm v_m, \qquad (33)$$

where $n' = 0, \pm 1, \pm 2, \ldots$ and where the plus sign in Eq. (33) is for L increasing and the minus sign for L decreasing.

Although the measurement of frequency shifts by this method determines only the shift modulo the free spectral range, it is often the case that for large v_{FSR} all but one of the possible shifts can be ruled out. The actual frequency shift can be confirmed by performing the AB scan sequence for a second free spectral range.

Wavelength measurements from these interferometer scans are not accurate enough to determine small shifts in frequency. They do, however, provide a further check on the assignment of the laser transition responsible for emission.

For the frequency-tuning curves presented here, the Stokes-frequency shifts were obtained by this method. The pump-frequency shifts were obtained in some cases from the grating positions and in some cases by interferometer measurement. The errors arise from several sources. For the grating-position measurement of the pump frequency, the unknown frequency-pulling effects contribute to the error. The determination of the grating position corresponding to the CH_3F absorption-line center or the

CO_2 gain maximum accounts for a common error of $\sim \pm 1$ GHz for both the pump and Stokes reference frequencies. Uncertainty in the peak positions from the interferometer scans contributes an error of $\sim \pm 0.5$ GHz for the FIR measurements and $\sim \pm 0.75$ GHz for the pump-frequency measurements. The total experimental error is presented with each specific set of data.

D. CH_3F FREQUENCY-TUNING DATA

Using the high-pressure CO_2 pump laser and far-infrared (FIR) waveguide laser, frequency-tuning experiments in $^{12}CH_3F$ were performed in the manner described. Far-infrared-emission frequency shifts were measured as a function of the pump-laser frequency shifts for P and R-branch Raman transitions in $^{12}CH_3F$.

The best demonstration of the generation of widely tunable FIR radiation from a CH_3F Raman laser occurs for the R branch (Mathieu and Izatt, 1981; Danly *et al.,* 1983). For these transitions the pump-laser power is high, and the threshold for Raman emission is low. Experimental results for R-branch frequency tuning are shown in Fig. 24. The FIR-emission frequency is plotted as a function of the pump-laser frequency for Raman emission on the $R(19) - R(22)$ transitions.

The experimental error is shown in the upper left; the pump frequency is determined from the CO_2-laser grating position. The theoretical tuning is shown (dashed lines), and the location of the line-center absorptions for each transition are indicated. The far-infrared emission was tunable from 33 to 39 cm^{-1} (256 – 300 μm). As predicted, the tuning is asymmetric about the absorption-line center, with negative pump offsets favored. As the pump frequency is decreased below a line-center frequency, the Raman emission occurs on a single transition and tunes linearly. However, as the pump frequency decreases further, the Raman process ceases operation on its present transition and shifts to operation on the next lower transition. This causes a discontinuous jump in the emission frequency.

These results demonstrate the basic tuning features predicted by the theory for R-branch transitions. They also constitute an important experimental advancement in the generation of high-power, tunable, far-infrared generation. Similar results for R-branch frequency tuning in $^{12}CH_3F$ have been obtained by Izatt and Mathieu (1981); their data further verify this tuning theory.

Experimental results for Raman tuning on the P-branch in $^{12}CH_3F$ have also been obtained; they are shown in Fig. 25. Because of the higher thresholds for P transitions at large J and the lower CO_2-laser output power on the 10-μm R branch, Raman-tuning data were obtained only on the $P(35)$ transition. The location of the $P(35)$ line-center absorption and the 10.2-μm

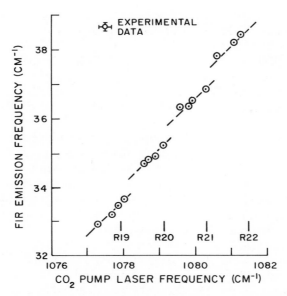

FIG. 24   Experimental results for *R*-branch tuning: ---, theoretical tuning.

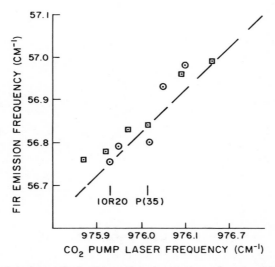

FIG. 25   Experimental results for *P*-branch tuning: O, data referred to CO_2-laser line-center frequency; □, data referred to *P*(35) absorption line center.

$R20$ CO_2-laser transition are also indicated in Fig. 25. Data shown as circles are referred to the CO_2-laser line-center frequency; the pump-frequency shifts were measured by the interferometer for this data. The data shown as squares are referenced to the $P(35)$ absorption-line center; the pump-frequency shifts were calculated from the grating position for these data. Experimental error for the Stokes frequencies is ± 1.5 GHz in both cases. Errors for the pump frequencies are ± 1.75 and ± 3 GHz for the circled and boxed data points, respectively.

Raman tuning by ± 4.5 GHz was obtained about line center. The pump-laser power was not sufficient to pump the Raman transitions any further off-resonance; consequently, the predicted tuning asymmetry for P transitions cannot be verified with these results. Nevertheless, these results serve to demonstrate the feasibility of frequency tuning on P transitions.

For the $R(20)$ Raman transition, data were also taken for a number of closely spaced pump frequencies in an attempt to measure ac Stark shifts. These data are shown in Fig. 26, where the $R(20)$ CH_3F absorption-line center and the 9.22-μm $R20$ CO_2-laser line center are also depicted. For the set of data denoted by circles, the frequency shifts were referenced to the $R(20)$ CH_3F line center. For the set of data denoted by squares, the shifts

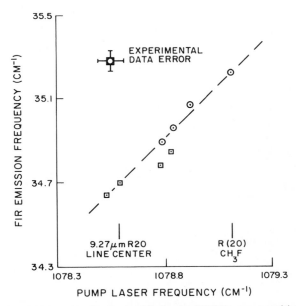

FIG. 26 Experimental study of ac Stark shifts for the $R(20)$ Raman transition: $\bigcirc$, frequency shifts referred to the $R(20)$ CH_3F line center; $\square$, frequency shifts referred to the CO_2-laser line center.

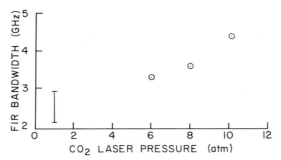

FIG. 27 The bandwidth of far-infrared emission versus CO_2 pump laser pressure. The vertical bar in lower left shows the typical range of FIR bandwidths for atmospheric-pressure CO_2 pump lasers.

were referenced to the CO_2-laser line center. Typical error bars are shown in the upper left.

For the experimental scans in Fig. 26, the measured pump intensity was ~ 4 MW/cm². This pump intensity corresponds to a $\beta_{13}/2\pi = 1.2$ GHz for the $R(20)$, $K = 3$ transition. The dipole moment in β_{13} has been spatially averaged. At these pump intensities, the expected ac Stark shift is smaller than the experimental error. Unfortunately, this precludes any determination of the relevance of the ac Stark shift to frequency-tuning curves. For the pump intensities ($I_p \lesssim 6$ MW/cm², $\beta_{13}/2\pi \lesssim 1.4$ GHz), pump bandwidth (4 GHz), and experimental error encountered in this investigation, the ac Stark shift, if present, is not significant. However, the effects of ac Stark shifts on the frequency-tuning curves may be important and measurable for high pump intensities or narrow-bandwidth, high-resolution studies of tunable FIR Raman lasers.

The bandwidth of the FIR emission depends on the pump-laser bandwidth and the gain. Bandwidth-narrowing effects in amplified spontaneous-emission lasers have been treated in detail by several authors (Allen and Peters, 1972; Casperson and Yariv, 1972; Yariv, 1975). The FIR-emission bandwidth is plotted as a function of the CO_2-laser pressure in Fig. 27. Typical values for atmospheric-pressure laser-pumped lasers are also shown. The pump-laser bandwidth is to some extent dependent on the gain bandwidth, which is pressure dependent. For higher pressures, the pump-laser bandwidth increases roughly linearly with pressure; this produces an increase in the FIR-emission bandwidth. At a pump-laser pressure of 10 atm, the measured pump bandwidth (FWHM) was typically $4-5$ GHz. This implies that gain narrowing of the FIR bandwidth is not significant.

A decrease in output power at the two-photon ($\nu_p + \nu_s$) absorption resonance has been observed by Mathieu and Izatt (1981). Their data are shown

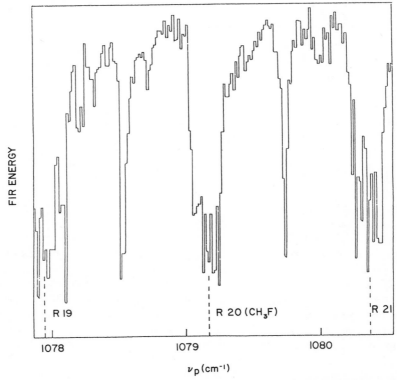

FIG. 28   Output power of a tunable far-infrared laser versus pump-laser frequency. (From Mathieu and Izatt, 1981.)

in Fig. 28. Far-infrared-laser output power is plotted as a function of pump-laser frequency. The decrease in FIR-laser power at the two-photon resonance is quite pronounced. The location of these two-photon absorption holes agrees with the theory presented here.

Insufficient pump-laser power at many frequencies made this study of tunable FIR lasers difficult. In particular, frequency-tuning data for Q-branch transitions in CH_3F were not obtained, because of the low power and inadequate tunability of the CO_2 pump laser on the 9-μm P branch. For the P transitions in CH_3F, the large thresholds (caused by small populations for high J) made it impossible to obtain an extensive set of tuning results with the available pump-laser power. Nevertheless, the experimental results presented here not only demonstrate Raman tuning on R and P branch transitions, but also demonstrate the feasibility of tunable far-infrared generation by stimulated Raman scattering in polyatomic molecules.

E. Suggested Experiments

There are several directions in which this study of tunable far-infrared lasers should be extended. A measurement of either the gain spectrum or the frequency dependence of the threshold pump intensity would provide a good quantitative check of the theoretical gain expression.

The use of gases other than $^{12}CH_3F$ would allow a greater coverage of the FIR region and provide additional tests of the general theory. In particular, Biron *et al.* (1981) have suggested the use of $^{13}CH_3F$ and NH_3; Izatt and Deka (1983) have presented results on frequency tuning in $^{13}CH_3F$. Other possibly useful gases include those known to operate well in conventional high-power FIR lasers: CH_3Cl, CH_3I, CD_3F.

One of the difficulties associated with tunable FIR lasers is the high intensity needed to drive the Raman transition far from resonance. The use of nanosecond or picosecond pump pulses would result in a large increase in peak power. The threshold pump intensities for tunable emission would then be more easily obtained over a much broader range of FIR frequencies and transitions.

A single-mode CO_2 pump laser would provide the narrow bandwidth necessary for high-resolution studies of both the frequency-tuning and ac Stark shifts. Far-infrared emission with a smaller bandwidth than that observed here could also be obtained. However, the K-level energy splittings in symmetric-top molecules will always contribute to large-bandwidth emission in a single-pass FIR laser.

The use of an oscillator for the FIR laser could result in lower pump-intensity thresholds and smaller-bandwidth emission. The optical feedback could be introduced by a periodic waveguide structure (Kneubuhl and Affolter, 1982) or by external mirrors. The problem of the propagation of radiation from the end of a dielectric waveguide to an external mirror and then back into the waveguide has been studied extensively (Abrams, 1972; Belland and Crenn, 1982, 1983; Degnan, 1973). For FIR laser oscillators, one of the most difficult problems remains that of the input and output coupling of the pump and FIR radiation.

IV. Efficiency Enhancement of Far-Infrared Lasers

The efficient, high-power generation of coherent far-infrared radiation remains an important problem for infrared physics. This problem and that of the generation of frequency-tunable FIR radiation are not unrelated. With the development of the previously described tunable FIR source, new approaches to efficiency enhancement can now be pursued experimentally as well as theoretically. The efficiency enhancement of FIR lasers by stimulated multiphoton-emission processes is discussed in this section.

A. BACKGROUND

The relatively low efficiency of optically pumped far-infrared lasers has been acknowledged for quite some time. Many techniques for improving the efficiency have been investigated (Walzer, 1983). Buffer gases have been used with some success in both pulsed and cw systems (Behn *et al.*, 1983; Chang and Lin, 1976; Lawandy and Koepf, 1980). In continuous-wave FIR lasers, the incorporation of waveguides into the FIR cavity leads to a higher wall-collision rate and faster relaxation of the laser final state to the ground state. FIR lasers pumped with circularly polarized pump beams have been shown to have increased power-conversion efficiency (Mansfield *et al.*, 1981; Petuchowski and DeTemple, 1981). It has been demonstrated both theoretically (Panock and Temkin, 1977) and experimentally (Woskoboini-kow *et al.*, 1979) that high-power FIR lasers operate more efficiently on Raman transitions than on line-center transitions.

The photon-conversion efficiency η is defined as the number of photons per unit volume and time emitted at ω_s, divided by the number of photons per unit volume and time absorbed at ω_p:

$$\eta = \frac{|dI(\omega_s)/dz|}{|dI(\omega_p)/dz|} \frac{\hbar\omega_p}{\hbar\omega_s}.$$

For Raman transitions, the conversion efficiency approaches unity in the limit of large Stokes fields; every pump photon is converted into an FIR photon (Panock and Temkin, 1977). The goal of all the aforementioned techniques is to maximize this conversion efficiency.

The overall efficiency is the product of this conversion efficiency and the quantum efficiency, $\eta_Q \equiv \omega_s/\omega_p$. Even for the maximum conversion efficiency ($\eta = 1$), the total (power) efficiency of an optically pumped laser is limited by the quantum efficiency.

The quantum efficiency appears to be the ultimate limit for optically pumped FIR lasers. It is in general quite low for FIR lasers; for a 10-μm pump photon and a 250-μm FIR photon, $\eta_Q = 0.04$. The pump-photon energy not converted into the FIR-photon energy is lost as thermal energy.

A renewed interest has developed in techniques to increase this limit to the overall laser efficiency (Danly *et al.*, 1981; DeLucia *et al.*, 1981; Thomas *et al.*, 1981, 1982). Thomas, Dasari, and Feld have focused on methods for recycling the energy $\hbar(\omega_p - \omega_s)$ remaining after the conversion of one pump photon into one FIR photon. This energy, normally wasted, would be converted back into photons at frequency ω_p.

An alternate method for increasing the quantum efficiency of FIR lasers is based on the stimulated emission of more than one FIR photon for each pump photon. The possibility of multiphoton transitions in atoms and

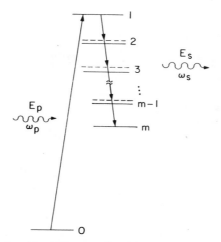

FIG. 29 A diagram of multiphoton emission.

molecules has been recognized for over fifty years (Goeppert-Meyer, 1929, 1931). The application of stimulated multiphoton-emission processes to lasers was first discussed by Sorokin and Braslau (1964) and Prokhorov (1965). However, they focused their attention on nondegenerate two-photon emission, in which the emission occurs at two incommensurate frequencies. The process of degenerate multiphoton emission is considered here; the photons emitted into the radiation field are at the same frequency. The application of this process to the efficiency enhancement of FIR molecular lasers has been discussed (Danly *et al.,* 1981; Thomas *et al.,* 1981, 1982).

In many FIR lasers, cascade and refilling transitions emit FIR photons in addition to those emitted by the optically pumped FIR transition. In these lasers, more than one FIR photon is obtained for each pump photon. However, these additional photons are at different frequencies and hence do not lead to an increased efficiency for generation of radiation at the desired frequency ω_s. In multiphoton emission, as is shown in Fig. 29, the laser emits several FIR photons at the same frequency, ω_s, for each pump photon at ω_p.

For quantum systems with harmonic energy-level spacing, the amplitude for multiphoton emission is quite large because of the simultaneous resonance of every transition with the radiation field. However, the amplitude for multiphoton absorption is even larger for a purely harmonic system. This tendency for a harmonic system to absorb photons from an incident radiation field rather than to emit photons into that field results from the difference in dipole moments. If the oscillator is in a state $|n\rangle$, then

$|\langle n+1|\mathrm{ex}|n\rangle|^2 > |\langle n-1|\mathrm{ex}|n\rangle|^2$. A small degree of anharmonicity is necessary for a net emission of radiation (Twiss, 1958).

Real physical systems exhibit varying degrees of anharmonicity. Lax and Mavroides (1960) and Lax (1961) have suggested the use of Landau levels in semiconductors as a slightly anharmonic system in which multiphoton transitions might be observed. The anharmonic-level spacings result from the nonparabolic energy–momentum bands. The vibrational states of molecules form an anharmonic ladder of states. The physics of multiphoton absorption and ionization in molecules is closely related to multiphoton emission and has been researched extensively (Allen and Stroud, 1982; Bakos, 1977; Bayfield, 1979; Cantrell *et al.*, 1980; Giacobino and Cagnac, 1980; Lambropoulos, 1976; Stenholm, 1979). Multiphoton absorption is of primary importance in spectroscopy and laser-isotope separation. In symmetric-top molecules the rotational-level spacing is also anharmonic; the energy of level J is given by $BJ + BJ^2$. For multiphoton transitions on rotational levels, only the first and last FIR photon begin and end on real molecular levels. All intermediate steps of the multiphoton transition occur between virtual levels. These virtual levels are resonantly enhanced by the adjacent rotational states.

Stimulated multiphoton emission is responsible for the operation of many devices, including klystrons, carcinotrons, and gyrotrons. The possibility of a laboratory maser was first proposed by Schneider (1959). In the quantum picture of the gyrotron or electron-cyclotron maser, electrons are initially prepared in a high-energy (momentum) state and injected into a cavity containing an axial magnetic field. The level spacings are anharmonic because of the relativistic correction to the electron kinetic energy. Sauter (1931) has calculated the allowed energies for a relativistic electron in a homogeneous magnetic field; the energy is given by

$$E = m_0 c^2 [1 + 2(n + \tfrac{1}{2})(\hbar \omega_c / m_0 c^2)]^{1/2} - m_0 c^2$$

for $n = 0, 1, 2, \ldots$. Each electron that is initially in a high-energy state is stimulated to emit many ($\sim 10^8$) photons as it transfers its energy to the electromagnetic field.

Since the first discussion of the application of multiphoton-emission processes to lasers, several theoretical treatments of the two-photon-emission laser have appeared in the literature (Bandilla and Voigt, 1982; Carman, 1975; Narducci *et al.*, 1977). However, these treatments have been limited to nondegenerate two-photon emission and nonresonantly enhanced two-photon emission. Nikolaus *et al.* (1981) have reported the observation of stimulated degenerate two-photon emission in atomic Li; however, their results have been questioned (Jackson and Wynne, 1982).

The many advantages of lasers that operate via multiphoton-emission processes necessitate the further investigation of this important problem.

With the stimulated emission of n photons at frequency ω_s, for each pump photon the quantum efficiency is increased by n over the single-photon quantum efficiency. In this section, the application of stimulated multi-photon-emission processes to optically pumped FIR lasers will be discussed.

B. MULTIPHOTON EMISSION IN SYMMETRIC-TOP MOLECULES

The case of degenerate multiphoton emission within the excited vibrational state of symmetric-top molecules is considered here, although the present discussion can be extended to other molecules. As shown in Fig. 29, a pump laser induces a rotational–vibrational transition in the molecule; this is followed by the simultaneous emission of n photons of frequency $v_{ny} = (E_1 - E_m)/nh$. For a resonant pump field, the multiphoton FIR transition begins on a rotational level J (level 1 in Fig. 29) and ends on a rotational level J_m. For n photon emission, $m = n + 1$ rotational levels are involved and $J_1 = J_m + n$. In the absence of ac Stark shifts, the emission frequency is given by

$$v_{ny} = B(2J_m + n + 1) + \delta'_p/n, \tag{34}$$

where we have included the possibility of a nonresonant pump laser: δ'_p is defined by

$$\delta'_p = v_p - (E_1 - E_0)/h = \delta_p/2\pi.$$

The assumption of negligible ac Stark shifts is somewhat tenuous. To stimulate multiphoton-emission processes, strong FIR fields are needed. In general, these strong fields will shift the position of the molecular levels. The multiphoton resonances derived here are strictly valid only in the limit of small FIR fields. However, in some cases the ac Stark shifts may cancel; this level analysis would be valid in those cases. Grischkowsky et al. (1975) calculated the susceptibility for resonantly enhanced two-photon transitions. Although no pump field was included in their treatment, the first-order Stark shifts that were obtained are appropriate to the problem considered here, provided the pump-laser field is weak. For strong two-photon fields, the upper and lower levels are Stark shifted by equal amounts. Consequently, the two-photon resonance frequency will be unchanged. The offset from the intermediate level will change, and the resonant pump frequency will shift relative to its zero-field value. Nevertheless, the level analysis presented here serves as a lowest-order estimate of the location of the multiphoton resonances.

In n-photon emission, the frequency offset of an intermediate virtual level from the adjacent rotational level J is given by

$$\Delta(J) = B[n(J - J_m) - (J - J_m)^2] + \delta'_p(J - J_m)/n. \tag{35}$$

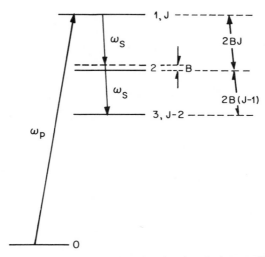

FIG. 30   A diagram of two-photon emission showing that the intermediate virtual level is off-resonance by the rotational constant B.

In two-photon emission with a resonant pump field ($\delta'_p = 0$), the emission frequency is $\nu_{2\gamma} = 2BJ_1 - B$. The intermediate virtual level is off-resonance by the rotational constant B (Fig. 30). This is a reasonably small offset considering FIR emission was obtained with comparable offsets in the frequency-tuning experiments. For off-resonant pumping, the virtual level is off resonance by $\Delta = B + \frac{1}{2}\delta'_p$; negative pump offsets will bring the two-photon-emission virtual level closer to resonance.

For degenerate three-photon emission, as shown in Fig. 31 with a resonant pump field, the emission frequency is given by $\nu_{3\gamma} = 2B(J_1 - 1) + \frac{1}{3}\delta'_p$. For $\delta'_p = 0$, the three-photon frequency $\nu_{3\gamma}$ exactly matches the line-center frequency for the middle rotational transition. In fact, for the emission of any odd number of degenerate photons, the emission frequency exactly equals the line-center frequency for the middle rotational transition. This can be quite advantageous. In multiphoton emission experiments, a laser field at the n-photon frequency must first be produced. This field can then be directed into the multiphoton oscillator or amplifier. When n is odd, the field at the n-photon frequency can be obtained by pumping a single-photon line-center transition in a gas that is the same molecular species as that used in the multiphoton amplifier.

The frequency offsets for three-photon emission are given by $\Delta(J_1 - 1) = 2B + \frac{2}{3}\delta'_p$ and $\Delta(J_1 - 2) = 2B + \frac{1}{3}\delta'_p$. For resonant pumping the offsets are equal. In general, for the emission of n photons the multiphoton frequency and offsets are given by Eqs. (34) and (35).

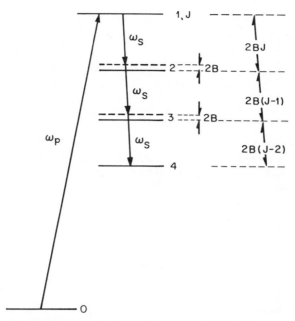

FIG. 31 A diagram of degenerate three-photon emission with a resonant pump field.

The maximum offset for n-photon emission, $\Delta_{\max}$, can be obtained by differentiation of Eq. (35) with respect to J. For n even and resonant pumping, the maximum offset occurs at level $J = J_m + \frac{1}{2}n$, where $\Delta_{\max} = \frac{1}{4}Bn^2$. For n odd and resonant pumping, the maximum offset occurs for two rotational levels, $J = J_m + \frac{1}{2}(n + 1)$ and $J = J_m + \frac{1}{2}(n - 1)$. The frequency offset is given by $\Delta_{\max} = \frac{1}{4}B(n^2 - 1)$. For off-resonant pumping, the maximum offsets occur at the following rotational levels:

$$J = J_m + \tfrac{1}{2}n, \qquad n \text{ even};$$

$$J = J_m + \tfrac{1}{2}(n + 1), \qquad n \text{ odd}, \quad 0 < \delta_p'/2B < 1;$$

$$J = J_m + \tfrac{1}{2}(n - 1), \qquad n \text{ odd}, \quad 0 > \delta_p'/2B > -1.$$

The corresponding frequency offsets are given, respectively, by

$$\Delta_{\max} = \tfrac{1}{4}Bn^2 + \tfrac{1}{2}\delta_p',$$

$$\Delta_{\max} = \tfrac{1}{4}B(n^2 - 1) + \delta_p'(n + 1)/2n,$$

$$\Delta_{\max} = \tfrac{1}{4}B(n^2 - 1) + \delta_p'(n - 1)/2n.$$

Note these offsets do not depend on the initial or final rotational level; they depend only on the rotational constant and the number of photons emitted.

Provided the emission-frequency offset from level 1 (δ_p') is not included, the arithmetic mean offset of the virtual levels from the adjacent rotational levels is given by

$$\langle \Delta \rangle = (n-1)^{-1} \sum_{J=J_m}^{J_m+n-1} \Delta(J) = (n-1)^{-1} \sum_{J'=0}^{n-1} \Delta(J'),$$

$$\langle \Delta \rangle = \tfrac{1}{6}Bn(n+1) + \tfrac{1}{2}\delta_p', \tag{36}$$

where $J' \equiv J_m - J$ and $\Delta(J)$ is given by Eq. (35).

To estimate the saturation intensity for n-photon emission, the geometric mean of the frequency offsets is required. It is well known that, in the perturbation limit, the process of n-photon absorption between an initial and a final state can be described in terms of a generalized Rabi frequency (Allen and Stroud, 1982; Larsen and Bloembergen, 1976). This generalized Rabi frequency is given by

$$\gamma_n = \beta_{12}\beta_{23} \cdots \beta_{n,\,n+1}/\Delta'(J_2)\Delta'(J_3) \cdots \Delta'(J_n),$$

where $\Delta'(J_i) = 2\pi\Delta(J_i)$. The $\Delta(J_i)$ are given by Eq. (35), and for $n = 1$, $\gamma_1 = \beta_{12}$. Systems undergoing multiphoton absorption behave as though they are two-level systems with a Rabi frequency γ_n, provided the $\Delta(J_i)$ are large. The saturation intensity for n-photon emission can be estimated from the condition $\gamma_n^2\tau^2 = 1$.

For $n = 2$, the condition $\gamma_n^2\tau^2 = 1$ agrees with results previously obtained. Grischkowsky *et al.* (1975) have calculated the susceptibility for degenerate two-photon absorption. They used an adiabatic following approximation and described coherent two-photon absorption in terms of the vector model originally developed by Feynman, Vernon, and Hellwarth (1957). The solution presented by Grischkowsky *et al.* is valid for large field strengths. The two-photon small-signal gain is reduced by a factor of two when the condition $\beta^2 = \Delta'/\tau$ is satisfied. This is equivalent to the saturation condition derived from $\gamma_n^2\tau^2 = 1$ for $n = 2$.

Because the dipole moments are approximately equal, the saturation condition for n-photon emission can be expressed as

$$\beta^{2n} = \left[\frac{1}{\tau} \prod_{i=2}^{n} \Delta'(J_i) \right]^2.$$

The n-photon saturation intensity I_{sat}^{ny} is given by

$$I_{\text{sat}}^{ny} = (\hbar^2 c/2\pi\mu^2)\overline{\Delta}^2, \tag{37}$$

where the mean offset $\overline{\Delta}$ is defined by

$$\overline{\Delta} = \left[\tau^{-1} \prod_{k=1}^{n-1} (2\pi Bk(n-k) + \delta_p k/n) \right]^{1/n}.$$

The offsets $\Delta'(J_i) = 2\pi\Delta(J_i)$ are obtained from Eq. (35). For a resonant pump field, this expression can be further simplified:

$$\overline{\Delta} = \{[(2\pi B)^{n-1}/\tau][(n-1)!]^2\}^{1/n},$$

$$\overline{\Delta} \simeq [(n-1)^2/e^2][2\pi(n-1)(2\pi B)^{n-1}/\tau]^{1/n}.$$

The second expression results from the use of Stirling's approximation, $n! = n^n e^{-n}\sqrt{2\pi n}$, and the assumption of large n.

The dipole moment in Eq. (37) is given by $\mu^2 \simeq \frac{1}{6}\mu_0^2$, where μ_0 is the permanent-dipole moment in the excited vibrational state. The n-photon saturation intensity is then given by

$$I_{sat}^{n\gamma} \simeq \frac{3\hbar^2 c}{\pi\mu_0^2}\left\{[(n-1)!]^2\frac{(2\pi B)^{n-1}}{\tau}\right\}^{2/n}. \tag{38}$$

An estimation of the saturation intensity for multiphoton emission in several symmetric top-molecules is shown in Table V for a pressure of $p = 20$ Torr and a homogeneous linewidth of 30 MHz/Torr. For large n, $I_{sat}^{n\gamma}$ varies weakly with the pressure as $p^{2/n}$. The operation of multiphoton amplifiers at the large pump intensities necessary for high conversion efficiency will require reasonably high gas pressures.

The saturation intensities are not unreasonable for high-peak-power FIR systems and $n \lesssim 4$. The use of molecular gases with smaller rotational constants can reduce the virtual-level frequency offsets and the saturation intensities significantly. The fundamental reason for this is that the anharmonic portion of the rotational energy is proportional to B. However, with smaller B the emission frequency for reasonable J values decreases. Consequently, for fixed n the quantum efficiency would also decrease.

In typical FIR laser systems, powers of 1–10 kW are obtained routinely. Far-infrared powers as high as 1.7 MW (5 J in 3 μsec) have been obtained

TABLE V

SATURATION INTENSITY FOR n-PHOTON EMISSION IN SYMMETRIC-TOP MOLECULES WITH RESONANT PUMP FIELD[a]

n	CH$_3$F $B = 25$ GHz $\mu_0 = 1.9$ D	CH$_3$Cl[b] $B = 13.3$ GHz $\mu_0 = 1.9$ D	CH$_3$Br[b] $B = 9.6$ GHz $\mu_0 = 1.8$ D	CH$_3$I[b] $B = 7.5$ GHz $\mu_0 = 1.65$ D
2	0.26	0.14	0.11	0.10
3	2.09	1.2	0.89	0.77
4	14	5.6	3.8	3.1
5	47	17	11	9.1

[a] Intensity values are given in megawatts per square centimeter.
[b] Molecular constants from Townes and Schalow (1955).

from a large-volume 385-μm D_2O laser (Semet *et al.*, 1983). The intensities necessary to saturate an n-photon transition for $n \lesssim 4$ are available with present-day laser systems. Future higher-power FIR sources may make multiphoton emission feasible for $n > 4$. Although multiphoton-emission processes in molecular gases require moderately large stimulating fields, the predicted efficiencies and saturation intensities compare favorably with other high-power FIR sources such as the free-electron laser.

C. DENSITY-MATRIX ANALYSIS OF TWO-PHOTON EMISSION

Two-photon emission in laser amplifiers is the simplest and most likely multiphoton-emission process. Since the first discussion of quantum amplifiers based on two-photon processes (Prokhorov, 1965; Sorokin and Braslau, 1964), there have been several theoretical treatments of the two-photon-emission problem. Carman (1975) employs a set of pulse-evolution-rate equations to describe stimulated nondegenerate two-photon emission in laser amplifiers. There is no intermediate state in his treatment, and Carman shows that, at least within the rate-equation approximation, anti-Stokes stimulated Raman scattering is a strong competitor to straight two-photon emission.

Narducci *et al.* (1977) addressed the problem of a degenerate two-photon laser amplifier with a self-consistent approach. The Bloch equations are used to derive the atomic polarization in terms of the atomic amplitudes for an active medium prepared in a state of inversion between two levels of the same parity. No intermediate state is present, and relaxation processes are not included.

For two-photon emission in optically pumped lasers, the pump-laser field must be included in the treatment of the problem. Furthermore, for multiphoton emission between rotational states in molecules, intermediate levels are present. They result in the important advantage of near-resonant enhancement and the disadvantage of dipole-allowed single-photon transitions.

A theory of two-photon emission in optically pumped lasers can be derived with the density-matrix formalism for arbitrary field strengths and offsets. The level notation is defined in Fig. 30; the nonzero dipole moments are taken as μ_{01}, μ_{12}, μ_{23}. The pump (Stokes) field has amplitude $E_p(E_s)$ and frequency $\omega_p(\omega_s)$. The electric field is written

$$E(t) = \tfrac{1}{2}E_p \exp(i\omega_p t) + \tfrac{1}{2}E_s \exp(i\omega_s t) + \text{c.c.}$$

The electric susceptibility $\chi''(\omega_s)$ is now given by

$$\chi''(\omega_s) = -\text{Im}(2/E_s^*)(\mu_{32}\lambda_{23} + \mu_{21}\lambda_{12}) \tag{39}$$

and includes both the single-photon gain and the two-photon $(2\nu_s)$ gain from levels $1-3$. We assume the same approximations emloyed in the previous discussion of the density matrix. From the density-matrix equations [Eq. (1)] and the ansatz

$$\rho_{12} = \lambda_{12} \exp(-i\omega_s t), \qquad \rho_{13} = \lambda_{13} \exp(-2i\omega_s t),$$

$$\rho_{23} = \lambda_{23} \exp(-i\omega_s t), \qquad \rho_{02} = \lambda_{02} \exp[i(\omega_p - \omega_s)t],$$

$$\rho_{01} = \lambda_{01} \exp(i\omega_p t), \qquad \rho_{03} = \lambda_{03} \exp[i(\omega_p - 2\omega_s)t],$$

the following equations are obtained in the rotating-wave approximation:

$$
\begin{aligned}
L_{12}\lambda_{12} &= \beta_{12}r_{12} + \beta_{32}\lambda_{13} - \beta_{10}\lambda_{02}, \\
L_{23}\lambda_{23} &= \beta_{23}r_{23} - \beta_{21}\lambda_{13}, \\
L_{02}\lambda_{02} &= \beta_{32}\lambda_{03} + \beta_{12}\lambda_{01} - \beta_{01}\lambda_{12}, \\
L_{13}\lambda_{13} &= \beta_{23}\lambda_{12} - \beta_{12}\lambda_{23} - \beta_{10}\lambda_{03}, \\
L_{03}\lambda_{03} &= \beta_{23}\lambda_{02} - \beta_{01}\lambda_{13}, \\
L_{01}\lambda_{01} &= \beta_{01}r_{01} + \beta_{21}\lambda_{02}.
\end{aligned}
\tag{40}
$$

The β_{ij} $(=\beta_{ji}^*)$ and L_{ij} are defined by

$$\beta_{01} \equiv \mu_{01}E_p/2\hbar, \qquad \beta_{21} \equiv \mu_{21}E_s/2\hbar, \qquad \beta_{32} \equiv \mu_{32}E_s/2\hbar,$$

and

$$
\begin{aligned}
L_{12} &\equiv \omega_s - \omega_{12} + i\tau^{-1}, & L_{13} &\equiv 2\omega_s - \omega_{13} + i\tau^{-1}, \\
L_{23} &\equiv \omega_s - \omega_{23} + i\tau^{-1}, & L_{03} &\equiv 2\omega_s - \omega_p + \omega_{30} + i\tau^{-1}, \\
L_{02} &\equiv \omega_s - \omega_p + \omega_{20} + i\tau^{-1}, & L_{01} &\equiv \omega_{10} - \omega_p + i\tau^{-1}.
\end{aligned}
$$

The population-difference equations in the steady-state limit are given by

$$
\begin{aligned}
r_{01} &= r_{01}^\circ + 4\tau \, \mathrm{Im}(\beta_{10}\lambda_{01}) - 2\tau \, \mathrm{Im}(\beta_{21}\lambda_{12}), \\
r_{12} &= r_{12}^\circ - 2\tau \, \mathrm{Im}(\beta_{10}\lambda_{01}) + 4\tau \, \mathrm{Im}(\beta_{21}\lambda_{12}) - 2\tau \, \mathrm{Im}(\beta_{32}\lambda_{23}), \\
r_{23} &= r_{23}^\circ - 2\tau \, \mathrm{Im}(\beta_{21}\lambda_{12}) + 4\tau \, \mathrm{Im}(\beta_{32}\lambda_{23}), \\
r_{13} &= r_{12} + r_{23}.
\end{aligned}
\tag{41}
$$

The general solution for the gain $\alpha(\omega_s)$ is obtained from $\chi''(\omega_s)$ and Eq. (40). The gain is given by

$$\alpha(\omega_s) = (4\pi\omega_s/c)\chi''(\omega_s),$$

where

$$\chi''(\omega_s) = -\text{Im}\,\frac{2}{E_s^*}\left\{\frac{\mu_{21}\beta_{12}}{L_{12}}r_{12} + \frac{\mu_{32}\beta_{23}}{L_{23}}r_{23} - \frac{\mu_{21}\beta_{12}\beta_{01}^2}{L_{12}L_{02}''}\right.$$

$$\times\left[\frac{r_{01}}{L_{01}} - \frac{r_{12}}{L_{12}} - \frac{\beta_{23}^2}{L_{13}'}\left(\frac{1}{L_{03}} + \frac{1}{L_{12}}\right)\left(\frac{r_{12}}{L_{12}} - \frac{r_{23}}{L_{23}}\right)\right]$$

$$+ \frac{\beta_{23}\beta_{12}}{L_{13}''}\left(\frac{\mu_{21}\beta_{32}}{L_{12}} - \frac{\mu_{32}\beta_{21}}{L_{23}}\right)\left[\frac{r_{12}}{L_{12}} - \frac{r_{23}}{L_{23}} - \frac{\beta_{01}^2}{L_{02}'}\right.$$

$$\left.\left.\times\left(\frac{1}{L_{03}} + \frac{1}{L_{12}}\right)\left(\frac{r_{01}}{L_{01}} - \frac{r_{12}}{L_{12}}\right)\right]\right\}$$

$$\equiv \chi_1'' + \chi_2'' + \chi_3'' + \chi_4'' \tag{42}$$

with the definitions

$$L_{13}' \equiv L_{13} - \frac{\beta_{01}^2}{L_{03}} - \frac{\beta_{12}^2}{L_{23}} - \frac{\beta_{23}^2}{L_{12}},$$

$$L_{02}' \equiv L_{02} - \frac{\beta_{01}^2}{L_{12}} - \frac{\beta_{12}^2}{L_{01}} - \frac{\beta_{23}^2}{L_{03}},$$

$$L_{13}'' \equiv L_{13}' - \frac{\beta_{01}^2\beta_{23}^2}{L_{02}'}\left(\frac{1}{L_{03}} + \frac{1}{L_{12}}\right)^2,$$

$$L_{02}'' \equiv L_{02}' - \frac{\beta_{01}^2\beta_{23}^2}{L_{13}'}\left(\frac{1}{L_{03}} + \frac{1}{L_{12}}\right)^2.$$

Although Eq. (42) is a complicated expression, it is the strong-field solution for the gain at frequency ω_s as a function of the populations.

The complete solution is obtained from Eq. (42) together with the strong-field solution for the population differences. We have not solved Eq. (41) for the r_{ij} in this limit. Nevertheless, several observations about the solution can be made. The term $\chi''(\omega_s)$ as given by Eq. (42) is written as the sum of four terms. The first term, χ_1'', corresponds to single-photon emission of the $1 \rightarrow 2$ transition. This term represents the incoherent two-step process: ω_p absorption followed by ω_s emission.

The second term corresponds to the contribution to $\chi''(\omega_s)$ from $2 \rightarrow 3$ transitions. It is nonzero only when $r_{23} \neq 0$ or when molecules are transferred into level 2. Both the coherent and two-step single-photon emission processes leave molecules in level 2. Consequently, this term represents the contributions of both two and three-step processes to two-photon emission at $\omega_s = \omega_{2\gamma}$.

The third term in Eq. (42) contributes to both single and two-photon emission. In the limit $\mu_{23} \rightarrow 0$, $\chi_1'' + \chi_3''$ constitute the exact single-photon-emission result for a three-level system (Panock and Temkin, 1977). The effect of level 3 on this part of χ_3'' is to produce additional Stark shifts of the single-photon gain.

The fourth term in Eq. (42) contributes exclusively to two-photon emission; in the limit $\mu_{23} \to 0$, $\chi_4'' \to 0$. The coherenet three-wave process and the two- and three-step processes are all included in χ_4''. For the coherent three-wave process, the molecule is transferred from level 0 to level 3 with no loss of phase information. The part of χ_4'' proportional to $\beta_{01}^2 r_{01}$ can be identified with this process. In the two-step process the molecule first undergoes the transition $0 \to 1$; it then undergoes the coherent two-photon-emission process corresponding to the transition $1 \to 3$. The three-step process involves the three separate transitions $0 \to 1$, $1 \to 2$, and $2 \to 3$ with a loss of phase at each stage.

A simpler expression for the susceptibility resulting from two-photon emission processes, $\chi_{2\gamma}''$, can be obtained in the limit of small laser fields ($\beta_{01}^2 \tau^2 \ll 1$, $\beta_{12}^2 \tau^2 \ll 1$, $\beta_{23}^2 \tau^2 \ll 1$). The lowest-order contribution to the two-photon emission susceptibility is fourth order in the fields. Inspection of Eq. (42) reveals that the population differences r_{12} and r_{23} are needed to second order, whereas r_{01} is needed to zeroth order in the fields. From Eq. (41), the population differences are given by

$$r_{01} = r_{01}^\circ + \mathcal{O}(\beta^2 \tau^2),$$

$$r_{12} = \frac{2\beta_{01}^2 \tau^2}{1 + \delta_p^2 \tau^2} r_{01}^\circ + \mathcal{O}(\beta^4 \tau^4),$$

$$r_{23} = \mathcal{O}(\beta^4 \tau^4),$$

where the upper three levels are assumed to have no population difference. This is a good approximation for most optically pumped FIR lasers.

The term χ_1'' also contributes to two-photon emission. However, even though $\chi_1'' \simeq \mathcal{O}(\beta^4 \tau^4)$, this term is of higher order than χ_4'' because of an additional $B^2 \tau^2$ in the denominator. The two-photon part of χ_3'' is of higher order than the lowest-order terms of χ_4''.

The small-field limit ($\beta_{01}^2 \tau^2 \ll 1$, $\beta_{12}^2 \tau^2 \ll 1$, $\beta_{23}^2 \tau^2 \ll 1$) of $\chi_{2\gamma}''(\omega_s)$ is then given by

$$
\begin{aligned}
\chi_{2\gamma}''(\omega_s) = \frac{4\hbar}{E_s^2} \frac{\beta_{01}^2 \beta_{12}^2 \beta_{23}^2 \tau^5 r_{01}^\circ}{(1 + z^2)[1 + (x + y)^2]} &\Big(2x'^2 y'(x - y) \\
\times \{(x + y)[2(1 - xy) - x^2] &+ x(1 - xy)\} \\
+ \{(xx' - yy')(xx' + tt') &+ (x' - y')(x' - t') \\
+ (x + y)[(x' - y')(xx' + tt') \\
+ (x' + t')(xx' - yy')]\} &\Big),
\end{aligned}
$$

(43)

where the following dimensionless resonance functions have been defined:

$$x \equiv \delta_{s1}\tau = (\omega_s - \omega_{12})\tau,$$

$$y \equiv \delta_{s2}\tau = (\omega_s - \omega_{23})\tau,$$

$$z \equiv \delta_p\tau = (\omega_p - \omega_{10})\tau,$$

$$t \equiv x + y - z,$$

$$x' \equiv (1 + x^2)^{-1},$$

$$y' \equiv (1 + y^2)^{-1},$$

$$t' \equiv [1 + (x + y - z)^2]^{-1}.$$

The first term of $\chi''_{2\gamma}(\omega_s)$ represents the two-step process leading to two-photon emission, and the second term represents the coherent three-wave contribution.

For the level configuration shown in Fig. 29 with $\delta_p = 0$ and $\delta_{s1} = -\delta_{s2} = -2\pi B$, the total susceptibility [Eq. (42)] at the two-photon frequency $\omega_s = \omega_{2\gamma}$ reduces in the small-field limit to

$$\chi''(\omega_{2\gamma}) = \frac{4\hbar}{E_s^2}\left[\frac{\beta_{01}^2\beta_{12}^2\tau}{(2\pi B)^2}r_{01}^\circ + \frac{6\beta_{01}^2\beta_{12}^2\beta_{23}^2\tau^3}{(2\pi B)^2}r_{01}^\circ\right.$$

$$\left. - \frac{4\beta_{01}^4\beta_{12}^2\tau^3}{(2\pi B)^2}r_{01}^\circ + \frac{\beta_{01}^2\beta_{12}^2}{(2\pi B)^4\tau}r_{01}^\circ + \mathcal{O}\left(\frac{\beta^6\tau}{B^4}, \frac{\beta^8\tau^3}{B^4}\right)\right]. \quad (44)$$

The second term is the two-photon emission contribution; this comes from Eq. (43) with the limits given here. The three remaining terms are single-photon-emission contributions to $\chi''(\omega_{2\gamma})$. These three terms can be obtained from Eq. (6). The last two terms are Stark shifts of the lowest-order term.

This result [Eq. (44)] has been verified by analytical computer calculation using the symbolic manipulation language MACSYMA developed by the Mathlab group at MIT.† This computer calculation serves as a check of the analytical solution leading to Eq. (44). Such a check is extremely valuable in a problem of this nature because there are a large number of algebraic manipulations required in its solution, and it is difficult to check the accuracy of all the calculations. The use of a symbolic-manipulation language should allow the solution of many difficult multiphoton-laser problems in the future.

† The Mathlab group is currently supported, in part, by NASA under grant NSG1323, by the ONR under grant N0014-77-C-0641, by the DOE under grant ET-78-C-02-4687, and by the U.S. Air Force under grant F49620-79-C-020.

D. SELF-EXCITATION AND COMPETING PROCESSES

Two important aspects of multiphoton-emission lasers are self-excitation and the competing single-photon-emission processes. The susceptibility as given by Eq. (44) contains two types of terms. The contribution to $\chi''(\omega_{2\gamma})$ from single-photon emission is proportional to I_p; the contribution from two-photon emission is proportional to $I_p I_s$. In general, the nonlinear susceptibility for n-photon emission (Fig. 29) is related to the fields, dipole moments, and offsets by

$$\chi''(\omega_{n\gamma}) \propto \frac{\tau}{E_s^2} \left| \frac{\beta_{01}\beta_{12}\beta_{23} \cdots \beta_{n,n+1}}{\delta_p \Delta(J_2)\Delta(J_3) \cdots \Delta(J_n)} \right|^2.$$

where $\delta_p \rightarrow 1/\tau$ for a resonant pump field and the $\Delta(J_i)$ are given by Eq. (35). For n-photon emission, $\chi''(\omega_{n\gamma})$ is proportional to $I_p I_s^{n-1}$ and the growth of the radiation field at the n-photon frequency is $dI_s/dz = \alpha I_s$:

$$dI_s/dz \propto (I_p I_s^{n-1}) I_s.$$

Because the gain is proportional to I_s^{n-1}, the radiation field at the multiphoton frequency must be present for the gain to be nonzero. Furthermore, for the n-photon gain to be larger than the single-photon gain, large FIR fields must be present. Multiphoton emission in lasers must be stimulated-emission process. We will therefore focus our attention on the application of stimulated multiphoton-emission processes to laser-power amplifiers and injection-locked oscillators.

One of the primary concerns for multiphoton lasers is the presence of lower-order competing processes. Single-photon emission is by far the most important of these. Unless a strong FIR field is present at the n-photon frequency, the emission of one photon at a line center or Raman frequency is more probable than the emission of n photons. For this reason, much of the work on this problem has focused on the emission of two photons between equal-parity atomic states (Carman, 1975; Narducci et al., 1977; Nikolaus et al., 1981). When single-photon emission is dipole forbidden, two-photon emission becomes one of the lowest-order noncollisional relaxation processes. Atoms transferred to the metastable upper state can then be de-excited by laser radiation at half the transition frequency.

Although the use of metastable states eliminates the competition from single-photon emission, the amplitude for the two-photon process is not resonantly enhanced by the presence of any intermediate quantum levels. For multiphoton FIR emission in molecules, the near-resonant intermediate rotational states increase the n-photon amplitude. However, the presence of intermediate levels of mixed parity also introduces the problem of competition with single-photon-emission processes.

In comparing the competition between single and multiphoton emission, we restrict the discussion to the case of two-photon emission. Similar arguments apply to the stimulated emission of more than two photons. The gain at the two-photon resonance frequency $\alpha(\omega_{2\gamma})$ is obtained from Eq. (44) for resonant pumping in the perturbation limit. This gain is written as $\alpha(\omega_{2\gamma}) = \alpha_1(\omega_{2\gamma}) + \alpha_2(\omega_{2\gamma})$, where $\alpha_1(\omega_{2\gamma})$ is the single-photon-emission contribution to the gain at the two-photon frequency and $\alpha_2(\omega_{2\gamma})$ is the two-photon contribution. For $\alpha_2(\omega_{2\gamma}) > \alpha_1(\omega_{2\gamma})$ or equivalently $6\beta_{23}^2\tau^2 > 1$, the two-photon process will contribute more to the total gain. Large FIR fields are needed for a quantum efficiency twice the single-photon value.

The gain for single-photon emission at the line center or one-photon frequency $\alpha(\omega_{1\gamma})$ is given in these limits by Eq. (44) with $(2\pi B)^2 \to 1/\tau^2$. For $\beta_{23}^2\tau^2 \ll 1$, $\alpha(\omega_{2\gamma}) \simeq \alpha_1(\omega_{2\gamma})$, and the ratio of gains is

$$\alpha(\omega_{2\gamma})/\alpha(\omega_{1\gamma}) = (2\pi B\tau)^{-2} \ll 1.$$

As expected, emission at $\omega_{1\gamma}$ dominates.

For $\beta_{23}^2\tau^2 \gtrsim 1$, the solution for two-photon gain is no longer strictly valid. However, useful estimates of the FIR intensity necessary for stimulated two-photon emission can still be obtained. For $6\beta_{23}^2\tau^2 \gtrsim 1$, $\alpha(\omega_{2\gamma}) \simeq \alpha_2(\omega_{2\gamma})$ and the gain ratio is given by

$$\frac{\alpha(\omega_{2\gamma})}{\alpha(\omega_{1\gamma})} = \frac{\chi''(\omega_{2\gamma})}{\chi''(\omega_{1\gamma})} \simeq \frac{6\beta_{23}^2}{(2\pi B)^2}.$$

When $\alpha(\omega_{2\gamma})/\alpha(\omega_{1\gamma}) > 1$, the stimulated two-photon process will dominate the emission. This condition reduces to a condition on the stimulating two-photon intensity:

$$I_2 \gtrsim 2\pi\hbar^2 c \left(\frac{B}{\mu_0}\right)^2,$$

$$I_2 \gtrsim 21 \left(\frac{B\,[\text{GHz}]}{\mu_0\,[\text{Debye}]}\right)^2 \quad [\text{kW/cm}^2].$$

In this expression the transition-dipole moment has been spatially averaged; μ_0 is the permanent-dipole moment. For CH_3F, CH_3Cl, CH_3Br, and CH_3I, this intensity is 3600, 1030, 600, and 430 kW/cm². These values are larger than the estimated saturation intensities.

A more rigorous treatment of the competition between single- and two-photon emission can be obtained from the evaluation of Eq. (42) at $\omega_s = \omega_{1\gamma}$ and $\omega_s = \omega_{2\gamma}$. The full-field solution for the populations is required.

This approximate treatment nevertheless indicates that the operation of high-power FIR amplifiers by stimulated multiphoton transitions is feasible. Furthermore, as with the single-photon Raman laser, the multiphoton gain of these amplifiers can be tuned to any desired frequency by varying the frequency of the amplifier pump laser.

There are several possible methods for suppressing single-photon emission. For near-resonant pumping, the emission pulse at line center is delayed relative to the pump pulse (Ahmed and Nicholson, 1983). Emission at the Raman frequency occurs simultaneously with the pump pulse. Consequently, provided both the amplifier pump pulse and the n-photon stimulating field temporally overlap, emission at the line center may not have time to build up.

Alternatively, off-resonant pumping of the amplifier will produce stimulated Raman scattering only if the pump intensity is larger than the threshold value. For large FIR fields and pump intensities below this threshold, stimulated multiphoton emission should be observable. A high-Q FIR oscillator injection-locked at the multiphoton frequency might also prove adequate to suppress the unwanted line-center or single-photon Raman emission. Another approach might be to use frequency-selective elements in the cavity, such as gratings, to reduce the single-photon gain. Nevertheless, amplified spontaneous emission will always cause some decrease in the multiphoton gain.

E. SUGGESTED EXPERIMENTS

The preceding discussion of multiphoton-emission processes in optically pumped far-infrared (FIR) lasers, although cursory, serves to outline several of the important aspects of this problem. These processes are important for both practical applications and fundamental research. Further theoretical and experimental work in this area would produce a better understanding of multiphoton-emission processes and a possible technique for FIR-laser efficiency enhancement. Several experiments are suggested.

Two- and three-photon emission experiments in a variety of molecular gases are possible with the high-power, tunable FIR laser discussed in Sections II and III. Two-photon experiments require smaller FIR laser intensities. For resonant pump fields, FIR radiation at the three-photon emission frequency is easily generated by line-center emission.

In two- or three-photon amplifier experiments, a pump field at frequency ω_p and an FIR field at frequency ω_s are both coupled into an amplifier section. Any of the well-known FIR laser gases such as $^{12}CH_3F$, $^{13}CH_3F$, or other methyl halides could be used in the amplifier. By measuring the amplifier gain as a function of ω_s and ω_p, one could look for peaks in the gain at the two- or three-photon resonance frequencies. The gain spectrum near the multiphoton resonances can be obtained by either of two methods. In the first case, the pump frequency is fixed; the stimulating field at frequency

ω_s, obtained from a tunable FIR laser, is swept through the resonance. Alternatively, the frequency of the stimulating FIR field is fixed and the amplifier pump frequency is swept through the multiphoton resonance frequency.

In an injection-locked FIR oscillator experiment, a high-Q FIR cavity can be tuned to the multiphoton resonance. Radiation at the multiphoton frequency is then injected into the oscillator to initiate the emission process. The spectral characteristics of the oscillator output will then depend on the cavity Q, the input frequencies ω_p and ω_s, and the input powers. The injected FIR intensity necessary to produce multiphoton emission should be significantly reduced by the presence of a cavity mode at that frequency.

The use of nanosecond and subnanosecond pump and FIR pulses would result in higher peak powers and larger amplitudes for these n-photon processes. However, the maximum obtainable peak powers may be limited by gas breakdown in air. Higher pressures for optimum conversion efficiency would be required. An additional possibility involves the use of high-pressure buffer gases. These buffer gases could collisionally broaden the molecular levels and effectively reduce the offset from resonance of the multiphoton virtual levels.

V. Summary

The theoretical analysis and experimental demonstration of a high-power, frequency-tunable far-infrared laser have been presented. The density-matrix theory of the double Raman process and the general analysis of frequency tuning in symmetric-top molecules have been described. Gain spectra and Raman tuning curves were calculated for P, Q, and R pump transitions.

Experimental results obtained with a frequency-tunable CO_2-laser-pumped CH_3F waveguide laser were presented for R and P-branch pump transitions. There is good agreement between theory and experiment. High peak power ($1-10$ kW/cm^2), frequency-tunable emission has been obtained in the $256-300$- and $175-176$-μm wavelength ranges on the R and P branches in CH_3F. The application of stimulated Raman scattering processes to the generation of continuously tunable radiation in the FIR region of the spectrum has been demonstrated; the straightforward extension of the principles to other FIR laser gases should produce tunable laser radiation at other FIR frequencies. In particular, $^{13}CH_3F$ is a promising candidate for the generation of tunable radiation at wavelengths up to 1.22 mm.

The problem of the efficiency enhancement of high-power FIR lasers has also been addressed. Stimulated multiphoton emission in optically pumped FIR lasers has been discussed. The saturation intensity for n-photon emission has been estimated, and a density-matrix treatment of two-photon emission in optically pumped lasers was presented. Several multi-photon experiments have been suggested. The application of tunable FIR lasers to the investigation of multiphoton processes in molecules is an important new direction for far-infrared laser research.

REFERENCES

Abrams, R. L. (1972). *IEEE J. Quantum Electron.* **QE-8**, 838.
Aggarwal, R. L., and Lax, B. (1977). *In* "Nonlinear Infrared Generation" (Y. R. Shen, ed.), pp. 19–80. Springer-Verlag, Berlin and New York.
Ahmed, H., and Nicholson, J. P. (1983). *IEEE J. Quantum Electron.* **QE-19**, 256.
Alcock, A. J., Leopold, K., and Richardson, M. C. (1973). *Appl. Phys. Lett.* **23**, 562.
Alcock, A. J., Fedosejevs, R., and Walker, A. C. (1975). *IEEE J. Quantum Electron.* **QE-11**, 767.
Allen, L., and Peters, G. I. (1972). *J. Phys. A* **5**, 695.
Allen, L., and Stroud, C. R. (1982). *Phys. Rept.* **91**, 1.
Arimondo, E., and Inguscio, M. (1979). *J. Mol. Spectrosc.* **75**, 81.
Bagratashvili, V. N., Knyazev, I. N., Letokhov, V. S., and Lobko, V. V. (1976). *Sov. J. Quantum Electron.* **6**, 541.
Bakos, J. S. (1977). *Phys. Rept.* **31**, 209.
Bandilla, A., and Voigt, H. (1982). *Opt. Commun.* **43**, 277.
Bayfield, J. E. (1979). *Phys. Rept.* **51**, 317.
Behn, R., *et al.* (1983). *Proc. 7th Int. Conf. Infrared and Millimeter Waves,* Marseille, p. 257. IEEE, New York.
Belland, P., and Crenn, J. P. (1982). *Appl. Opt.* **21**, 522.
Belland, P., and Crenn, J. P. (1983). *Opt. Commun.* **45**, 165.
Bernard, P., Mathieu, P., and Izatt, J. R. (1981). *Opt. Commun.* **37**, 285.
Berry, A. J., Hanna, D. C., and Hearn, D. B. (1982). *Opt. Commun.* **43**, 229.
Biron, D. G. (1980). Ph. D. Thesis, Physics Dept. Massachusetts Institute of Technology, Cambridge, Massachusetts.
Biron, D. G., Danly, B. G., Temkin, R. J., and Lax, B. (1981). *IEEE J. Quantum Electron.* **QE-17**, 2146.
Bloembergen, N. (1965). "Nonlinear Optics." Benjamin, Reading, Massachusetts.
Bloembergen, N. (1967). *Am. J. Phys.* **35**, 989.
Brewer, R. G., and Hahn, E. L. (1975). *Phys. Rev. A* **11**, 1641.
Cantrell, C. D., Letokhov, V. S., and Makarov, A. A. (1980). *In* "Coherent Nonlinear Optics: Recent Advances" (M. S. Feld and V. S. Letokhov, eds.). Springer-Verlag, Berlin and New York.
Carman, R. L. (1975). *Phys. Rev. A* **12**, 1048.
Casperson, L. W., and Yariv, A. (1972). *IEEE J. Quantum Electron.* **QE-8**, 80.
Chang, T. Y. (1977). *In* "Nonlinear Infrared Generation" (Y. R. Shen, ed.), pp. 215–272. Springer-Verlag, Berlin and New York.
Chang, T. Y., and Bridges, T. J. (1970). *Opt. Commun.* **1**, 423.
Chang, T. Y., and Lin, C. (1976). *J. Opt. Soc. Am.* **66**, 362.
Cotter, D., Hanna, D. C., and Wyatt, R. (1975). *Appl. Phys.* **8**, 333.
Danly, B. G., and Temkin, R. J. (1980). *IEEE J. Quantum Electron.* **QE-16**, 587.
Danly, B. G., Temkin, R. J., and Lax, B. (1981). *Proc. 6th Int. Conf. Infrared and Millimeter Waves,* Miami Beach, TALK F-4-1, IEEE, New York.
Danly, B. G., Evangelides, S. G., Temkin, R. J., and Lax, B. (1983). *Proc. 7th Int. Conf. Infrared and Millimeter Waves,* Marseille, France. p. 292

Degnan, J. J. (1973). *Appl. Opt.* **12**, 1026.
Degnan, J. J. (1976). *Appl. Phys.* **11**, 1.
Deka, B. K., Dyer, P. E., and Winfield, R. J. (1981). *Opt. Commun.* **39**, 255.
DeLucia, F. C., Herbst, E., Feld, M. S., and Happer, W. (1981). *IEEE J. Quantum Electron.* **QE-17**, 2171.
DeMartino, A., Frey, R., and Pradere, F. (1978). *Opt. Commun.* **27**, 262.
DeMartino, A., Frey, R., and Pradere, F. (1980). *IEEE J. Quantum Electron.* **QE-16**, 1184.
DeTemple, T. A. (1979). *In* "Infrared and Millimeter Waves," vol. 1 (K. J. Button, ed.), pp. 129–184. Academic Press, New York.
Drozdowicz, Z. (1978). Ph. D. Thesis, Physics Dept., Massachusetts Institute of Technology, Cambridge, Massachusetts.
Drozdowicz, Z., Temkin, R. J., and Lax, B. (1979a). *IEEE J. Quantum Electron.* **QE-15**, 170.
Drozdowicz, Z., Temkin, R. J., and Lax, B. (1979b). *IEEE J. Quantum Electron.* **QE-15**, 865.
Ducuing, J., Frey, R., Pradere, F. (1976). *Proc. Loen Conf. Tunable Lasers and Applications* (A. Mooradian, T. Jaeger, and P. Stokseth, eds.). Pp. 81–87. Springer, Berlin.
Dupertuis, M. A., Salomaa, R., and Siegrist, M. R. (1983a). *Proc. 7th Int. Conf. Infrared and Millimeter Waves,* Marseille, France. IEEE, New York.
Dupertuis, M. A., Salomaa, R., and Siegrist, M. R. (1983b). *Proc. 8th Int. Conf. Infrared and Millimeter Waves,* Miami Beach, Florida. Talk TH2.6, IEEE, New York.
Dupertuis, M. A., Rainer, R., Salomaa, R., and Siegrist, M. R. (1984). *IEEE J. Quantum Electron* **QE-20**, 440.
Fetterman, H. R., *et al.* (1979). *Appl. Phys. Lett.* **34**, 123.
Feynman, R. P., Vernon, Jr., F. L., and Hellwarth, R. W. (1957). *J. Appl. Phys.* **28**, 49.
Finkelstein, V. Yu. (1982). *Phys. Rev. A* **27**, 961.
Freund, S. M., Duxbury, G., Romheld, M., Tiedje, J. T., and Oka, T. (1974). *J. Mol. Spectrosc.* **52**, 38.
Frey, R., Pradere, F., and Ducuing, J. (1977). *Opt. Comm.* **23**, 65.
Giacobino, E., and Cagnac, B. (1980). *In* "Progress in Optics," vol. 17 (E. Wolf, ed.) pp. 87–161. North-Holland Publ., Amsterdam.
Gibson, R. B. (1979). Ph. D. Thesis, Physics Department, Massachusetts Institute of Technology, Cambridge, Massachusetts.
Gibson, R. B., Javan, A., and Boyer, K. (1978). *Appl. Phys. Lett.* **32**, 726.
Goeppert-Mayer, M. (1929). *Naturwissenschaften* **17**, 932.
Goeppert-Mayer, M. (1931). *Ann. Phys. (Leipzig)* **9**, 273.
Gondhalekar, A., Heckenberg, N. R., and Holzhauer, E. (1975). *IEEE J. Quantum Electron.* **QE-11**, 103.
Gordy, W. (1960). *In* "Microwave Research Institute Symposia," vol. 9 (J. Fox, ed.), pp. 1–23. Brooklyn Polytech. Press, Brooklyn, New York.
Grischkowsky, D., Loy, M. T. T., and Liao, P. F. (1975). *Phys. Rev. A* **12**, 2514.
Hall, D. R., Gorton, E. K., Jenkins, R. M. (1976). *J. Appl. Phys.* **48**, 1212.
Hartig, W., and Schmidt, W. (1979). *Appl. Phys.* **18**, 235.
Henningsen, J. O. (1977). *IEEE J. Quantum Electron.* **QE-13**, 435.
Hertzberg, G. (1945). "Infrared and Raman Spectra." Van Nostrand, Princeton, New Jersey.
Hirshfield, J. L. (1979). *In* "Infrared and Millimeter Waves," vol. 1 (K. J. Button, ed.) pp. 1–54. Academic Press, New York.
Hodges, D. T., Tucker, J. R., and Hartwick, T. S. (1976). *Infrared Phys.* **16**, 175.
Izatt, J. R., and Deka, B. K. (1983). *Proc. 7th Int. Conf. Infrared and Millimeter Waves,* Marseille, France.
Izatt, J. R., and Mathieu, P. (1981). *Proc. 6th Int. Conf. Infrared and Millimeter Waves,* Miami Beach, Florida.

Jackson, D. I., and Wynne, J. J. (1982). *Proc. 12th Int. Quantum Electron.Conf.,* Munich, Federal Republic of Germany, p. 238.

Jassby, D. L., Cohn, D. R., Lax, B., and Halverson, W. (1974). *Nucl. Fusion* **14**, 745.

Javan, A. (1957). *Phys. Rev.* **107**, 1579.

Johns, J. W. C., McKellar, A. R. W., Oka, T., and Romheld, M. (1975). *J. Chem. Phys.* **62**, 1488.

Kantorowicz, G., and Palluel, P. (1979). *In* "Infrared and Millimeter Waves," vol. 1 (K. J. Button, ed.), pp. 185–212. Academic Press, New York.

Kneubuhl, F. K., and Affolter, E. (1982). *In* "Infrared and Millimeter Waves," vol. 5 (K. J. Button, ed.), pp. 305–337. Academic Press, New York.

Lambropoulos, P. (1976). *In* "Advances in Atomic and Molecular Physics," vol. 12 (D. R. Bates and B. Bederson, eds.), pp. 87–164. Academic Press, New York.

Larsen, D. M., and Bloembergen, N. (1976). *Opt. Commun.* **17**, 254.

Lawandy, N. M., and Koepf, G. A. (1980). *Opt. Lett.* **5**, 336.

Lax, B. (1961). *Proc. 2nd Int. Conf. Quantum Electron.,* Berkeley, California (J. R. Singer, ed.). Columbia Univ. Press, New York.

Lax, B. (1982). *In* "Infrared and Millimeter Waves," vol. 5 (K. J. Button, ed.), pp. 1–28. Academic Press, New York.

Lax, B., and Mavroides, J. G. (1960). *Solid State Phys.* **11**, 261.

Lin, Y., and Gong, D. (1982). *Acta Optica Sinica* **2**, 63.

Loudon, R. (1973). "The Quantum Theory of Light." Oxford Univ. Press, London.

Loy, M. M. T., and Roland, P. (1977). *Rev. Sci. Instrum.* **48**, 554.

Mansfield, D. K., Tesauro, G. J., Johnson, L. C., and Semet, A. (1981). *Opt. Lett.* **6**, 230.

Marcatili, E. A. J., and Schmeltzer, R. A. (1964). *Bell Syst. Tech. J.* **43**, 1783.

Marcuse, D. (1980). "Principles of Quantum Electronics." Academic Press, New York.

Mathieu, P., and Izatt, J. R. (1981). *Opt. Lett.* **6**, 369.

Mizuno, K., and Ono, S. (1979). *In* "Infrared and Millimeter Waves," vol. 1 (K. J. Button, ed.) pp. 213–233. Academic Press, New York.

Morikawa, E. (1977). *J. Appl. Phys.* **48**, 1229.

Narducci, L. M., Eidson, W. W., Furcinitti, P., and Eteson, D. C. (1977). *Phys. Rev. A.* **16**, 1665.

Nikolaus, B., Zhang, D. Z., and Toschek, P. E. (1981). *Phys. Rev. Lett.* **47**, 171.

Oka, T. (1973). *In* "Advances in Atomic and Molecular Physics," vol. 9 (D. R. Bates and I. Estermann, eds.), pp. 127–206. Academic Press, New York.

Osche, G. R. (1978). *J. Opt. Soc. Am.* **68**, 1293.

Panock, R. L., and Temkin, R. J. (1977). *IEEE J. Quantum Electron.* **QE-13**, 425.

Peebles, W. A., Brower, D. L., Luhmann, N. C., and Danielewicz, E. J. (1980). *IEEE J. Quantum Electron.* **QE-16**, 505.

Petuchowski, S. J., and DeTemple, T. A. (1981). *Opt. Lett.* **6**, 227.

Polanyi, J. C., and Woodall, K. B. (1972). *J. Chem. Phys.* **56**, 1563.

Prokhorov, A. M. (1965). *Science* **149**, 828.

Rogowski, W. (1941). *In* "Gaseous Conductors: Theory and Engineering Applications" (J. D. Cobine, ed.), pp. 177–181. Dover, New York.

Rolland, C., Reid, J., Garside, B. K., Jessop, P. E., and Morrison, H. D. (1983). *Opt. Lett.* **8**, 36.

Rosenbluh, M., Temkin, R. J., and Lax, B. (1976). *Appl. Opt.* **15**, 2635.

Sauter, F. (1931). *Z. Phys.* **69**, 742.

Schneider, J. (1959). *Phys. Rev. Lett.* **2**, 504.

Semet, A., Johnson, L. C., and Mansfield, D. K. (1983). Princeton Plasma Physics Laboratory Rept. PPPL-1741.

Shen, Y. R., and Bloembergen, N. (1965). *Phys. Rev.* **137**, A1787.

Shoemaker, R. L., Stenholm, S., and Brewer, R. G. (1974). *Phys. Rev. A* **10**, 2037.

Siegrist, M. R., Morgan, P. D., and Green, M. R. (1978). *J. Appl. Phys.* **49**, 3699.

Smith, P. W. (1972). *Proc. IEEE* **60**, 422.

Sorokin, P. P., and Braslau, N. (1964). *IBM J. Res. Dev.* **8**, 177.

Sorokin, P. P., Wynne, J. J., and Lankard, J. R. (1974). *Appl. Phys. Lett.* **33**, 1183.

Sprangle, P., Smith, R. A., and Granatstein, V. L. (1979). *In* "Infrared and Millimeter Waves," vol. 1 (K. J. Button, ed.), pp. 279–327. Academic Press, New York.

Stenholm, S. (1979). *Contemp. Phys.* **20**, 37.

Strumia, F., and Inguscio, M. (1981). *In* "Infrared and Millimeter Waves," vol. 5 (K. J. Button, ed.), pp. 130–213. Academic Press, New York.

Takami, M. (1976a). *Japan J. Appl. Phys.* **15**, 1063.

Takami, M. (1976b). *Japan J. Appl. Phys.* **15**, 1889.

Temkin, R. J., and Cohn, D. R. (1976). *Opt. Commun.* **16**, 213.

Thomas, J. E., Dasari, R. R., Feld, M. S. (1981). *Proc. 6th Int. Conf. Infrared and Millimeter Waves,* Miami Beach, Florida, *IEEE* Cat. No. 81 CH 1645-1 MTT.

Thomas, J. E., Dasari, R. R., and Feld, M. S. (1982). *Int. J. Infrared Submil. Waves* **3**, 137.

Townes, C. H., and Schalow, A. L. (1955). "Microwave Spectroscopy." McGraw-Hill, New York.

Trappeniers, N. J., and Elenbaas-Bunshoten, E. W. A. (1979). *Chem. Phys. Lett.* **64**, 205.

Twiss, R. Q. (1958). *Aust. J. Phys.* **11**, 564.

Vass, A., Davis, B. W., Firth, W. J., and Pidgeon, C. R. (1982). *Appl. Phys. B* **29**, 131.

Walzer, K. (1983). *In* "Infrared and Millimeter Waves," vol. 7 (K. J. Button, ed.), pp. 119–163. Academic Press, New York.

Weitz, E., Flynn, G., and Ronn, A. M. (1972). *J. Chem. Phys.* **56**, 6060.

Wiggins, J. D., Drozdowicz, Z., and Temkin, R. J. (1978). *IEEE J. Quantum Electron.* **QE-14**, 23.

Wood, O. R. (1974). *Proc. IEEE* **62**, 355.

Woskoboinokow, P., Praddaude, H. C., Mulligan, W. J., and Lax, B. (1979). *J. Appl. Phys.* **50**, 1125.

Woskoboinokow, P., *et al.* (1981). *Proc. 6th Int. Conf. Infrared and Millimeter Waves,* Miami Beach, Florida, *IEEE* Cat. No. 81 CH 1645-1 MTT.

Wynne, J. J., and Sorokin, P. P. (1977). *In* "Nonlinear Infrared Generation" (Y. R. Shen, ed.), pp. 159–214. Springer-Verlag, Berlin and New York.

Yariv, A. (1975). "Quantum Electronics." Wiley, New York.

CHAPTER 6

Far-Infrared Laser Scanner for High-Voltage Cable Inspection*

P. K. Cheo

United Technologies Research Center
East Hartford, Connecticut

I. Introduction

Extruded high-molecular-weight (HMW) and cross-linked (XL) polyethylene are solid dielectric materials commonly used for insulating the conductor employed in the underground electric-power distribution network. The intrinsic dielectric strength of polyethylene has been rated as high as 800 kV/mm. However, the value specified for in-service cables has been greatly derated by the industry to enhance system reliability. Typical operating stresses are limited to values ranging from 1 to 3 kV/mm. Even at these levels, the number of power outages caused by in-service cable failure is increasing at an alarming rate. It has been well established that the catastrophic breakdown of underground systems develops during an aging process known as "treeing." The growth of a tree inside the insulation originates at a

* This work is funded by the Electric Power Research Institute under Contract RP794.

site of an imperfection such as a void or a contaminant. In time, the tree branches will extend across the insulation wall. The useful life of the cable is dictated by the quality of insulation material and operating parameters as well as by environmental condition. To detect these defects while manufacturing the cable, we have introduced a laser-scattering technique (Cheo, 1978; Cantor *et al.*, 1981) that provides a signature of the defect inside the insulation in real time. Through Mie scattering, the signature of a spherical void can be correlated to the size of the void. To demonstrate this scheme, we chose the 118.8-μm CH_3OH laser for three reasons: at 118.8 μm, the lattice absorption in polyethylene is significantly reduced; a defect as small as $\lambda/4$ can be detected without difficulty; and the wavelength is long enough to avoid diffuse scattering from crystalline boundaries.

This chapter describes a cable-inspection system that is a continuation of our previous work (Cheo, 1978; Cantor *et al.*, 1981) and that has been developed to be suitable for use in a cable factory. The system design criteria are based on two important considerations: reliability and practicality. The maximum speed of extruding solid-dielectric cable insulation is about 90 ft/min. To provide a continuous and complete inspection with a full coverage of the cable insulation, a high-speed optical scanner has been formulated that produces a 360° rotation of a far-infrared laser beam around a moving cable without being obstructed by the central conductor. Using a focused beam with a diameter $w_0 = 3$ mm, it is necessary to rotate the laser beam at an angular speed of 9000 rpm to provide at least 50% overlap of the beam around the cable insulation moving at the maximum speed. The laser inspection data are collected by high-speed tracking optics synchronized with the illuminating optics and are detected by liquid-helium-cooled gallium-doped germanium photoconductor. High-speed data acquisition, signal processing, logging, and reduction are carried out in real time by a dedicated microprocessor. System reliability is ultimately dependent on the performance of the far-infrared laser. For production-line use, a fully automated, hands-off operation of the laser is required. Therefore, a laser structure must be designed in such a way that its power output and frequency will not fluctuate when it is subjected to large temperature variations (40–120°F) and the noisy environment of the factory floor. In addition, the minimum time between major servicing must be greater than 2000 h.

System practicality is established by the actual performance of a computer-interactive system that can be operated by untrained personnel and that can be used conveniently to inspect all sizes of high-voltage cables. This chapter presents a detailed description of the far-infrared laser structure, the stabilization technique, and the high-speed optical scanner. It also briefly describes the mechanical and electronic control mechanisms that are responsible for obtaining an extremely stable passive and active laser cavity, which is absolutely essential for use in cable factories.

II. System Requirements and Performance Estimates

The specifications for HMW and XLPE insulated power cables rated from 5 through 69 KV are established and reviewed each year by the Association of Edison Illuminating Companies (AEIC). Presently, the specifications call for nearly perfect cable insulation that must be free from voids larger than 50 μm and contaminants larger than 250 μm. Either a short piece or a complete reel of the cable must be subjected to a variety of electrical tests. At present, visual inspection of voids and contaminants is also required. This is done by slicing two ends of a reel of cable into wafers and examining these wafers with a microscope. The laser inspection system offers a unique capability to perform nondestructive and complete inspection during manufacturing. With this method users of polyethylene-insulated cables may be assured that the quality over the entire length of the cable truly complies to the AEIC specifications.

The laser system, as shown in Fig. 1, is designed to perform real-time, on-line inspection of complete distribution-type cables of all sizes. To provide complete coverage of the insulation during extrusion, the laser beam must illuminate the entire insulation. Because the scattered signal from an inclusion is usually very weak, it is necessary to maintain a focused laser beam over the entire path length inside the insulation. For a spot size of about 3 mm, it is necessary to produce a 360° rotation of the beam around the cable at a maximum angular speed of 9000 rpm, to completely illuminate the cable insulation moving at 90 ft/min. It is possible to rotate mirrors at this speed, provided that the mirror module is properly designed so that it can be dynamically balanced at high speeds. As the conductor size increases, the extrusion speed decreases. Therefore the scanner is capable of inspecting the entire class of underground power-distribution cables. Major features of the cable inspection system required by the utility industry are listed here:

(1) fully automatic, hands-off operation;
(2) reliable for use in cable-factory environment;
(3) industrial safety certification;
(4) greater than 95% of cable insulation at a maximum extrusion speed of 90 ft/min;
(5) applicable to all HMW and XLPE distribution cables;
(6) detects voids, contaminants, protrusions, and surface irregularities;
(7) minimum detectable size 50 μm;
(8) real-time signal processing including data acquisition, computation, recording, presentation, and storage.

The size of an insulation defect can be accurately determined from the measurement of scattered laser power (Cantor *et al.*, 1981). Scattered sig-

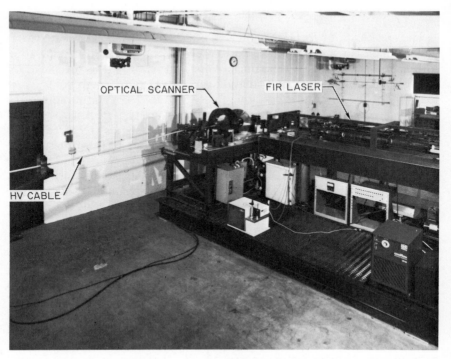

FIG. 1 Photograph showing the prototype far-infrared (FIR) laser scanner system under construction.

nals from spherical voids have spherical symmetry about the axis of propagation. Therefore, a distinction between a void and a contaminant can be made by a symmetry test. However, the present system does not have this capability. To perform this task, a multiple-detection-system configuration is required. The system in its present form can only determine the presence of an insulation defect during manufacturing, the exact size of a void or the average cross section of a contaminant.

For cable factory applications, the far-infrared laser and scanner system must be fully automated by using electromechanical devices and microprocessors to control and command its operation. The system must also comply with industrial safety codes. Detailed descriptions of various system components will be given in the following sections. In this section, the description of the system as a whole will first be introduced. Figure 2 is a block diagram of the entire system. It consists primarily of three subsystems: laser module, scanner module, and processor and control module.

To start up the laser system, the operator initiates a command sequence to

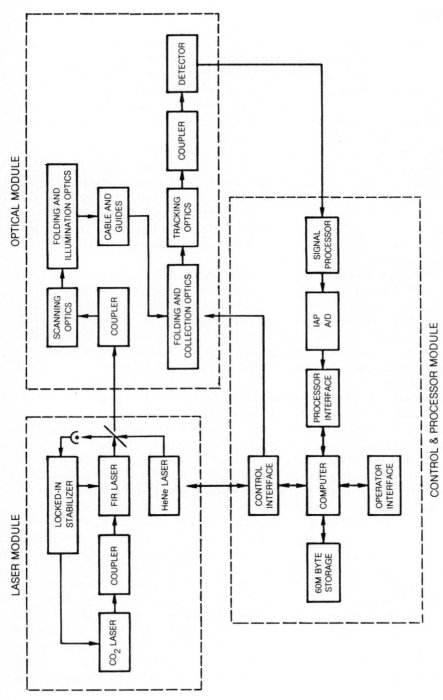

Fig. 2 A block diagram of a far-infrared laser system.

the electronic controller, which turns on various electronics, vacuum pumps, gas mixtures, water cooling, and high voltage. System parameters are monitored continuously by placing sensors at several crucial locations. If any malfunction occurs, the system is protected by instantly shutting off the high voltage and the gas mixture. System safety is provided by using interlocking devices that are programmed with a time-sequence logic to avoid operator error. During extrusion, laser-scattering signals detected by the liquid-helium-cooled Ga:Ge photoconductor are logged in conjunction with measurements of cable length, extrusion speed, and scanner speed. The signal processor will precondition the signal waveform by filtering and discriminating the signal from the noise. The computer will perform real-time analysis of signals to provide information on the size and number density of the defects. The acquired data can be stored, statistically analyzed, and presented in a variety of ways in accordance with the manufacturer's requirements.

The system performance can be estimated with respect to its specific configuration and the existing component characteristics. The analysis is made by taking into account the following factors: far-infrared laser intensity I_0, watts per square centimeter; optical configuration; environmental conditions; cable dimensions ℓ, in centimeters; differential scattering cross section $\sigma(\theta, \varphi)$, in square centimeters per steradian; material absorption $\alpha(\lambda, \ell)$; surface scattering; detector noise and frequency response; and probability of false alarm.

Because this system is based on a simple method utilizing direct detection of forward-scattered signals, it is necessary to consider only the optical paths for the incident and the scattered far-infrared laser radiation. Figure 3 depicts optical paths of an over-simplified system. The laser beam is focused on a rotating-mirror module, which is attached to one end of a drive shaft of a high-speed motor. The beam is offset by reflecting to another mirror, which is rotating at the same speed. This offset mirror is located at a distance that is dependent on the cable dimension and the illumination optics. Details of the optical scanner will be presented in Section III. The rotating beam is directed to the cable by means of two flat-folding mirrors, and then illuminates the cable with a focused spot inside the cable insulation by means of an ellipsoidal mirror. The arrangement of the collection optics is very similar to that of the illumination optics. The collection ellipsoid relays a focused beam of scattered light at an angle about 15° off the forward-beam direction onto a rotating tracker located at the other end of the drive shaft of the high-speed motor. The tracker relays the scattered signal to the Ga:Ge photoconductor with the help of additional optics (not shown in Figure 3). For a 3-mm beam rotating at 9000 rpm, this system provides complete coverage of the cable moving at a maximum extrusion speed of

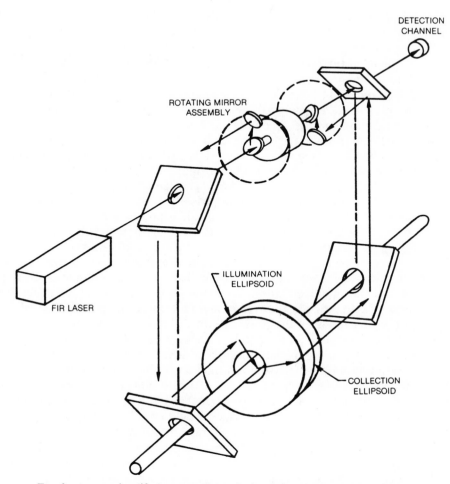

FIG. 3   An oversimplified system schematic showing optical paths of the scanner.

about 90 ft/min. Most power-distribution cables are extruded at speeds lower than this because of the large conductor size.

The laser power P_s scattered from a defect into a solid angle Ω subtended by the collection aperture can be written as

$$P_s = \int_\Omega I_0 \sigma(\theta, \varphi) \, d\Omega. \tag{1}$$

In Eq. (1) the integration can be carried out numerically for various values of σ that have been calculated (Cantor *et al.,* 1981) for spherical voids by using the approach suggested by Deirmendjian (Deirmendjian, 1969), as given in the appendix. According to Mie-scattering code, the optimum scattering occurs primarily in the forward direction at $\lambda \simeq d$, where d is the diameter of the void. The scattered signal power calculated from Eq. (1) must be modified to account for losses of various components employed in the system. The analysis must also include the losses caused by absorption of the far-infrared radiation by polyethylene and the by-products, such as acetophenone, moisture, cumyl alcohol, and methane.

The system noise consists of the detector's noise and background noise. The latter is caused by scattering from the microporosity and from the surface roughness of the insulation material. The detector used in this system is a 3×3-mm gallium-doped germanium crystal cooled to 4 K in a liquid-helium-cooled Dewar with properly designed thermal and radiative isolation. The measured noise equivalent power (NEP) is less than 10^{-11} W/Hz$^{1/2}$, and the frequency response is greater than 80 kHz. Although this crystal is very small, it has an effective aperture of 12.5 mm, which is obtained by placing the crystal inside a cavity at the vertex of a parabolic horn. This horn has an opening 12.5 mm in diameter and a gradual taper, so that all radiation confined within a solid angle defined by $f/3.3$ can reach the detector. In practice, the system noise is always background limited and is primarily caused by the roughness of the cable surface. The system performance estimate for a far-infrared laser power of 100 mW at 118.8 μm is shown in Fig. 4, in which the calculated signal-to-noise (S/N) ratio is plotted as a function of cable-insulation thickness. Curve a represents the anticipated S/N from a 50-μm void. A 2-dB uncertainty of this calculated value is attributed to the variation in by-product absorption. Curve b represents the background scattering from surface roughness. For a very smooth surface with an rms height $\bar{h} \leq 0.5\,\mu$m and rms length $\bar{\ell} \leq 6\,\mu$m, the measured S/N is about 19 dB for a 1/0 25-kV dry, cured, cable insulation. A part of this may be contributed by microvoids in the insulation. For the case of a very rough surface ($\bar{h} \simeq 10\,\mu$m, $\bar{\ell} \simeq 50\,\mu$m), the measured S/N for a 1/0 25-kV cable is about 33 dB, which is only 3 dB below that of the 50-μm voids. Figure 4 also includes S/N measurements of a 100- and a 250-μm void in a 1/0 25-kV cable with an o.d. $\simeq 2.6$ cm. The measured and the calculated values are in excellent agreement. For a highly reliable system, a design S/N of 17 dB is desirable. This value corresponds to a detection probability of 0.999, which is equivalent to an average of one false alarm in a period of about 5.5 h. In other words, it is posible to reliably detect insulation defects as small as 50 μm with an error probability of 10^{-9}, provided that all distribution cables are manufactured with very smooth surfaces. It is still possible to detect 50-μm defects in poorly made cables, however, the ability to distinguish the insulation defects from the surface defects is greatly decreased.

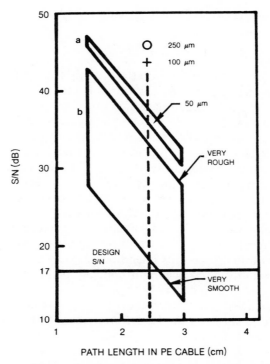

FIG. 4 Performance estimate of a cable inspection system: curve a, S/N for 50-μm void; curve b, background scattering from surface roughness for 1/0 25-kV cable.

III. Component Development

This section describes three key components used in the cable inspection system: the laser module, the optical-scanner module, the electronics-control module. The far-infrared laser is pumped by a continuous-wave CO_2 laser oscillating on a P(36) line in the 9.6-μm band. This far-infrared laser, which emits at 119 μm, is one of the most powerful that can be generated by optical pumping. With an 80-W single-mode, single-frequency CO_2 laser tuned on the CH_3OH molecular line center, it is possible to generate over 400 mW far-infrared laser power at 119 μm. The rotating optical module produces a 360° far-infrared laser beam that illuminates the cable insulation and collects the scattered radiation from the cable insulation to form a focused spot on a far-infrared photodetector. These components are designed for a specific purpose: they must meet the criteria and specifications described in Section II. In addition, they can be integrated into a system that must be evaluated through a series of reliability and safety tests to be qualified for industrial applications.

A. LASER MODULE

There are many factors that govern the stability and spectral purity of the output of the optically pumped far-infrared laser. Most important is the thermal and mechanical structure, which must be rigid and isolated from external disturbances so that the resonant frequency of the CO_2 laser pump can be locked onto the CH_3OH absorption line center by a closed-loop electronic feedback circuit. In this way, all energy emitted from the CO_2 molecules in a single mode can be transferred to the CH_3OH molecules. Because there exists a slight difference in frequency between the two laser lines, the CO_2 laser frequency must be offset from its peak to obtain the maximum far-infrared laser output. Consequently, in actual operation the CO_2 laser power must be lowered by about 10–15% from its peak value to achieve the optimum pumping. Another important consideration in the design is the commonly observed phenomenon known as optical feedback. Amplitude instability often occurs in optically pumped laser systems, in which a frequency-pulling effect of the pump laser can cause pulsations in the output when a very small amount of optical power is reflected back into the pump-laser cavity from the far-infrared-laser cavity. It is imperative that a reliable way be found to eliminate such optical cross-talk between the two laser cavities. Therefore, extreme care must be exercised to control the CO_2 laser frequency and to isolate the feedback. Finally, the far-infrared laser cavity must also be constructed so to avoid thermal and mechanical fluctuations.

1. Structure

The structures of both the CO_2 and the far-infrared laser cavities are very similar. The cavity length is about 2 m. The cavity mirrors and the gain tubes of both lasers are water-cooled and supported by super-invar rods that are freely expandable along their length through a series of split-ring supports. All parts are made with high precision so that very little angular adjustments are needed for the cavity mirrors. Once they are aligned, they are hardened and require no further adjustment. The length variation of super-invar rods over a finite temperature range varied from 50 to 120°F, which is extremely small. Nevertheless it is compensated by a reverse expansion of one of the mirror holders. There are eight split-ring supports for stiffening the cavity. They are anchored on a 5-cm-thick machined-stock aluminum plate having a length of 2 m. The cavity assembly is fastened to a 12-in.-thick honeycomb table top. The structure, which includes the table supports, has been analyzed by using a computer model for a typical noise spectrum obtained by measuring the disturbance from the floor in an industrial environment. The noise disturbance is allowed to be transmitted to the laser structure through the supporting legs without active pneumatic damping. The maximum rms values of the structural deformations in length $\Delta \ell$ and tilt $\Delta \theta$ are calculated to be

$$(\Delta \ell)_{rms} \leq 3 \times 10^{-8} \quad \mu m,$$

$$(\Delta \theta)_{rms} \leq 7 \times 10^{-13} \quad rad, \tag{2}$$

$$\Delta \ell / \Delta T \leq 2.4 \times 10^{-5} \quad \mu m/^\circ F.$$

These results are insignificant in comparison with laser wavelengths. The key point here is that the structure has been designed so stiff that the resonant frequencies of the normal modes are far above the random-noise spectrum of the floor. As a result, the structure can easily be stabilized by using a piezoelectric transducer (PZT) to track any slow drift in cavity length. Two PZTs that have 10 and 75 $\mu m/kV$ are used for the CO_2 and CH_3OH lasers, respectively. Figure 5 is a photograph showing PZT-controlled mirrors of both the CO_2 pump and the far-infrared lasers.

2. CO_2 Pump Laser

The CO_2 pump laser is the most important part of the far-infrared laser system, because of the close resonant and high-threshold requirements, which demand extremely high spectral and spatial purity not readily available in commercial lasers. The CO_2 laser developed for cable-inspection application is a windowless system with internal mirrors. The laser tube is made of pyrex, 2-m long, having an i.d. of 12 mm, in which a series of wire rings of 250-μm o.d. are inserted for mode discrimination. Experimentally, we found that removing the ZnSe Brewster window, commonly used in externally mirrored cavities, significantly increases not only the laser output power but also the reliability and stability. This can be attributed to the high circulation of optical power inside the cavity, which can cause a slight change in the optical path through the window material because of thermal effects. The removal of these windows necessitates the isolation of one of the cavity mirrors from the high voltage required for the gas discharge. Care must also be taken to protect the cavity mirrors from deposition and contamination. We have performed a continuous test on a laser system of this type over a very long period of time (≥ 2000 h) without any measurable degradation in the output power.

The cavity mirrors consist of one flat grating, which has 125 lines/mm and is blazed at 9.6 μm, and one dielectrically coated ZnSe spherical mirror, which has a radius of curvature of 20 m with a reflection coefficient of 70% at 9.6 μm. These mirrors are mounted on holders made of heat-treated metal,

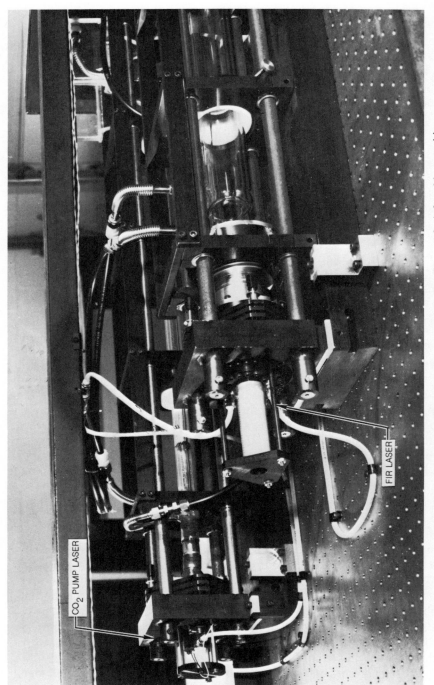

FIG. 5 A photograph showing PZT-controlled mirror assemblies for CO_2 and far-infrared laser cavities.

machined into a U-shape spring structure that allows only one degree of angular adjustment. All parts of the system are water-cooled to near room-temperature. As indicated by Eq. (2), the mechanical deformation is expected to be insignificant compared with the thermal deformation. Great care has been taken to reduce the thermal instability. Experimentally, we have found that additional external disturbances can couple into the laser structure through noise sources such as pulsations in the cooling-water and vacuum-pump lines. Through the use of proper baffling and isolation, these have been eliminated. At a very slow gas-flow rate (≤ 100 cc/min at STP), the CO_2 laser output power is about 90 W at 9.6 μm and 110 W at 10.6 μm. Spectrum analysis of the output confirms that there exists only one transverse mode. By increasing the gas-flow rate, it is expected (Cheo, 1967) that the output will increase almost linearly with the flow rate over a wide range of values.

3. Far-Infrared Laser

The far-infrared laser structure is very similar to that of the CO_2 pump laser. The water-cooled laser tube is also pyrex, 2 m long, having an i.d. of 48 mm. The CH_3OH gas is allowed to flow very slowly through the tube so that the tube life will not be limited by the gas degradation. Such a large bore size was chosen to reduce the wall interference to the pump-laser beam, which is injected into the far-infrared cavity from an off-axis hole. We have considered the possibility of using an optical isolator to reject the cross talk between the two lasers. Although there are several types of optical isolators for 10 μm, they are not commerically available for powers in excess of 10 W and would require extensive research and development. For this reason we have selected the off-axis injection scheme over the isolator for reliability. The far-infrared laser cavity consists of a concave copper mirror having a radius of curvature of 6 m and a flat dielectric-coated silicon mirror. The off-axis pumping is accomplished by coupling the CO_2 pump beam through a 5-mm hole in the curved copper mirrors. The position of this hole with respect to the center of the mirror is critical for efficient coupling and is determined by the geometry of the cavity. The flat dielectric-coated silicon mirror is chosen for maximum reflection at 9.6 μm to allow the pump beam to be reflected back and forth in the cavity several times. Between the coating and the silicon substrate, this mirror has a gold coating everywhere except for a 10-mm hole in the center that provides the necessary output coupling for the far-infrared light.

With 80-W CO_2 pump power tuned to the peak of the CH_3OH absorption line center, we obtain 400-mW far-infrared laser output, which is measured by an uncalibrated calorimeter (Scientech Inc.) in the lowest-order mode. The output profile, as shown in Fig. 6, is obtained by scanning the beam with

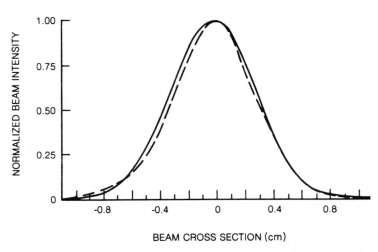

FIG. 6   Far-infrared laser-beam profile: ——, Gaussian; — —, measured.

a pyroelectric detector having a very small sensitive area at a distance about one meter away from the far-infrared output mirror. The solid curve is the best fit for a Gaussian. The output power can be increased by selecting the higher-order modes. This can be done by increasing the size of the hole coupler. However, the penalty is an exchange of poor beam quality for higher power output.

4.   Stabilization

As mentioned before, the cavity spacing of these lasers can easily be kept on the peak of the CH_3OH molecular-resonance line if the cavity structure is passively stable. Two stabilization schemes have been investigated. Figure 7a shows a technique known as optoacoustic spectrophone, which utilizes the resonance absorption of the pump radiation by CH_3OH molecules in a cell. This interaction causes a change in the cell pressure, which can be used as the error signal for establishing a phase-sensitive detection for the feedback loop. Within the absorption cell, a cylindrical miniature microphone is placed coaxially with the optical path. As the dithered CO_2 laser power passes through this cell, an audio signal is generated by the microphone and has a characteristic response correlated with the line profile of the CO_2 pump laser. A null in the acoustic response corresponds to the absorption peak of the CH_3OH molecule. Therefore it is very convenient to use this optoacoustic null for locking the CO_2 laser. This technique works well except that the microphone is too susceptible to microphonic noise. It is difficult to isolate the microphone completely from the external noise, which could

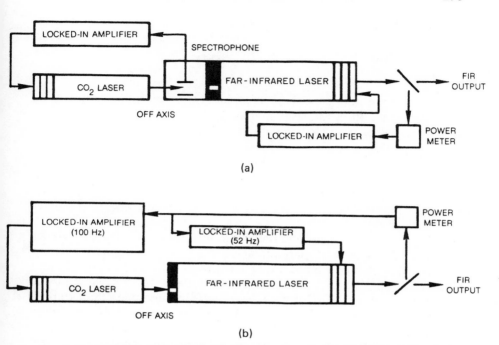

FIG. 7 Laser stabilization methods: (a) using an optoacoustic cell and (b) monitoring the far-infrared laser output.

cause instabilities in the laser system. A more reliable and simpler approach is shown in Fig. 7b, in which the dithered far-infrared laser is used as the error signal for locking both the CO_2 pump laser and the far-infrared laser. The feedback signal is fed into two synchronous detection channels, one at 52 Hz and the other at 100 Hz, as shown in Fig. 7b. This scheme works very well to maintain a maximum far-infrared laser output against the long-term drift.

B. OPTICAL-SCANNER MODULE

A unique optical scanner has been developed specifically for cable-inspection application. Basically, a 360° rotating laser beam is directed tangent to the inner conductor, and the detector looks at only a small portion of the scattered beam along an axis oriented at about 15° from the forward-transmitted beam direction. It is convenient to describe the optical system by dividing it into two parts, illumination optics and collection optics. One unique feature common to both cases is the way that a stationary laser beam is projected onto a cable in a 360° rotation and, conversely, a rotating target inside the cable is brought onto a stationary detector. This is accomplished by using two folding mirrors, as shown in Fig. 3. With these two flats, it is easy to visualize that the laser, scanner motor, ellipsoid, and cable axes can all be made colinear.

1. Illumination Optics

Figure 8 is a simplified scheme to explain the illumination optics. The scanner mirrors offset the laser source so that, as the mirrors rotate, the source appears to originate from a ring with a fixed radius centered on the cable axis. The ellipsoidal mirror that is located concentrically with the cable is represented by a lens in Fig. 8 and images the laser source to a corresponding locus of points offset from the cable center. The magnification factor of the ellipse determines the ratio of the scanner offset to the conductor radius. Therefore, for different conductor size, different scanner offset is required.

Owing to physical limitations, the minimum distance between the scanner mirror and the ellipsoid is approximately 1 m. Over this distance, the beam will expand to about 4 cm because of diffraction. A focal distance of 20 cm is needed to generate an $f/5$ focal cone with the desired spot size in the cable insulation. For a detailed system design, a computer-aided three-dimensional ray-tracing program (ACCOS V) has been extensively utilized to determine various parameters, including various mirror-clear apertures, tilt angles, depth of field inside the insulation, beam cross sections, etc.

The actual mechanical layout for the optical module is shown in Fig. 9. The laser beam enters the module through a hole and is imaged on the illumination scanner mirror, which is attached to the shaft of a high-speed motor. The scanner optics are configured in a conical shape, as shown in Fig. 9, and because of high rotating speed (~ 9000 rpm) must be carefully designed so that the mirrors will not be deformed and the structure can be

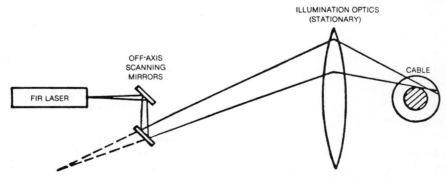

FIG. 8   A simplified schematic to illustrate the cable-illumination optics.

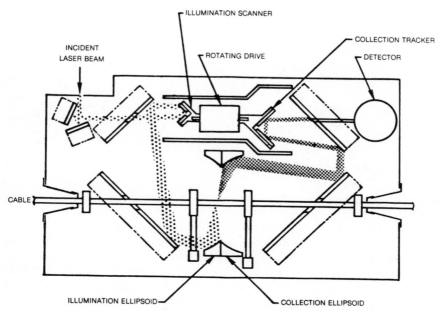

FIG. 9 The mechanical layout for the optical module.

dynamically balanced. To analyze such a complex structure, another computer-aided design program is employed to determine mass properties of the rotating optics. This information is used as input for the critical-speed analysis that provides natural frequencies of the system. These results have been used to predict the vibrational amplitudes and stress distribution and thereby to allow a proper selection of material and structural configuration.

2. Collection Optics

The mechanical layout for the collection optical system is very similar in many respects to that of the illumination optical system. However, its design is under considerably more constraints than the illumination optics because of the finite size of the detector aperture and its restricted field of view. For these reasons, and also for a maximum collection of scattered radiation, larger optics are required on the collection side. Figure 10 is a simplified schematic to illustrate this system. Arbitrarily, we have selected three points within the illumination cone in the cable for use as scattering sources for ray tracing, to evaluate prospective collection-system optical configurations. The first source point is at a location where the central ray of the illumination cone is perpendicular to a diameter of the cable. The

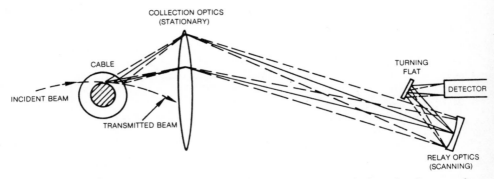

Fig. 10   A simplified schematic to illustrate the collection optics for laser signals scattered from insulation defects.

second is near the cable exit point of the central ray of the illumination cone, and the third is located midway between the first two.   Hence, only far-infrared radiation from those scattering centers between the illumination midpoint and the exit point are considered as signal sources.   All other points along the illumination path will be examined from different orientations of the illumination scan.   In this manner, the whole cable insulation is examined only once per 360° revolution of the illumination cycle.

The primary collection aperture resides on the collection ellipsoid that is located back to back with the illumination ellipsoid.   The forward-transmitted beam returns toward the source after reflection from the illumination ellipsoid.   As shown in Fig. 10, a relay mirror is used to image the collection aperture (at the ellipsoid) onto the entrance aperture of the detector.   By this technique, all of the radiation incident on the collection aperture is transmitted to the detector.   The amount of scattered radiation collected is limited by the size of the collection aperture, which is constrained by the size and field of view of the detector, as well as the depth of field of the radiation source.   All of these considerations were taken into account in designing the collection-system optics.

The tracking feature of the collection system is accomplished in a manner analogous to the illumination system, by two turning flats in a cone that is physically located on the opposite end of the shaft holding the illumination scanning mirrors.   This can also be seen in Fig. 9.   This arrangement ensures that the illumination and collection optics remain synchronous while maintaining the correct phase relationship for efficient signal collection. The collection system also utilizes two turning flats to image the scanner axis to the collection-side cable axis.   This serves the same function as the illumination configuration, namely, to be able to physically locate the scanner axis remote from the cable axis.

The detailed design for the collection optics is also made by using the ACCOS V program, which specifies all optical elements in terms of all six mechanical degrees of freedom and gives precisely the curvatures of either spherical or aspheric surfaces. The present design of the optical collection system utilizes a 10-cm-diameter collection aperture, a 7-cm-diameter relay mirror, and a 1.8-cm-diameter detector aperture with a ±7° field of view. This configuration is compatible with commercially available detector size and field of view.

Figure 11 shows the beam profile within the cable insulation. The results, in general, show that the beam is distorted and deviates in the worst case from the 2-mm focused Gaussian spot to an irregular shape of roughly 3 × 5 mm. The surfaces 1, 2, and 3, as shown in Fig. 11, are planes perpendicular to the forward beam direction and containing the three points, as shown in Fig. 10. The beam distortion is caused in part by the cylindrical cable surface and in part by the off-axis nature of the imaging optics. Similar beam-profile analyses at the tracking optics are obtained for the purpose of

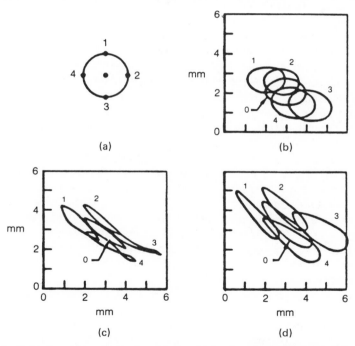

FIG. 11 Far-infrared laser-beam aberrations calculated at three different locations inside cable insulation: (a) source (1-cm diameter) with object points used in ray tracing, (b) surface 1 near the conductor, (c) surface 2 at the center of the insulation, and (d) surface 3 near the outer surface of the insulation.

choosing a proper collection aperture. Clearly, beam distortion reduces the scattered signal-to-noise ratio. However, because of the increase in the beam size, the scanner and tracker speed can be correspondingly reduced. A factor-of-two reduction in rotating speed can considerably relax the difficulties encountered in the mechanical design of the high-speed optics. A reduction in rotating speed also reduces the required system bandwidth, which helps to recover some of the loss in the signal-to-noise ratio.

C. ELECTRONIC-CONTROL MODULE

The operation of this far-infrared laser and optical-scanner system requires many electronic, thermal, and mechanical devices. A central control module, which consists of an electronics package and a dedicated computer, has been developed to provide power distribution, control and command, and interfacing for all subsystems, as well as to perform signal processing of the scattered far-infrared laser waveforms. The processed signals are correlated with the detected insulation defects in real time during manufacturing.

The electronic-controller hardware package contains many TTL circuits that are interactive with all electromechanical devices for specific purposes, such as to turn on or turn off vacuum pump, gas and water valves, and power supplies, and to monitor various safety interlocks. Figure 12 is a block diagram showing the cable-inspection system-control interface. Figure 13 is a block diagram showing the dedicated computer hardware employed in the system. Details of the performance of this computer-interactive control and real-time signal processing of inspection data will be published elsewhere. In this chapter, we shall briefly describe only the controller, with regard to its hardware and functions.

The control unit primarily serves as a logic or gate for various operational functions involved in running the two lasers. Both local (panel) and remote (computer) controls are implemented but are not mutually exclusive. Aside from several sequencing and safety requirements, electrically switched devices can be actuated asynchronously from either the local panel toggles or the computer keyboard. All control functions utilize straight-through logic that merely interfaces computer digital outputs and toggle switches to solid-state relay drivers. In turn, the relays apply line power directly to the functional hardware.

Interlock and time-sequence logic are used to control two important events: application of high voltage and positioning of the interlaser coupling mirror. Laser discharge is dependent on high voltage at the electrodes. For safety reasons, the power supply must be in the off position if protective covers over the laser area are improperly placed or the laser-optics cooling water is not flowing. Both requirements must be physically satisfied to authorize power supply output. However, these two interlocks can be

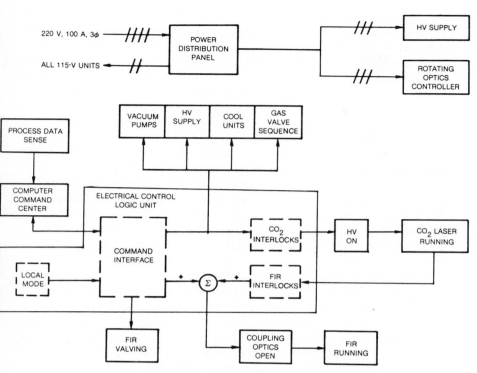

FIG. 12   A block diagram of the electronic-control interface.

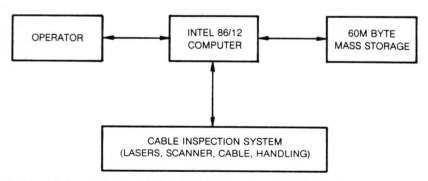

FIG. 13   A dedicated computer for the cable inspection system.   The Intel 86/12 computer uses the following: 8086 microprocessor, 8087 math co-processor, 512K-byte memory, CPM 86 operating system, FORTRAN, and Assembly language.

jumped for self test only. In addition, to protect the laser, CH_3OH vapor-inlet-valve logic must be in the valve-open state before the coupling mirror can be positioned.

Both the mirror and the high-voltage output have sequential restrictions as well, and each circuit protects against operator error. Following line-voltage excitation to the power supply, a brief transient-settling period ensues. Regardless of the state of the high-voltage turn-on signal (computer bit or toggle switch), circuit-logic delays high-voltage authorization for 1.5 sec after line-voltage application. The "standby" and "HV output" signals to the high-voltage unit are one-time pulses triggered by input leading edges. The standby command will override when the HV-output instruction is initiated prematurely, and it will be necessary to reactivate the sequence once the line voltage and interlocks are satisfied. A simpler situation exists with the coupling mirror. One and one-half seconds after the CH_3OH vapor inlet valve-open command, the coupling mirror can be positioned. Should the mirror-actuation signal occur prematurely, the mirror will respond automatically following the gas-valve command and built-in delay.

EMI – RFI power-line filters are used to reduce noise coupling into sensitive electronic sections of the overall system, which includes the computer assembly, the controller, and a general instrumentation bus. A digital signal interface that utilizes fiber optic links to provide electrical isolation is inserted between computer and controller. In the controller, Schmitt-type input gates are used to reduce the susceptibility to noise. As a special case, constraints of the high-voltage control circuits dictate very clean switching signals. A latching flip-flop serves to condition that toggle input. As noted previously, the momentary high-voltage-ON and standby circuit is designed to give precedence to the standby state. Access to the controller from the computer interface is via a multiconductor shielded cable. Terminal strips on the controller rare panel provide connections to pallet hardware, such as valves, through solid-state relays. Remote relay placement adjacent to hardware components can further reduce noise coupling. The entire module has been specially packaged for use in a cable-factory environment.

IV. Conclusion

An automated and rugged far-infrared laser scanner has been built for production-line use to inspect the quality of high-voltage cable insulation. This far-infrared laser system scans the cables extruded at speeds up to 90 ft/min. A complete inspection of the cable insulation is accomplished by using a rotating mirror system, which provides a laser beam to rotate around the cable at a maximum angular speed of 9000 rpm. The system is capable of detecting all types of insulation defects, including voids, contaminants, and protrusions, and also surface roughness as small as 50 μm in diameter. The first prototype has been installed at Essex Cables in Lafayette, Indiana, a Division of United Technologies Corporation, in April 1984 for factory evaluation of its usefulness. This far-infrared laser can also be used for plasma diagnostics and studies of physical and chemical properties of polymeric materials.

Appendix: Mie Scattering from Voids

A. CROSS SECTIONS

When a collimated laser beam of intensity I_0 (W/cm^2) is incident on a single particle at the origin of a laboratory coordinate system, the amount of power scattered (P_s) and the amount absorbed (P_a) are proportional to I_0, so that it is possible to define scattering and absorption cross sections σ_s and σ_a by the simple relations

$$P_s = \sigma_s I_0, \tag{3}$$

$$P_a = \sigma_a I_0. \tag{4}$$

Because the total power removed from the incident beam is $P_e = P_s + P_a$, one can also define an extinction cross section

$$\sigma_e = \sigma_s + \sigma_a. \tag{5}$$

The σ have the dimensions of an area (square centimeters) and are clearly related to the geometrical cross-sectional area of the particle, which is πr^2 for a sphere, and the effectiveness of the interaction. The efficiency factors K_{sc}, K_{ab}, K_{ex} are simply defined by

$$\sigma_s = \pi r^2 K_{sc}, \quad \text{etc.}, \tag{6}$$

and the dimensionless K can only depend on the ratio r/λ and the relevant indices of refraction m and m_0 of the particles and the background medium, respectively.

The absorbed power goes into heating the particle; it is assumed that the index of refraction is not thereby altered. The scattered power is scattered in all possible directions. Imagine a spherical coordinate system (r, θ, ϕ) with the direction of the propagation vector of the incident plane wave defining the polar axis. Then the scattered power in the element $d\Omega$ of solid angle

$$d\Omega = \sin \theta \, d\theta \, d\phi \tag{7}$$

is

$$dP/d\Omega = I_0 \, d\sigma/d\Omega. \tag{8}$$

The notation

$$d\sigma/d\Omega = \sigma(\theta, \phi) \qquad (9)$$

is sometimes used for this differential cross section to simplify equations.

The forward direction $\theta = 0°$ of the incident beam, together with the direction θ, ϕ of the scattered ray, defines a "scattering plane." Depending on whether the initial polarization is perpendicular (1) or parallel (2) to this plane, one gets unique scattering cross sections $\sigma_1 = \sigma_1(\theta, \phi)$, $\sigma_2 = \sigma_2(\theta, \phi)$ from which, unfortunately, the results for any other polarization cannot in general be determined. For unpolarized incident light, however, the correct result is the average of these two numbers. The treatment of arbitrary polarization requires a knowledge of the scattering amplitudes A_1 and A_2.

B. Scattering Amplitudes

The exact Mie solution requires an expansion of the incident wave, the scattered wave, and the field inside the sphere in terms of spherical-coordinate solutions of the electromagnetic-wave equations. The expansion coefficients (a's and b's) are determined by satisfying boundary conditions at the surface of the scatterer, and at infinity. The scattered fields at a distance from the scatterer behave like outgoing spherical waves, with an additional factor, the scattering function or scattering amplitude, which gives the angular dependence of the scattered waves.

Two scattering amplitudes A_1 and A_2—or their dimensionless equivalents S_1 and S_2—are required for a complete solution. In terms of the standard angular functions $\pi_n(\theta)$ and $\tau_n(\theta)$, the two amplitudes are given by

$$S_1(\theta) = kA_1(\theta) = \sum_{n=1}^{\infty} \frac{2n+1}{n(n+1)}(a_n\pi_n + b_n\tau_n); \qquad (10)$$

$$S_2(\theta) = kA_2(\theta) = \sum_{n=1}^{\infty} \frac{2n+1}{n(n+1)}(b_n\pi_n + a_n\tau_n). \qquad (11)$$

These amplitudes are complex numbers because the a_n and b_n are complex; $k = 2\pi/\lambda$.

Two amplitudes are needed to give all of the polarization properties of the scattered field. Suppose the incident plane wave propagates along the $+z$ direction, which is also the polar axis of spherical polar coordinates (r, θ, ϕ), and is polarized so that its electric field is along the $+x$ axis (from which the angle ϕ is measured). The incident direction ($\theta_0 = 0$, $\phi_0 =$ arbitrary) and the scattering direction (θ, ϕ) define the so-called scattering plane. With respect to this plane, the incident field has components perpendicular and parallel to it given by

$$E_0(1) = \sin \phi \exp(i\omega t - ikz), \qquad (12)$$

$$E_0(2) = \cos \phi \exp(i\omega t - ikz), \qquad (13)$$

assuming a unit amplitude initially.   The corresponding components of the scattered fields are

$$E_s(1) = \sin \phi S_1(\theta) \exp(i\omega t - ikr)/ikr, \qquad (14)$$

$$E_s(2) = \cos \phi S_2(\theta) \exp(i\omega t - ikr)/ikr. \qquad (15)$$

The magnitudes of S_1 and S_2 determine the intensity of the scattered fields; the phases of S_1 and S_2 determine the state of polarization of the scattered field.

From S_1 and S_2, or rather from A_1 and A_2, the following four quantities with the dimensions of differential scattering cross sections can be constructed:

$$\sigma_1(\theta) = A_1^* A_1 = |A_1|^2, \qquad (16)$$

$$\sigma_2(\theta) = A_2^* A_2 = |A_2|^2, \qquad (17)$$

$$\sigma_3(\theta) = \mathrm{Re}(A_2^* A_1), \qquad (18)$$

$$\sigma_4(\theta) = -\mathrm{Im}(A_2^* A_1). \qquad (19)$$

Thus, the magnitude of the scattered electric field is

$$\mathbf{E}^* \cdot \mathbf{E} = |E_s(1)|^2 + |E_s(2)|^2$$

$$= [\sin^2 \phi \sigma_1(\theta) + \cos^2 \phi \sigma_2(\theta)]/r^2, \qquad (20)$$

which shows that σ_1 and σ_2 are the differential cross sections for radiation polarized perpendicularly and parallel, respectively, to the scattering plane, and that

$$\sigma_\phi = (\sin^2 \phi)\sigma_1 + (\cos^2 \phi)\sigma_2 \qquad (21)$$

is the correct result for arbitrary linear polarization.   For unpolarized light one obtains

$$\sigma = \tfrac{1}{2}(\sigma_1 + \sigma_2) \qquad (22)$$

by averaging Eq. (21) over all possible values of the angle ϕ.

In general, the Mie-scattered light is fully polarized (i.e., not partially polarized or unpolarized) if the incident light is fully polarized.   However, there can be a rotation of the polarization so that the plane-polarized light will scatter into circularly or elliptically polarized light.   If the complex amplitudes are written

$$A_1 = |A_1| \exp(-i\epsilon_1), \qquad (23)$$

$$A_2 = |A_2| \exp(-i\epsilon_2), \qquad (24)$$

then the additional phase angle produced in the scattering is

$$\delta = \epsilon_1 - \epsilon_2 = \tan^{-1}(\sigma_4/\sigma_3). \qquad (25)$$

If $\delta = 0$, the incident polarization angle is unchanged. However, because σ_1 and σ_2 are not necessarily equal, circular polarization will become elliptical, in general.

C. ASYMPTOTIC LIMITS

Instead of r/λ, it is customary to use the dimensionless parameter

$$x = kr = 2\pi r/\lambda, \qquad (26)$$

which occurs naturally in theoretical formulations. In the limit $x \to \infty$, i.e., the so-called geometrical optics limit of small wavelength and large spheres, the efficiency for extinction approaches the limiting value

$$K_{ex} = 2 \qquad (27)$$

for a particle in air, independent of its index of refraction (Chylek, 1975).

The index of refraction of the particle m is a complex number, in general $m = n - i\kappa$. When m is real, there can be no absorption ($K_{ab} = 0$), and $K_{ex} = K_{sc}$ for this case. The apparently paradoxical result that the effective cross sectional area is $2\pi r^2$ instead of the geometrical area πr^2 is explained by the interesting fact that the pure diffraction contribution, which is half the total value, actually survives into the geometrical optics limit. However, this diffracted part is strongly concentrated in the forward direction, and it is often experimentally difficult to separate it from the unattenuated beam. A good discussion of this paradox is to be found in Middleton's book (Middleton, 1952, sections 3.3.1 and 3.3.2).

For a bubble, the same asymptotic value $K_{ex} = K_{sc} = 2$ must be valid because only the relative index of refraction can cause scattering. For a real index, the background medium merely causes an apparent change in the incoming wavelength. Relative to this new wavelength, the index of refraction of the bubble appears to be less than unity. So, if λ_0 is the wavelength in vacuum, the two important scale parameters are $x = 2\pi r(n_0/\lambda_0)$ in the background medium, and $y = 2\pi r(n/\lambda_0) = (n/n_0)x$ in the particle. For a particle in air, $n_0 = 1$ and $n > 1$. For a bubble, $n = 1$ and $n_0 > 1$. Hence, a bubble may be regarded as a particle with an index less than unity relative to its surroundings. Consequently, published results for the Mie-scattering coefficients are not generally applicable to bubbles. Optically speaking, a particle (like a water droplet) acts like a positive (focusing) lens, whereas a bubble acts like a negative (defocusing) lens.

When the scatterer is absorbing in the geometrical optics limit ($x \to \infty$), the scattering efficiency has the value (Chylek, 1975)

$$K_{sc} = 1 + |(m - 1)/(m + 1)|^2 \qquad (28)$$

which, however, does not give the correct result for a nonabsorbing sphere (for which $K_{sc} = K_{ex} = 2$). For metallic spheres, i.e., those with large values of κ, the value of Eq. (28) is closely attained for quite small x. However, for weakly absorbing spheres one would have to go to very large x before the asymptotic limit would prevail.

Another asymptotic limit that is believed to be exact (Deirmendjian, 1969), but which has not yet been proven, involves the backscatter coefficient (also called the "radar" coefficient)

$$\sigma_1(180°) = \sigma_2(180°) = (1/4\pi)\pi r^2 |(m - 1)/(m + 1)|^2 \qquad (29)$$

in the limit $x \to \infty$ for absorbing spheres. A limit for nonabsorbing spheres is believed not to exist. The factor $|(m - 1)/(m + 1)|^2$ is understandable as the reflection coefficient at a plane air–liquid interface at normal incidence.

D. RAYLEIGH SCATTERING

For small nonabsorbing particles where $x = 2\pi r/\lambda \ll 1$ and $y = xn \ll 1$, the leading term in the exact solution gives

$$S_1(\theta) = ix^3(n^2 - 1)/(n^2 + 2), \qquad (30)$$

$$S_2(\theta) = S_1(\theta) \cos \theta. \qquad (31)$$

Hence,

$$\sigma_1 = k^4 r^6 (n^2 - 1)^2/(n^2 + 2)^2, \qquad (32)$$

$$\sigma_2 = \sigma_1 \cos^2 \theta, \qquad (33)$$

$$\delta = 0. \qquad (34)$$

Small particles are thus poor scatterers ($\sigma \simeq a^6/\lambda^4$) but have very simple polarization and angular properties. The perpendicular polarization scatters isotropically. The parallel polarization scatters as $\cos^2 \theta$, i.e., equally in forward and backward directions, and zero at right angles. There is no rotation of the phase angle.

The total-extinction cross sections follow (by a theorem) from the values of S_1 and S_2 in the forward direction ($\theta = 0°$):

$$K_{ex}^{(i)} = (4/x^2) \operatorname{Re} S_i(0°). \qquad (35)$$

Now, in the forward direction,

$$\pi_n(0) = \tau_n(0) = n(n + 1)/2, \tag{36}$$

which leads to the general result

$$K_{ex}^{(i)} = (2/x^2) \sum_{n=1}^{\infty} (2n + 1) \, \text{Re}\{a_n + b_n\}, \tag{37}$$

which is thus the same for either polarization, as it must be. Upon calculating S_i to a higher order than in Eqs. (31) and (32) it follows that

$$K_{sc} = K_{ex} = \tfrac{8}{3}x^4[(n^2 - 1)/(n^2 + 2)]^2. \tag{38}$$

The use of a complex index of refraction alters these results. In S_1 and S_2 we merely replace n by m and find

$$K_{ex} = -\text{Im}\left\{4x\left(\frac{m^2 - 1}{m^2 + 2}\right) + \frac{4}{15}x^3\left(\frac{m^2 - 1}{m^2 + 2}\right)\left(\frac{m^4 + 27m^2 + 38}{2m^2 + 3}\right)\right\}$$

$$+ x^4 \, \text{Re}\left\{\frac{8}{3}\left(\frac{m^2 - 1}{m^2 + 2}\right)^2\right\} + \cdots, \tag{39}$$

where all terms up to order x^4 are shown for completeness. The scattering efficiency must be calculated separately. *It is not* the last term of Eq. (39) except for the case of a real index. The correct result is

$$K_{sc} = \tfrac{8}{3}x^4|(m^2 - 1)/(m^2 + 2)|^2. \tag{40}$$

The absorption efficiency is defined [recall Eqs. (5) and (6)] as the difference between K_{ex} and K_{sc}.

The leading term in absorption in the Rayleigh limit gives

$$K_{ab} = 24n\kappa x/(n^2 + 2)^2. \tag{41}$$

The standard material absorption coefficient is

$$\alpha \, (\text{cm}^{-1}) = 2k\kappa. \tag{42}$$

Because $x = kr$,

$$K_{ab} = 12nr\alpha/(n^2 + 2)^2, \tag{43}$$

and thus

$$\sigma_{ab} = (4\pi r^3/3) \, \alpha[9n/(n^2 + 2)^2], \tag{44}$$

showing that in the limit of a weakly absorbing, nonrefracting sphere $(n \to 1)$, the absorption cross section results from just the volume absorption. For a random collection of these spheres, with number density N

(spheres/cm³), the effective path absorption becomes

$$\alpha_{\text{eff}} = \sigma_{\text{ab}} N. \tag{45}$$

The Rayleigh scattering functions are useful for $x \leq 0.6$, i.e., $r \leq 0.1\lambda$, for typically low values of the index of refraction. Beyond this limit it pays to use the correct Mie-scattering computation. There are, however, a number of other useful theoretical results that give some important insights into the nature of Mie scattering, and which even permit some accurate estimates of the location of maxima and minima in the angular scattering functions, as well as of the scattering efficiencies. These methods will be discussed here.

As $m \rightarrow \infty$ (either through the real or imaginary parts) the sphere becomes totally reflecting. A formula for this limiting case that has been known since before the Mie theory is given here:

$$S_1 = ix^3(1 - \tfrac{1}{2} \cos \theta); \tag{46}$$

$$S_2 = ix^3(\cos \theta - \tfrac{1}{2}). \tag{47}$$

Here, $x \ll 1$, but obviously $xm \ll 1$ would be impossible. The distribution of scattering is predominantly backwards. Detailed calculations (Diermendjian, 1969) show that realistic reflecting particles do not have large enough values of m for Eqs. (46) and (47) to be of much use.

E. RAYLEIGH–GANS SCATTERING

When the index of refraction of a particle, or of the medium surrounding a void, is close to unity, i.e., $|m - 1| \ll 1$, an approximate formula exists valid for a larger range of x values than the Rayleigh formula, but limited by $x|m - 1| \ll 1$. The result is

$$S_1 = i\tfrac{2}{3}x^3(m - 1) \, G(2x \sin \tfrac{1}{2}); \tag{48}$$

$$S_2 = S_1 \cos \theta; \tag{49}$$

$$\delta = 0; \tag{50}$$

where the new function

$$G(u) = 3(\sin u - u \cos u)/u^3 = 3j_1(u)/u \tag{51}$$

is related to the spherical Bessel function j_1. The differential cross sections are

$$\sigma_1 = \tfrac{4}{9}k^4r^6|m - 1|^2 \, G^2(2 \sin \tfrac{1}{2}); \tag{52}$$

$$\sigma_2 = \sigma_1 \cos^2 \theta. \tag{53}$$

The new feature of the Rayleigh–Gans scattering formulas is the prediction of a lobe structure in the radiation pattern. In particular, in addition to the 90° null in the parallel polarization component, both σ_1 and σ_2 have new nulls at the same locations determined by the zeros of $j_1(u)$. These occur at the approximate values (Abramowitz and Stegun, 1970) $u_1 = 4.4934$, $u_2 = 7.7253$, $u_3 = 10.904$, $u_4 = 14.066$, $u_5 = 17.221$, etc. Thus a null, or a sharp dip, is expected in the angular scattering functions at the angles ($i = 1$, $2, 3, \ldots$)

$$\theta_i = 2 \sin^{-1}(u_i/2x). \tag{54}$$

Actually, the nulls are sensitive to the index of refraction. A generalization of Eq. (52) that works well will be discussed here.

The total scattering efficiency for polarized or unpolarized light can be found by integrating Eqs. (52) and (53) over all solid angles. The result is

$$K_{sc} = |m - 1|^2 \{ \tfrac{3}{2} + 2x^2 - (\sin 4x)/4x - 7(1 - \cos 4x)/16x^2 \\ + (1 - 4x^2)[\gamma + \log 4x - \text{Ci}(4x)]/2x^2 \}, \tag{55}$$

where $\gamma = 0.577 \ldots$ is Euler's constant, and Ci is the cosine-integral function:

$$\text{Ci}(y) = -\int_y^\infty \cos u/u \, du. \tag{56}$$

F.   VAN DE HULST SCATTERING

van de Hulst (1957) first derived a remarkable formula for the extinction efficiency valid in the limit that the complex index of refraction $m = n - i\kappa$ goes to unity, i.e., $|m - 1| \rightarrow 0$, but suitable for large x. Using the two parameters

$$\rho = 2x(n - 1) \tag{57}$$

and

$$g = \tan^{-1}[\kappa/(n - 1)], \tag{58}$$

van de Hulst's results are

$$K_{ex} = 2 - (4 \cos g/\rho) \, e^{-\rho \tan g} \sin(\rho - g) \\ + 4(\cos g/\rho)^2 [\cos 2g - e^{-\rho \tan g} \cos(\rho - 2g)]. \tag{59}$$

When $m = n$ is real ($\kappa = 0$), this simplifies to

$$K_{sc} = k_{ex} = 2 - 4 \sin \rho/\rho + 4(1 - \cos \rho)/\rho^2, \tag{60}$$

which has the correct limit $K_{ex} = 2$ as $\rho \rightarrow \infty$, but the wrong limit

$$K_{sc} = \rho^2/2 = 2(n - 1)^2 x^2 \tag{61}$$

as $\rho \rightarrow 0$. Thus, the van de Hulst formula applies only to intermediate and large x values.

There are no simple results for the angular-scattering functions in the van de Hulst theory.

G. HART–MONTROLL SCATTERING

Hart and Montroll (1951) have developed a formula (Eq. 37 of Hart and Montroll, 1951) which contains the van de Hulst extinction result in the limit of $n \to 1$ but which is somewhat better in that it is still fairly accurate for n as large as 1.5. However, it departs from the correct asymptotic limit as n departs from unity (though by only 25% at $n = 1.5$), while at the same time behaving better in the small x (Rayleigh) limit, i.e., $K_{sc} = (32/27)(n-1)^2 x^4$.

It is worth noting that Hart and Montroll's formula for the total-scattering cross section is based on an approximate integration of their underlying results for the angular scattering function, which may be written (Hart and Montroll, 1951, eq. A-35b):

$$\sigma_1^{H/M}(\theta) = \frac{4n^2(n-1)^2 r^2 x^4}{1 - 2\epsilon^2 \cos 4nx + \epsilon^4} \left| \frac{j_1(px)}{px} - \epsilon \, e^{2inx} \frac{j_1(qx)}{qx} \right|^2, \tag{62}$$

where

$$\epsilon = (n-1)/(n+1), \tag{63}$$

$$p = \sqrt{1 - 2n \cos \theta + n^2}, \tag{64}$$

$$q = \sqrt{1 + 2n \cos \theta + n^2}. \tag{65}$$

This more accurate result improves the Rayleigh limit to $K_{sc} = (\frac{8}{27})(n^2 - 1)^2 x^4$ and may improve the asymptotic limit too. What is known for sure is that Eq. (62) gives an excellent approximation to the normalized angular-scattering functions (i.e., the differential cross section normalized to unity at $\theta = 0°$). Much of that improvement is already included in the leading term $(j_1(px)/px)^2$. For instance, the locations of the nulls are now given by

$$\theta_i = \cos^{-1}\left(\frac{1 + n^2 - u_i^2/x^2}{2n}\right). \tag{66}$$

Later examples will show the improved performance of Eq. (66) compared with Eq. (54) and also the performance of Eq. (62).

H. COMPARISON OF VOID AND PARTICLE SCATTERING

A spherical particle of index n in air and a spherical void of the same size in a material of index $n_0 = n$ scatter very much alike provided that n is close to 1. In Fig. 14, parts (a) and (b) show the angular scattering patterns for a particle and a void, respectively, when $n = 1.1$. The particle scatters more in total (about 22% more), somewhat less in the forward direction, and more to

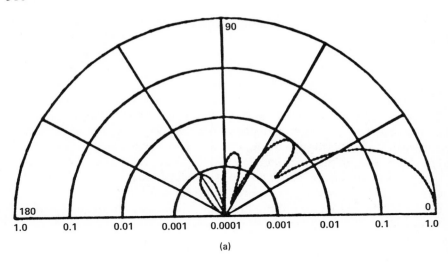

(a)

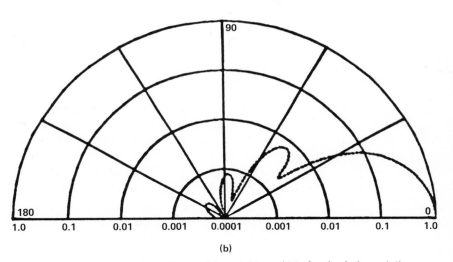

(b)

Fig. 14   Scattering from (a) a particle and (b) a void (refractive index = 1.1).

the sides and backward.   But the bulk of the scattered power (recall that these are logarithmic plots) lies in the same forward lobes.

When the index differs significantly from unity the same trends are seen, but they are exaggerated.   In Fig. 15, parts (a) and (b) are for a particle and a void, respectively, with $n = 1.5$, $n_0 = 1.5$.   Again the particle scatters more, this time by 40%.   The void, however, scatters more than twice as much (per unit solid angle) in the forward direction, but the particle more than makes

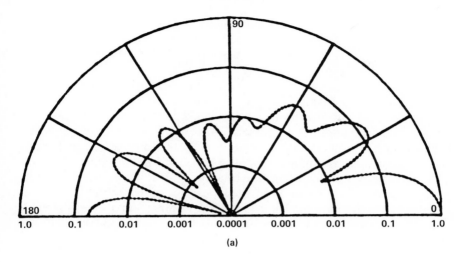

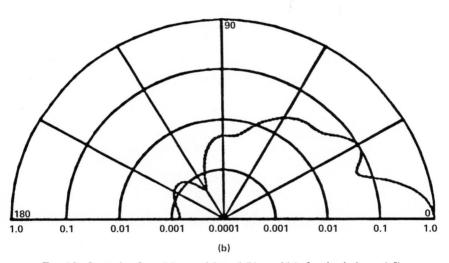

FIG. 15 Scattering from (a) a particle and (b) a void (refractive index = 1.5).

up for this in the backward lobes. Despite these differences, the patterns in the forward hemisphere bear a distinct resemblance to one another, as is most readily seen by superimposing the two patterns. For example, the scattering angles at which the major dips occur are nearly the same.

The particular case illustrated in Fig. 15a is a spherical void in polyethylene plastic. The wavelength (118.8 μm) is a strong far-infrared line in the output of a methyl-alcohol laser (pumped by a CO_2 laser). It is necessary to

P. K. CHEO

use far-infrared light to penetrate the plastic. The bulk absorption coefficient is $\alpha \simeq 1/cm$. This implies an imaginary term in the index of refraction $K = \lambda\alpha/4\pi \simeq 1 \times 10^{-3}$, which has a negligible effect on the scattering.

These remarks apply to the parallel polarization as well, although the referenced illustrations all involve the perpendicular polarization. In comparing the two different polarizations, the first dip in the angular distribution (nearest the forward direction) appears to occur at the same angle. The other dips are more difficult to assess. According to Rayleigh–Gans and Hart–Montroll, nulls should occur at exactly the same angles in the two polarizations, apart from the extra $\cos^2 \theta$ null at 90° in the parallel polarization. In practice very few of the dips qualify as true nulls, but it is of interest to compare the predicted and observed locations of these minima.

Table I gives the computed positions of the minima for particles and voids and the prediction of the Rayleigh–Gans formula, Eq. (54), and of the Hart–Montroll formula, Eq. (66), for three index values. Although it is obvious that Rayleigh–Gans predicts the same angles for particles and voids, it is less obvious that Hart–Montroll does so too. To see this, recall that

$$p(n, \theta) \, x = u_i \qquad (67)$$

gave the position of the angles in the case of a particle. Now, for a void, as we argued in Section II, the effective wavelength in the medium is $\lambda_e = \lambda/n$, and

TABLE I

MINIMA OF THE DIFFERENTIAL CROSS SECTION AT
VARIOUS ANGLES θ (deg)[a]

Type			Minima	
$n = 1.1$				
Rayleigh–Gans	40	72	111	—
Hart–Montroll	37	67	103(104)	(149)
Mie { Particle	38	67	100	156
Mie { Void	38	70	113	—
$n = 1.3$				
Rayleigh–Gans	40	72	111	—
Hart–Montroll	31	59	91	135(138)
Mie { Particle	32	59	86	129
Mie { Void	—	—	78	116
$n = 1.5$				
Rayleigh–Gans	40	72	111	—
Hart–Montroll	22	51	80(81)	115(119)(160)
Mie { Particle	21	49	74/93/114/138/165	
Mie { Void	20	—	—	120

[a] $\lambda_0 = 118.8 \ \mu m$; $r = 125 \ \mu m$.

the apparent index in the void (as seen by the effective wavelength) is $n_e = 1/n$. Hence, the effective x value is $x_e = 2\pi r/\lambda_e = xn$, and the effective p function is

$$p_e = [1 + (1/n^2) - (2/n) \cos \theta]^{1/2}. \tag{68}$$

Thus, it is easy to see that

$$p_e x_e = px = u_i, \tag{69}$$

which implies the same values of θ_i. In fact, because $q_e x_e = qx$ also, the minima predicted by the more complete Hart–Montroll formula, Eq. (62), are also at the same angles for particles and for voids. When these values differ from the predictions of the simpler formula they are shown in parenthesis.

It is clear from Table I that as n deviates from unity not all of the observed wiggles can be explained by the simple theory. The Rayleigh–Gans predictions are a very rough approximation to the correct minima. The Hart–Montroll predictions, however, are quite good, especially in the forward hemisphere, but also in the back hemisphere. The more correct Hart–Montroll formula, Eq. (62), is needed for the back hemisphere; it can even predict minima that the simpler formula, Eq. (54), cannot.

The predictions for particles are better than for voids. Some of the expected minima for voids do not even occur; at best, they are suggested by plateaus or knees in the angular function. On the other hand, some of the dips in the curves for particles are unexpected.

In summary, the Mie-scattering code can be used to evaluate the scattering features of spherical particles and voids in considerable detail, and rapidly. The broad features of the computed results are understandable through the available analytical results, especially in the case of particles. However, use of the code is recommended when detailed information or accurate numerical results are required.

ACKNOWLEDGMENTS

The author would like to acknowledge the contributions of a number of his colleagues. He especially wants to thank M. C. Foster and R. W. Gagnon for their contributions to the design of the optical scanner, E. J. McComb for his effort in the design of laser structures, H. M. Packard for the design of the electronic package, J. D. Farina and R. O. Decker for their effort in the design of a dedicated computer, and A. J. Cantor for his contribution to developing the Mie-scattering code. Finally, the author would like to thank J. B. Austin, Jr. for packaging the electronic controller, V. Failla for assembling the laser structures, and R. P. Muth for his excellent technical assistance throughout this project.

REFERENCES

Abramowitz, M., and Stegun, I. A. (1970). "Handbook of Mathematical Functions," AMS55, National Bureau of Standards, Washington, D.C.

Cantor, A. J., Cheo, P. K., Foster, M. C., and Newman, L. A. (1981). *IEEE J. Quantum Electron.* **QE-17,** 477.

Cheo, P. K. (1978). *Opt. Lett.* **2,** 42.

Cheo, P. K. (1967). *IEEE J. Quantum Electron.* **QE-3,** 683.

Chylek, P. (1975). *J. Opt. Soc. Am.* **65,** 1316.

Deirmendjian, D. (1969). "Electromagnetic Scattering on Spherical Polydispersions." Elsevier, New York.

Hart, R. W., and Montroll, E. W. (1951). *J. Appl. Phys.* **22,** 376, 1278.

Middleton, W. E. K. (1952). "Vision Through the Atmosphere." Univ. of Toronto Press. Toronto, Ontario.

van de Hulst, H. C. (1957). "Light Scattering by Small Particles." Wiley, New York.

Wolfe, W. L., ed. (1965). "Handbook of Military Infrared Technology." Naval Research Laboratory, Washington, D.C.

INDEX

A

Absorption coefficient
  dispersive Fourier transform spectroscopy,
    7, 9
  at millimeter wavelengths, 2–4, 6–7, 9,
    18–19, 24, 26, 28, 30–33, 35–37, 39
Absorption processes, tunable lasers, 228–234
Acoustic-phonon scattering, in high magnetic
    fields, 105–106
ac Stark effect
  in lasers, 207–208, 213, 228–229,
    234–235, 254–255, 257
  multiphoton emission in symmetric-top
    molecules, 261
Aircraft, atmospheric infrared emission
    studies, 159–160, 165
Albumin, far-infrared studies, 64
Alfvén wave, 111, 119–127
Alumina
  dielectric properties, 16, 18, 20
    absorption coefficients, 4, 26
    dielectric permittivity, 27–28
    loss-tangent spectra, 28
    refractive index, 26
Anharmonicity, 260
Anisotropic material, dielectric measurements,
    17
Anomalous water-vapor absorption, *see*
    Water-vapor continuum absorption
Atmosphere, spectral thermal infrared
    emission, 145–193
  broadband measurement, 168–169
  conclusion, 188–190
  electronics, 162–163
  experimental results, 165–180
    airplane and radiosonde-balloon flights,
      165–168
    continuous spectra of thermal emission,
      169–172

experimental setup, 147, 158–165
  aircraft, 159–160
  calibration, 162–165
  measurement techniques, 158–159
  radiometers, 160–162
height profiles of thermal emission at
    discrete wavelengths, 173–180
introduction, 145–148
LOWTRAN model, 148–154
water-vapor continuum absorption,
    146–147, 154–158
  interpretation, 180–188

B

Balloon flights, atmospheric infrared emission
    studies, 165–168
Beam splitter
  Michelson interferometer, 10
  polarizing interferometer, 10–11
Beryllia
  dielectric properties, 18, 20
    absorption coefficients, 4, 28
    dielectric permittivity, 29
    loss-tangent spectra, 30
    refractive index, 29
Beryllium–copper magnet, 78
Bismuth
  Alfvén wave propagation, 120
  cyclotron resonance, 118
  dielectric constant and magnetic field,
    122–123
  Landau levels, 123–124
  magnetoreflection, 112, 116–119, 139
  magnetotransmission, 120, 139
  mass density and magnetic field, 121–122
  semimetal–semiconductor transition,
    124–126
  Shubnikov–de Haas effect, 123–125

315